Maritime Structural Fitter

Level Three

Trainee Guide

PEARSON

Boston Columbus Indianapolis New York San Francisco Amsterdam
Cape Town Dubai London Madrid Milan Munich Paris Montreal Toronto Delhi
Mexico City Sao Paulo Sydney Hong Kong Seoul Singapore Taipei Tokyo

NCCER

President: Don Whyte
Director of Product Development: Daniele Dixon
Maritime Structural Fitter Project Manager: Chris Wilson
Senior Manager: Tim Davis
Quality Assurance Coordinator: Debie Ness
Desktop Publishing Coordinator: James McKay
Permissions Specialist: Adrienne Payne
Production Specialist: Adrienne Payne
Editor: Karyn Payne

Writing and development services provided by Topaz Publications, Liverpool, NY

Lead Writer/Project Manager: Tom Burke
Desktop Publisher: Joanne Hart
Art Director: Alison Richmond
Permissions Editors: Toni Burke, Andrea LaBarge
Writers: Tom Burke, Troy Staton, Veronica Westfall

Pearson

Director, Global Employability Solutions: Jonell Sanchez
Head of Global Certifications/Associations: Andrew Taylor
Associate Director: Sasha Jiwani
Editorial Assistant: Collin Lamothe
Program Manager: Alexandrina B. Wolf
Operations Supervisor: Deidra M. Skahill
Art Director: Diane Ernsberger
Digital Studio Project Managers: Heather Darby, Tanika Henderson
Directors of Marketing: David Gesell, Margaret Waples
Field Marketer: Brian Hoehl

Composition: NCCER
Printer/Binder: RR Donnelley
Cover Printer: RR Donnelley
Text Fonts: Palatino and Univers

Credits and acknowledgments for content borrowed from other sources and reproduced, with permission, in this textbook appear at the end of each module.

ScoutAutomatedPrintCode

PEARSON

Perfect bound ISBN-13: 978-0-13-375044-7
ISBN-10: 0-13-375044-2

Preface

To the Trainee

Congratulations! If you're training through an NCCER-Accredited Training Sponsor, you have likely completed *Maritime Structural Fitter Level One* and *Level Two*. Now you are well on your way to more specific skills and training in *Maritime Structural Fitter Level Three*.

In *Maritime Structural Fitter Level Three*, you will learn how to interpret structural drawings from the highest level general arrangement drawing to the lowest level piece-part drawing. You will prepare for the ship construction process, from laying out locations of structural members and equipment to installation, and then to erection of the vessel itself. Finally, you will learn welding techniques, metallurgy, and the importance of crew leadership.

The maritime industry is facing an aging workforce and shortage of skilled workers. This shortage extends to the structural fitter trade. Skilled structural fitters are in high demand in almost every shipyard. By completing this training you are gaining the abilities needed to help fill that demand. We hope you continue your training beyond this textbook to further your knowledge and skills in this field.

If you're training through an NCCER-Accredited Training Program Sponsor and successfully pass the module exams and performance tests in this course, you may be eligible for credentialing through the NCCER Registry. Check with your instructor or local program sponsor to find out. To learn more, go to **www.nccer.org** or contact us at 1.888.622.3720.

We invite you to visit the NCCER website at **www.nccer.org** for information on the latest product releases and training, as well as online versions of the *Cornerstone* magazine and Pearson's NCCER product catalog.

Your feedback is welcome. You may email your comments to **curriculum@nccer.org** or send general comments and inquiries to **info@nccer.org**.

NCCER Standardized Curricula

NCCER is a not-for-profit 501(c)(3) education foundation established in 1996 by the world's largest and most progressive construction companies and national construction associations. It was founded to address the severe workforce shortage facing the industry and to develop a standardized training process and curricula. Today, NCCER is supported by hundreds of leading construction and maintenance companies, manufacturers, and national associations. The NCCER Standardized Curricula was developed by NCCER in partnership with Pearson, the world's largest educational publisher.

Some features of the NCCER Standardized Curricula are as follows:

- An industry-proven record of success
- Curricula developed by the industry for the industry
- National standardization providing portability of learned job skills and educational credits
- Compliance with the Office of Apprenticeship requirements for related classroom training (*CFR 29:29*)
- Well-illustrated, up-to-date, and practical information

NCCER also maintains a Registry that provides transcripts, certificates, and wallet cards to individuals who have successfully completed a level of training within a craft in NCCER's Curricula. *Training programs must be delivered by an NCCER Accredited Training Sponsor in order to receive these credentials.*

Special Features

In an effort to provide a comprehensive, user-friendly training resource, we have incorporated many different features for your use. Whether you are a visual or hands-on learner, this book will provide you with the proper tools to get started in the Structural Fitter trade.

Introduction

This page is found at the beginning of each module and lists the Objectives, Performance Tasks, and Trade Terms for that module. The Objectives list the skills and knowledge you will need in order to complete the module successfully. The Performance Tasks give you an opportunity to apply your knowledge to the real-world duties that structural fitters perform. The list of Trade Terms identifies important terms you will need to know by the end of the module.

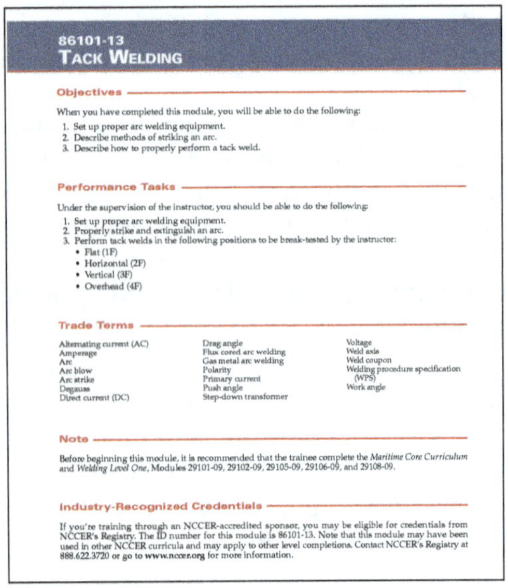
86101-13
TACK WELDING

Objectives

When you have completed this module, you will be able to do the following:

1. Set up proper arc welding equipment.
2. Describe methods of striking an arc.
3. Describe how to properly perform a tack weld.

Performance Tasks

Under the supervision of the instructor, you should be able to do the following:

1. Set up proper arc welding equipment.
2. Properly strike and extinguish an arc.
3. Perform tack welds in the following positions to be break-tested by the instructor:
   - Flat (1F)
   - Horizontal (2F)
   - Vertical (3F)
   - Overhead (4F)

Trade Terms

Alternating current (AC)	Drag angle	Voltage
Amperage	Flux cored arc welding	Weld axis
Arc	Gas metal arc welding	Weld coupon
Arc blow	Polarity	Welding procedure specification (WPS)
Arc strike	Primary current	Work angle
Degauss	Push angle	
Direct current (DC)	Step-down transformer	

Note

Before beginning this module, it is recommended that the trainee complete the *Maritime Core Curriculum* and *Welding Level One*, Modules 29101-09, 29102-09, 29105-09, 29106-09, and 29108-09.

Industry-Recognized Credentials

If you're training through an NCCER-accredited sponsor, you may be eligible for credentials from NCCER's Registry. The ID number for this module is 86101-13. Note that this module may have been used in other NCCER curricula and may apply to other level completions. Contact NCCER's Registry at 888.622.3720 or go to www.nccer.org for more information.

Special Features

Features provide a head start for those entering the Structural Fitter field by presenting technical tips and professional practices from craftworkers in various disciplines. These features often include real-life scenarios similar to those you might encounter on the job site.

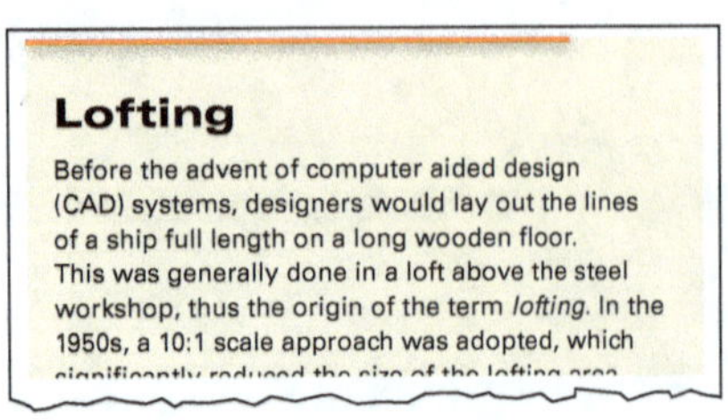
Lofting

Before the advent of computer aided design (CAD) systems, designers would lay out the lines of a ship full length on a long wooden floor. This was generally done in a loft above the steel workshop, thus the origin of the term *lofting*. In the 1950s, a 10:1 scale approach was adopted, which

Color Illustrations and Photographs

Full-color illustrations and photographs are used throughout each module to provide vivid detail. These figures highlight important concepts from the text and provide clarity for complex instructions. Each figure reference is denoted in the text in italics for easy reference.

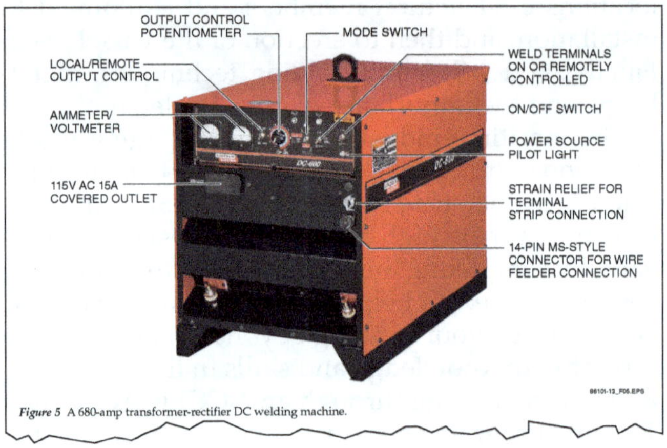

Figure 5 A 680-amp transformer-rectifier DC welding machine.

Notes, Cautions, and Warnings

Safety features are set off from the main text in highlighted boxes and are organized into three categories based on the potential danger of the issue being addressed. Notes simply provide additional information on the topic area. Cautions alert you of a danger that does not present potential injury but may cause damage to equipment. Warnings stress a potentially dangerous situation that may cause injury to you or a co-worker.

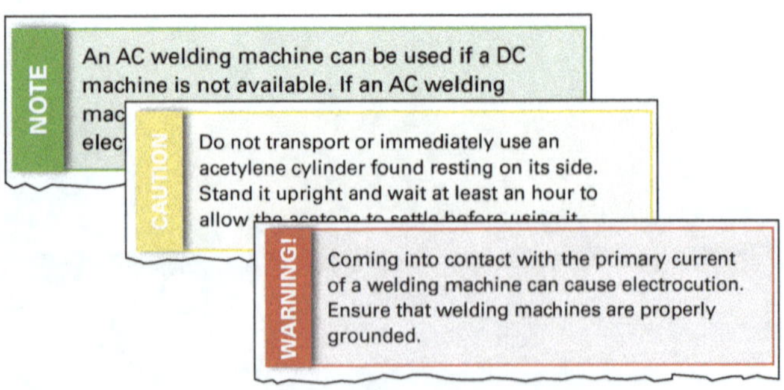
NOTE

An AC welding machine can be used if a DC machine is not available. If an AC welding mac… elec…

CAUTION

Do not transport or immediately use an acetylene cylinder found resting on its side. Stand it upright and wait at least an hour to allow …

WARNING!

Coming into contact with the primary current of a welding machine can cause electrocution. Ensure that welding machines are properly grounded.

Going Green

Going Green looks at ways to preserve the environment, save energy, and make good choices regarding the health of the planet. Through the introduction of new construction practices and products, you will see how the "greening of America" has already taken root.

> GOING GREEN
>
> **Test Coupons**
>
> Test coupons are routinely taken for testing in the welding industry. When those test coupons are no longer needed, they should be properly recycled, to save the company money and to keep salvageable items out of the landfills.

Step-by-Step Instructions

Step-by-step instructions are used throughout to guide you through technical procedures and tasks from start to finish. These steps show you not only how to perform a task but also how to do it safely and efficiently.

> Follow these steps to make a vertical-up fillet tack weld:
>
> *Step 1* In the 1F position, tack two plates together to form a T-joint for the fillet tack weld coupon.
>
> *Step 2* Clamp or tack-weld the coupon in the vertical position.
>
> *Step 3* Run a number of ½" to ¾" long tack welds along the root of the joint (starting at the bottom) using an electrode (work) angle of

Trade Terms

Each module presents a list of Trade Terms that are discussed within the text and defined in the Glossary at the end of the module. These terms are denoted in the text with **bold, blue type** upon their first occurrence. To make searches for key information easier, a comprehensive Glossary of Trade Terms from all modules is located at the back of this book.

> Welding machines that produce only DC welding voltage and current can be further classified by the characteristics of the welding current they produce. If the welding current varies between the higher open-circuit voltage and the lower operating voltage, it is classified as a variable-voltage constant-current power source. This type of welding machine is used for SMAW. If open-circuit and operating voltages are nearly the same, the machine is classified as a constant-voltage or constant-potential DC welding machine. This type of machine is used for **gas metal arc welding (GMAW)** and **flux cored arc welding (FCAW)**.

Review Questions

Review Questions reinforce the knowledge you have gained and are a useful tool for measuring what you have learned.

> **Review Questions**
>
> 1. SMAW is often referred to as ____.
>   a. AC welding
>   b. stick welding
>   c. gap welding
>   d. resistive welding
> 2. When working in a confined space, the oxygen and acetylene tanks must be left outside the space.
>   a. True
>   b. False
> 3. During SMAW welding, current jumps across the ____.
>   a. workpiece clamp from the tip of the electrode to the surface of the workpiece
>   b. gap from the tip of the electrode to the surface of the workpiece
>   c. electrode holder from the tip of the electrode to the surface of the workpiece
>   d. molten metal from the tip of the electrode to the surface of the workpiece
> 4. In AC current, the number of cycles completed in one second is called the ____.
>   a. polarity
>   b. travel
>   c. frequency
>   d. current value
> 5. Electrical current that has no frequency is known as ____.
>   a. DC current
>   b. AC current
>   c. AC voltage
>   d. 120VAC
> 6. Which of the following only applies to DC current?
>   a. Frequency
>   b. Polarity
>   c. Amperage
>   d. Current flow
> 7. The voltage in a welding machine is reduced or stepped down by a(n) ____.
>   a. limiter
>   b. generator
>   c. alternator
>   d. transformer
> 8. The duty cycle of a welding machine is based on a ____.
>   a. 10-minute period
>   b. 15-minute period
>   c. 20-minute period
>   d. 30-minute period
> 9. To prevent the lead from pulling loose from the welding machine terminal, welding cable lugs are generally ____.
>   a. slotted
>   b. soldered in place
>   c. ringed
>   d. insulated
> 10. When preparing to practice welding of fillet tack welds, the metal should be cut into ____.
>   a. 3" × 12" rectangles for the base
>   b. 4" × 12" rectangles for the base
>   c. 4" × 4" squares for the base
>   d. 12" × 12" squares for the base
> 11. The easiest method of striking an arc is the ____.
>   a. scratching method
>   b. grounding method
>   c. weaving method
>   d. tapping method
> 12. Arc blow is caused by ____.
>   a. strong winds
>   b. magnetic fields
>   c. fluctuating weld current
>   d. damaged electrodes
>
> 29101-13 Tack Welding Module Six 25

NCCER Standardized Curricula

NCCER's training programs comprise more than 80 construction, maintenance, pipeline, and utility areas and include skills assessments, safety training, and management education.

Boilermaking
Cabinetmaking
Carpentry
Concrete Finishing
Construction Craft Laborer
Construction Technology
Core Curriculum: Introductory Craft Skills
Drywall
Electrical
Electronic Systems Technician
Heating, Ventilating, and Air Conditioning
Heavy Equipment Operations
Highway/Heavy Construction
Hydroblasting
Industrial Coating and Lining Application Specialist
Industrial Maintenance Electrical and Instrumentation Technician
Industrial Maintenance Mechanic
Instrumentation
Insulating
Ironworking
Masonry
Millwright
Mobile Crane Operations
Painting
Painting, Industrial
Pipefitting
Pipelayer
Plumbing
Reinforcing Ironwork
Rigging
Scaffolding
Sheet Metal
Signal Person
Site Layout
Sprinkler Fitting
Tower Crane Operator
Welding

Maritime

Maritime Industry Fundamentals
Maritime Pipefitting
Maritime Structural Fitter

Green/Sustainable Construction

Building Auditor
Fundamentals of Weatherization
Introduction to Weatherization
Sustainable Construction Supervisor
Weatherization Crew Chief
Weatherization Technician
Your Role in the Green Environment

Energy

Alternative Energy
Introduction to the Power Industry
Introduction to Solar Photovoltaics
Introduction to Wind Energy
Power Industry Fundamentals
Power Generation Maintenance Electrician
Power Generation I&C Maintenance Technician
Power Generation Maintenance Mechanic
Power Line Worker
Power Line Worker: Distribution
Power Line Worker: Substation
Power Line Worker: Transmission
Solar Photovoltaic Systems Installer
Wind Turbine Maintenance Technician

Pipeline

Control Center Operations, Liquid
Corrosion Control
Electrical and Instrumentation
Field Operations, Liquid
Field Operations, Gas
Maintenance
Mechanical

Safety

Field Safety
Safety Orientation
Safety Technology

Supplemental Titles

Applied Construction Math
Tools for Success

Management

Fundamentals of Crew Leadership
Project Management
Project Supervision

Spanish Titles

Acabado de concreto: nivel uno
Aislamiento: nivel uno
Albañilería: nivel uno
Andamios
Carpintería: Formas para carpintería, nivel tres
Currículo básico: habilidades introductorias del oficio
Electricidad: nivel uno
Herrería: nivel uno
Herrería de refuerzo: nivel uno
Instalación de rociadores: nivel uno
Instalación de tuberías: nivel uno
Instrumentación: nivel uno, nivel dos, nivel tres, nivel cuatro
Orientación de seguridad
Paneles de yeso: nivel uno
Seguridad de campo

Acknowledgments

This curriculum was revised as a result of the farsightedness and leadership of the following sponsors:

Bollinger Shipyards, Inc.
Huntington Ingalls
Signal International
Tecnico Corporation

This curriculum would not exist were it not for the dedication and unselfish energy of those volun-teers who served on the Authoring Team. A sincere thanks is extended to the following:

Cornelious L. Davis IV
Brian Folse
Ken Inabinett
Brad Knight
Ronnie Parrish
Terry Stead

NCCER Partners

American Fire Sprinkler Association
Associated Builders and Contractors, Inc.
Associated General Contractors of America
Association for Career and Technical Education
Association for Skilled and Technical Sciences
Construction Industry Institute
Construction Users Roundtable
Design Build Institute of America
GSSC – Gulf States Shipbuilders Consortium
ISN
Manufacturing Institute
Mason Contractors Association of America
Merit Contractors Association of Canada
NACE International
National Association of Women in Construction
National Insulation Association
National Technical Honor Society
NAWIC Education Foundation
North American Crane Bureau
North American Technician Excellence
Pearson
Prov
SkillsUSA®
Steel Erectors Association of America
U.S. Army Corps of Engineers
University of Florida, M. E. Rinker, Sr. School of Construction Management
Women Construction Owners & Executives, USA

NCCER Business Partners

Contents

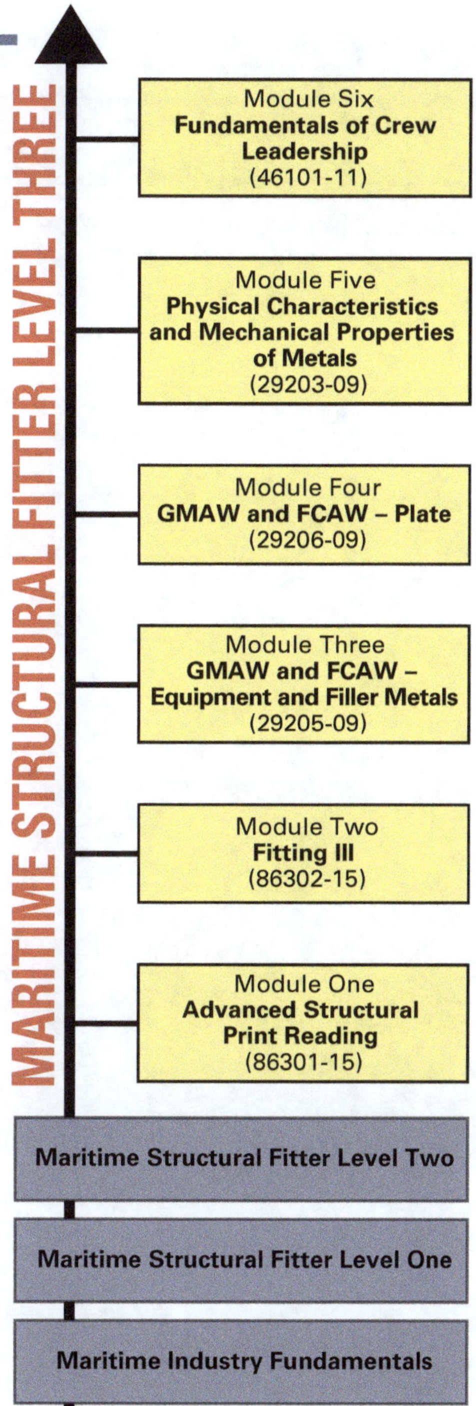

This course map shows all of the modules in *Maritime Structural Fitter Level Three*. The suggested training order begins at the bottom and proceeds up. Skill levels increase as you advance on the course map. The local Training Program Sponsor may adjust the training order.

86301-15

Advanced Structural Print Reading

Module One

Trainees with successful module completions may be eligible for credentialing through the NCCER Registry. To learn more, go to **www.nccer.org** or contact us at 1.888.622.3720. Our website has information on the latest product releases and training, as well as online versions of our *Cornerstone* magazine and Pearson's product catalog.

Your feedback is welcome. You may email your comments to **curriculum@nccer.org**, send general comments and inquiries to **info@nccer.org**, or fill in the User Update form at the back of this module.

This information is general in nature and intended for training purposes only. Actual performance of activities described in this manual requires compliance with all applicable operating, service, maintenance, and safety procedures under the direction of qualified personnel. References in this manual to patented or proprietary devices do not constitute a recommendation of their use.

86301-15

Advanced Structural Print Reading

Objectives

When you have completed this module, you will be able to do the following:

1. Describe how drawings for a vessel are structured.
2. Interpret the types of drawings commonly used by structural fitters.
   a. Assembly drawings
   b. Mechanical drawings
   c. Foundation drawings
3. Explain the drawing change process.

Performance Task

Under the supervision of your instructor, you should be able to do the following:

1. Use a drawing to establish the work lines for an individual structure.

Trade Terms

Floor
Piece part
Retrofit
Stanchion

Industry Recognized Credentials

If you are training through an NCCER-accredited sponsor, you may be eligible for credentials from NCCER's Registry. The ID number for this module is 86301-15. Note that this module may have been used in other NCCER curricula and may apply to other level completions. Contact NCCER's Registry at 888.622.3720 or go to **www.nccer.org** for more information.

Note

This module refers to drawings included in the supplemental drawing package provided with this module, including the following:

- General Arrangement Drawings (A1)
- Assembly Drawings (A2, A3, A4)
- Mechanical Drawings (A5, A6)

Contents

Topics to be presented in this module include:

Figures

The following figures are part of the supplemental drawing package provided with this module:

1.0.0 Introduction

Drawings define the parts, subassemblies, and assemblies to be used in a structure, where they are to be placed, and how they are assembled. On any given day, your foreman may hand you a package of prints and instruct you to carry out the task represented on the prints. Therefore, the ability to read engineering drawings is a critical skill that every fitter must master.

Drawings are developed in a top-down structure in which the vessel (hull, platform, etc.) itself is the top drawing and all other elements descend from it in a hierarchy. The hull drawing for a ship typically identifies four major divisions: systems, mechanicals, structural elements, and habitability (quality of life). See *Figure 1*.

- *Systems* include radar, sonar, and electrical services.
- *Mechanicals* include propulsion, along with heating, ventilation, and air conditioning (HVAC).
- *Structural elements* include modules (blocks), units, assemblies, subassemblies, and piece parts.
- *Habitability* (quality of life) drawings include berthing and living areas for crew and (if applicable) passengers.

The structural division is the ship fitter's area of interest. In a typical structural drawing arrangement, the vessel is organized into modules, or blocks, which are the major structures assembled in the drydock. The modules consist of units that are, in turn, made up of assemblies. Subassemblies are the smallest grouping of structural components (piece parts).

1.1.0 General Arrangement (GA) Drawings

General arrangement (GA) drawings are usually the first drawings developed by ship designers. They are used initially to show the layout of the vessel and the locations of spaces, equipment, and structures. As the design firms up, the GA drawings become increasingly detailed. In creating these drawings, designers are able to verify that all the features specified by the contractor can be accommodated. The GA drawing also allows designers and owners to arrange spaces for maximum efficiency and convenience and helps designers of one area avoid conflicts with other areas.

The top-level GA drawing may include both elevation views of the vessel and plan views of each deck or level. *Figure A1* is a GA drawing for an offshore supply vessel. This is a fairly simple drawing because most of this ship consists of cargo holds and tanks. In contrast, the GA drawing for a warship or passenger liner that might have 10 or more deck levels would require many sheets.

Elevation views of the ship are used to show a deck-by-deck view with identification of each space and numbering of frames. The elevation views are divided into port and starboard outboard profiles, as well as inboard profiles referenced from the ship center line.

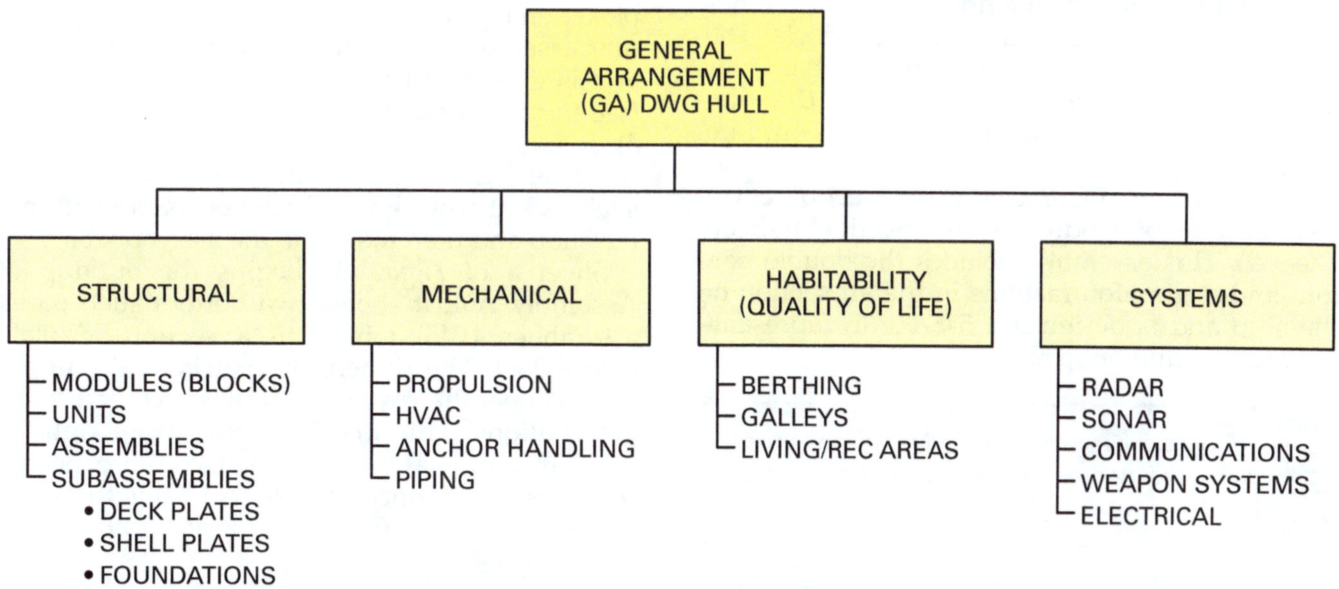

Figure1 Typical drawing structure for a vessel.

A plan view of each deck and major equipment space would be used to show the layout of the spaces on the decks, along with the locations of bulkheads, passageways, and stairs. Separate GA drawings of key spaces are also prepared. *Figure 2* shows a simplified version of a GA drawing for one deck of a vessel, with details of the engine room. In *Figure 3*, the engine room GA drawing is further broken into plan, elevation, and section views.

2.0.0 Working With Drawings

The drawings used in this module come from a specific shipyard and thus follow the drawing practices used by that yard. These drawings are provided to give you an understanding of basic drawing relationships and an opportunity to practice interpreting drawings. The drawing practices may differ significantly from those used by your company, so your instructor should conduct this lesson using drawings that are common to your company.

2.1.0 Assembly Drawings

Assembly drawings are typically provided for all multi-part structures, from the lowest subassembly to the largest module. Assembly drawings are the primary tools used by structural fitters in assembling these structures. This section uses drawings for a supply vessel to illustrate how drawings are used in the ship assembly process.

Figure A3 is a General Arrangement or Module Breakout drawing for this vessel. Sheet 1 of the drawing provides a plan view and an elevation view of the ship and identifies the four modules that will be erected in the drydock. The profile view includes the numbering scheme for the transverse frames. Sheet 2 of the drawing shows how the four modules are broken out into assemblies.

The drawings referenced in this section cover Assembly 61 of Module 60 (see Section 15-B on Sheet 2). This assembly includes the double bottom and engine foundations in the aft section of the ship and is designated 545/60/61, representing hull/module/assembly.

NOTE

This vessel has only four modules despite the gap in the module numbering (jump from 30 to 60).

Lofting

Before the advent of computer aided design (CAD) systems, designers would lay out the lines of a ship full length on a long wooden floor. This was generally done in a loft above the steel workshop, thus the origin of the term *lofting*. In the 1950s, a 10:1 scale approach was adopted, which significantly reduced the size of the lofting area required. CAD systems began replacing physical lofting at the larger shipyards in the 1960s.

Source: *Ship Construction*

The assembly drawing for assembly A61 is represented in *Figure A4*. This is the drawing that fitters working in the assembly bay would use to assemble A61. The bill of materials uses "P" and "K" designations to identify assemblies and subassemblies. *Figure A2* identifies the panel assemblies ("P" designations) such as tank top panels, bulkheads, double bottom floors, shell plates, and girders, along with structural assemblies ("K" designations) such as foundations, bulwarks, and tank ballasts.

In total, A61 contains 53 assemblies, excluding padeyes, along with approximately 160 individual piece parts. These assemblies and parts would need to be accumulated at the shipyard assembly bay in order to fabricate the assembly. Individual parts are marked with their drawing numbers.

If two or more items are attached, it is considered an assembly or subassembly and is given a drawing number. *Figure 4* is an example of a simple assembly consisting of two parts. This particular assembly is one of the double bottom floors and is designated P01 on the assembly drawing (see the upper right-hand area of *Figure A4*, Sheet 5). It is also represented on the isometric view on Sheet 5 (lower right-hand corner). Assemblies such as P01 would most likely be assembled in a fab shop and then moved to the assembly bay.

Sheet 4 of *Figure A4* begins the buildup of Assembly A61. It shows two bottom shell panel assemblies (P45) joined to a section of plate (60A61-280). The dimension matrix at the upper right shows the distances between circled letter designations. This matrix is read from right to left. Note, for example, that the two P45 panel assemblies are identical in size (A to B is the same as C to D and H to G is the same as E to F).

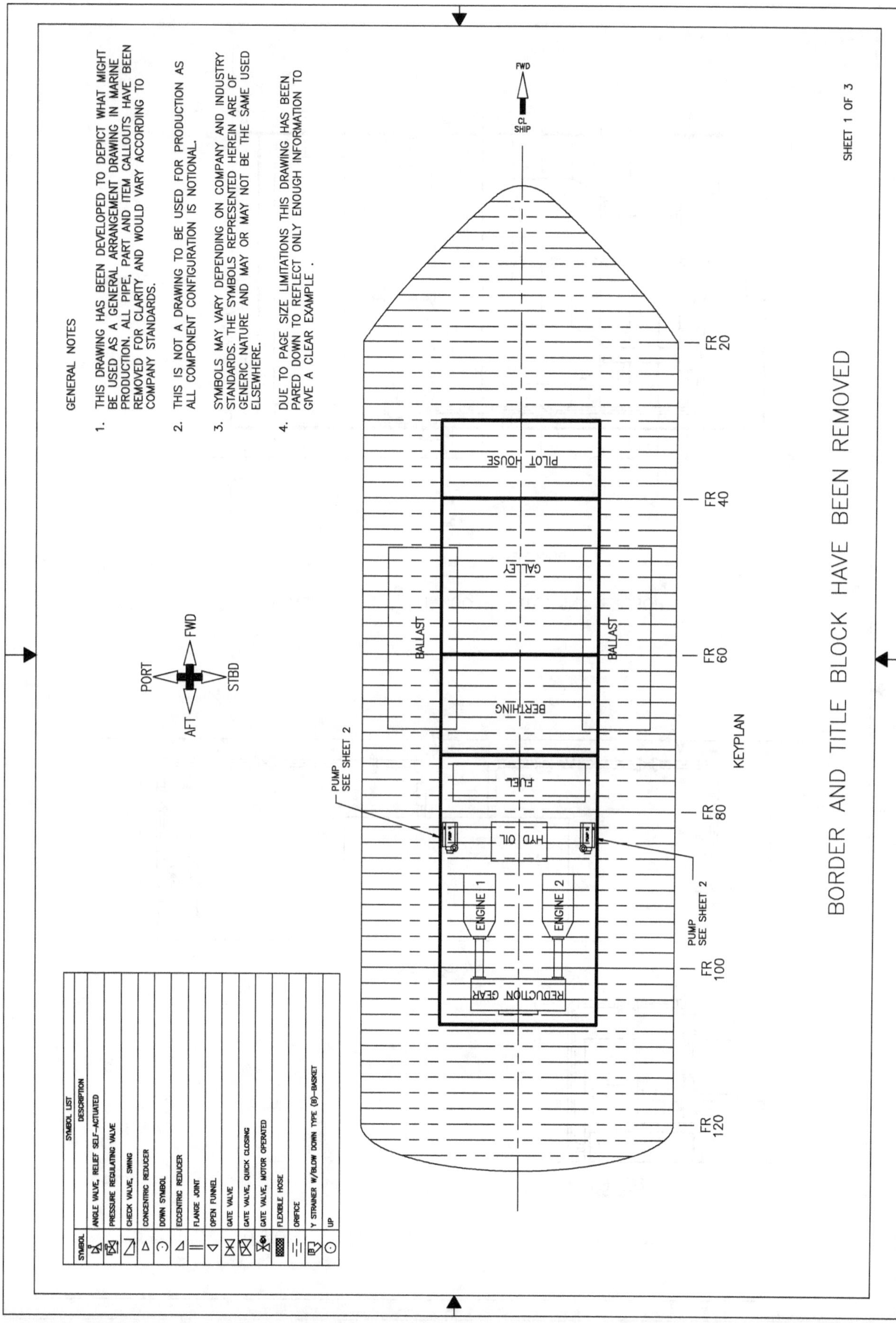

Figure 2 Simplified plan view of a deck.

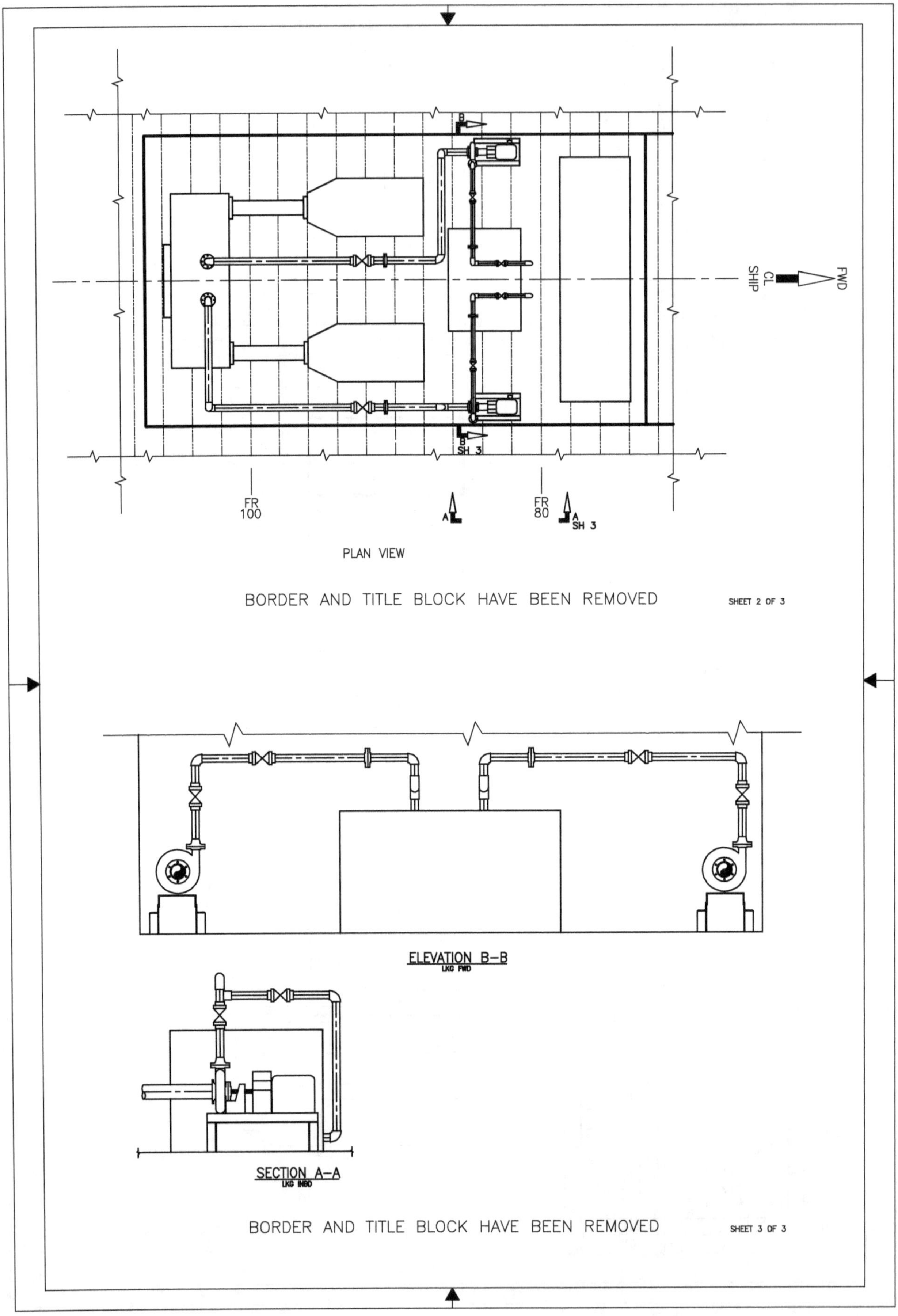

Figure 3 GA drawings of an engine room.

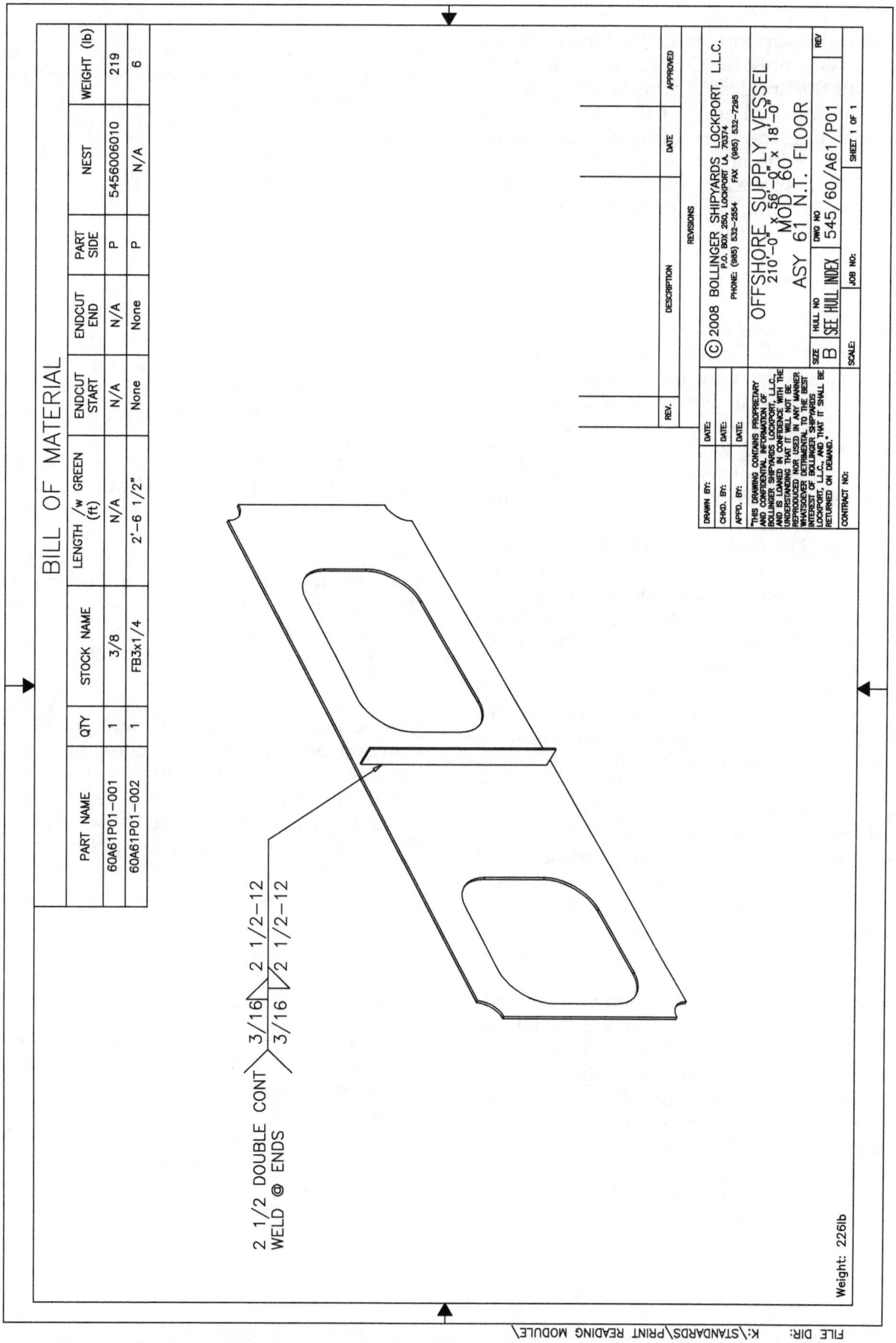

Figure 4 A simple assembly.

Sheet 5 of *Figure A4* shows the engine support girders and the support structure for the double bottom of Assembly 61. On the left side of the sheet are all the applicable dimensions. Looking aft from the forward edge of the assembly, you can see that the first lateral frame is positioned 1'-11" from the edge of the panel. From then on, the frames are spaced 2' apart. At every other frame, there is a double bottom floor assembly. The longitudinal positioning of the various stiffeners and floor assemblies, as well as that of the engine girders, is measured from the ship center line.

Sheet 6 of *Figure A4* provides a three-dimensional perspective of the area covered on Sheet 5 to more clearly indicate how the pieces fit together. The Section at Frame 70 shown on the right-hand side of Sheet 6 represents the support structure at Frame 70, which is the first frame aft of the forward edge of Assembly 61. It shows one side of that structure starting from the ship center line and moving to port. The starboard side is a mirror image. The section includes the longitudinal bulkheads that run the length of the assembly fore and aft, as well as the engine girders and double bottom floors. The drawings referenced at the bottom of Sheet 6 are the lifting details showing how to build the padeyes.

As shown on Sheet 5, the engine girders are made from a combination of fabricated assemblies and individual structural cross members. *Figure 5* is a drawing of one of the four engine girder assemblies, in this case P37. This girder would likely have been assembled in a fab shop and sent from there to the assembly bay.

Sheet 7 of *Figure A4* serves two purposes. First, it shows Assembly 61 with the tank top plates in place. Second, it shows the parts of the engine foundation that are installed above the tank top. The tank top consists of fabricated assemblies, as well as individual sections of deck plate. The plate assemblies (P15, P16) are fitted with stiffeners that serve to back up heavy foundations. *Figure 6* shows foundation backers installed on the underside of a deck plate assembly.

Sheet 8 of *Figure A4* identifies the engine girder support brackets and shows the positioning of the brackets in relation to the transverse frames. Section A-A shows a typical bracket installation. The individual parts numbered 60A61-94, 96, 103, and 104 are doubler plates for the stanchions.

2.2.0 Foundation Drawings

Main engines, boilers, pumps, fans, generators, and other items of heavy equipment are mounted on foundations, which carry and distribute the load. *Figure 7* shows an example of a foundation with a pump and motor bolted to it. The equipment is bolted to the foundation so that it can be removed if necessary. Foundations are typically made using steel plate, flat bar, and angle. Foundations for large equipment such as engines may require the use of I-beams or T-beams. Fitters have the responsibility to fabricate the foundations, lay out their locations, and install them.

Foundations are generally fabricated in the shop using drawings such as the one shown in *Figure 8*, and are then sent to the assembly bays where they are installed. Extra stock is usually included in the construction of the vertical section so that the foundation can be trimmed in order to achieve a perfect level during installation. This is necessary because the surface on which the foundation is installed may not be perfectly level. *Figure 9* shows a fitter leveling a foundation. Once the foundation is level, the fitter will scribe the foundation to the deck plate in order to determine how much material to remove. Trimming is typically done with a cutting torch.

Sheet 9 of *Figure A4* provides the locations for all the foundations in Assembly 61. These assemblies are designated with a "K" on the drawing and the bill of materials. The engine girders are also shown. Fitters will use this sheet of the drawing to lay out and install the foundations. Locations in the transverse plane (port to starboard) are referenced to the forward and aft edges of the assembly. In the longitudinal plane (fore to aft), the locations are referenced to the ship center line.

Blueprints

The blueprint method of copying was invented in 1842 by British astronomer John Herschel. This method of reproducing drawings was in common use for the next hundred years until it was gradually replaced by xerography and other methods. Even though the method is obsolete, it is still common for people to call drawings "blueprints".

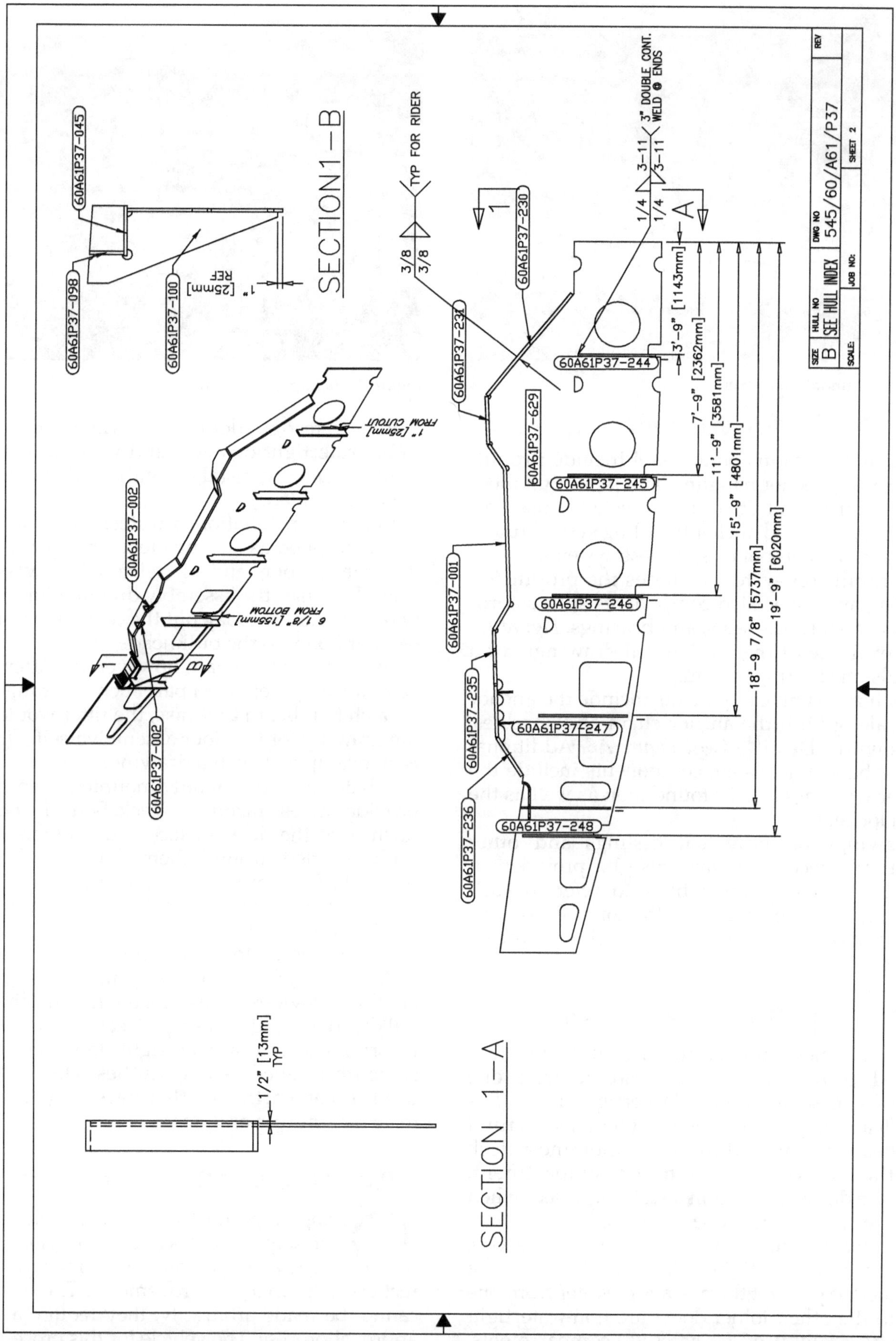

Figure 5 Assembly drawing for an engine girder.

Figure 6 Foundation backers.

Figure 7 Example of a foundation.

2.3.0 Mechanical Drawings

Mechanical systems on a vessel include propulsion, HVAC, anchor handling, and piping. Of these systems, structural fitters are expected to use propulsion system and anchor handling system drawings to install components of those systems.

The propulsion system includes the propulsion shafting and associated components such as propellers, stern tubes, seals, and bearings. *Figure A5* is an example of the mechanical drawings for a propulsion shafting system.

Anchor system components include the anchor itself, along with the anchor chain and shackles, as shown in Detail 3-G of *Figure A6.* Additional anchor handling system components include the anchor windlass and its foundation as well as the chain lockers.

Drawings of individual castings and other structural piece parts may also be provided. It may be necessary for a fitter to refer to such drawings for information on the composition of a casting in order to determine the welding process to be used.

2.4.0 Doors, Scuttles, and Hatches

Doors, hatches, and scuttles are similar in that they all provide access for personnel from one part of a vessel to another. Watertight doors (*Figure 10*) are used for passage through a structural bulkhead, and to and from the uppermost deck (weather deck) and an interior compartment. They are designed to form a watertight seal when they are closed and properly secured.

Non-watertight doors are used for access through non-watertight bulkheads. They are not designed to prevent the passage of water from one side to the other. Joiner doors are non-watertight doors typically used for access to compartments. Unlike watertight doors, joiner doors do not have to be watertight or withstand water pressure, so they are usually installed with hardware instead of being welded.

In a new installation, a watertight door is generally supplied as a complete assembly, including the frame, door, and all related hardware. The fitter then uses the assembly drawing for the next higher assembly to lay out the door assembly and weld or bolt it to the bulkhead.

In the case of a repair, however, the fitter might be required to replace a part such as a hinge, dog, or latch handle. In that case, the fitter would need the drawing for the door assembly itself. *Figure 11* is an example of such a drawing.

A hatch is a horizontally-mounted opening that provides access through a deck. Some hatches are flush with the deck surface, while others have a frame welded around them that extends above the deck. The frame extension, known as a coaming, prevents water from leaking in when the hatch is opened. *Figure 12* shows an example of a hatch commonly found on ships.

A scuttle (*Figure 13*) is a small opening with a lid that provides access through a hatch, deck, bulkhead, or hull of a ship. A scuttle may also be incorporated into a watertight door. Scuttles are sometimes called escape scuttles. They are often used in compartments that have only one other means of ingress or egress.

3.0.0 Drawing Change Control

During ship construction, especially for the first ship in a class, it is common to find that drawings need to be revised to correct errors or make improvements. These changes cannot be made arbitrarily; they require a formal approval process. The vehicle for this process is the

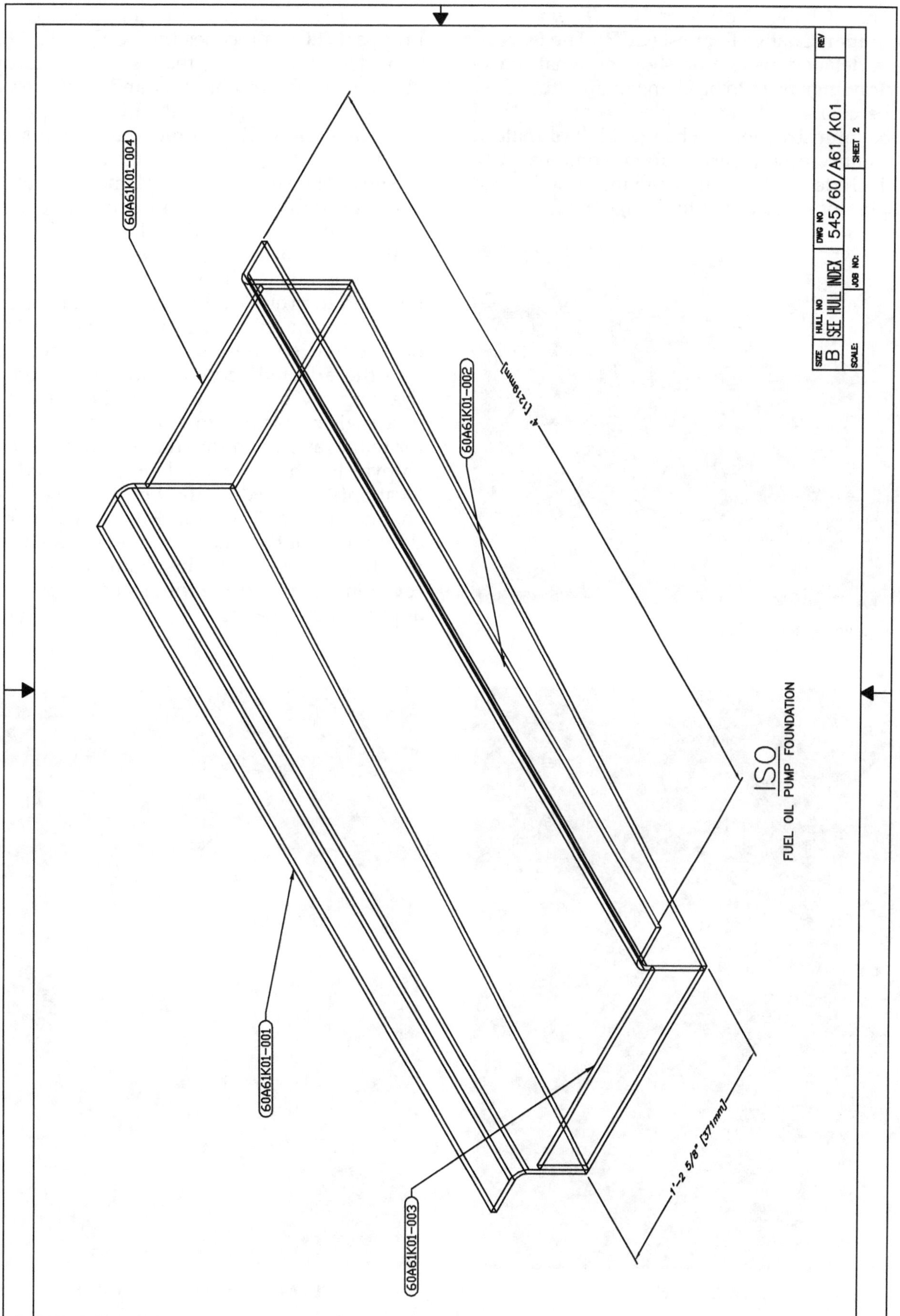

Figure 8 Simple foundation drawing.

Engineering Change Notice, or ECN (*Figure 14*) or Configuration Change Request (CCR). The ECN or CCR is a detailed instruction showing what drawing or drawings need to be changed and how they are to be changed. In its simplest form, the ECN might be an instruction to change a bill of materials to identify a new part. A more complex ECN would include a marked-up drawing showing the CAD operator exactly how to change the drawing.

Figure 9 Leveling a foundation.

In some instances, an Engineering Change Proposal (ECP) precedes the ECN/CCR. The ECP is presented to project management and owners in order to obtain approval and necessary funding for the change. Once the ECP is approved, an ECN or CCR will be prepared and distributed and the drawing will be changed.

The change cannot be implemented until it has gone through a review and approval cycle. This is done to make sure that the proposed change will not have an adverse impact. Once all approvals are obtained, the ECN is sent to all parties involved to provide advance notice of the change.

At the same time, the ECN is sent to the CAD department so the change can be incorporated into the affected drawings and recorded in the drawing revision block. *Figure 15* is an example of a revision block showing the revision history for that drawing. In the drawing systems of some shipyards, the revision letter is placed at the location(s) on the drawing that are affected by the change. The revision block typically includes a description of the change(s) made to each sheet of the drawing, the date the change was made, the revision letter, and the initials of the person who approved the changes.

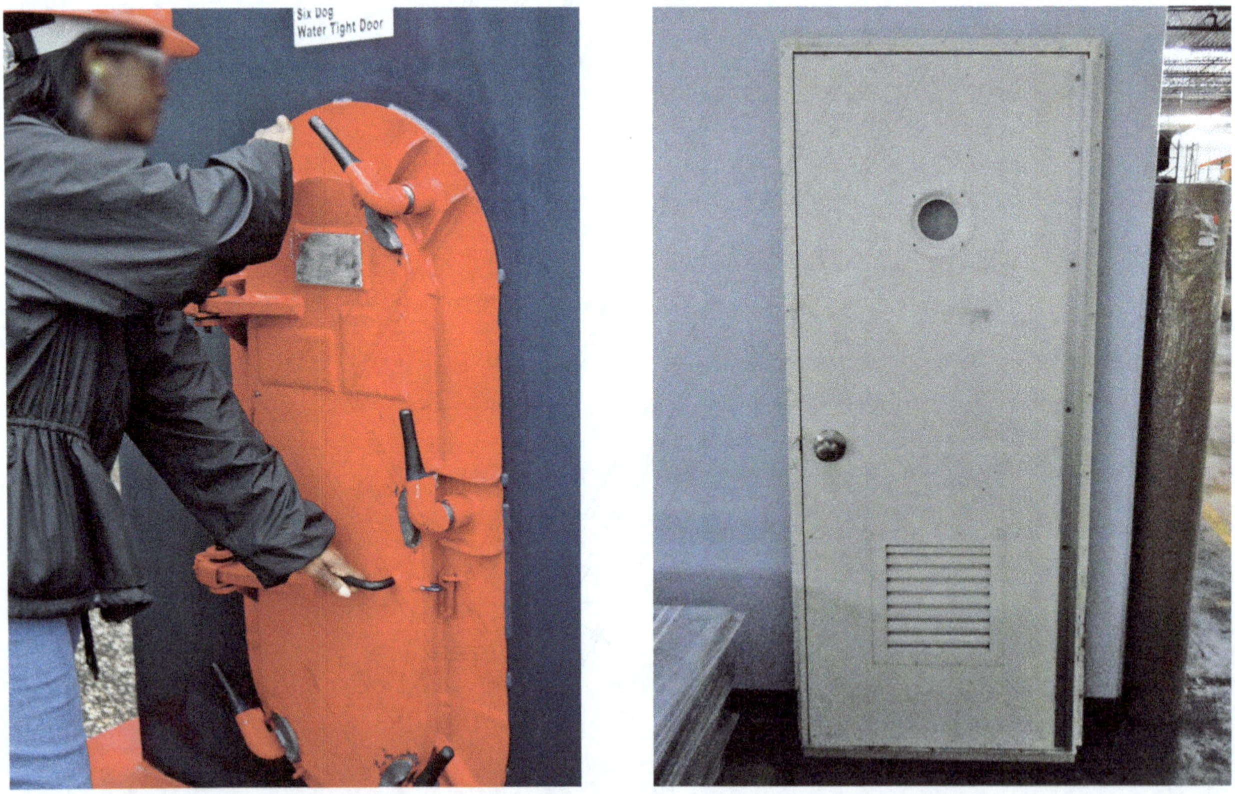

(A) WATERTIGHT DOOR

(B) NON-WATERTIGHT DOOR

Figure 10 Watertight and non-watertight doors.

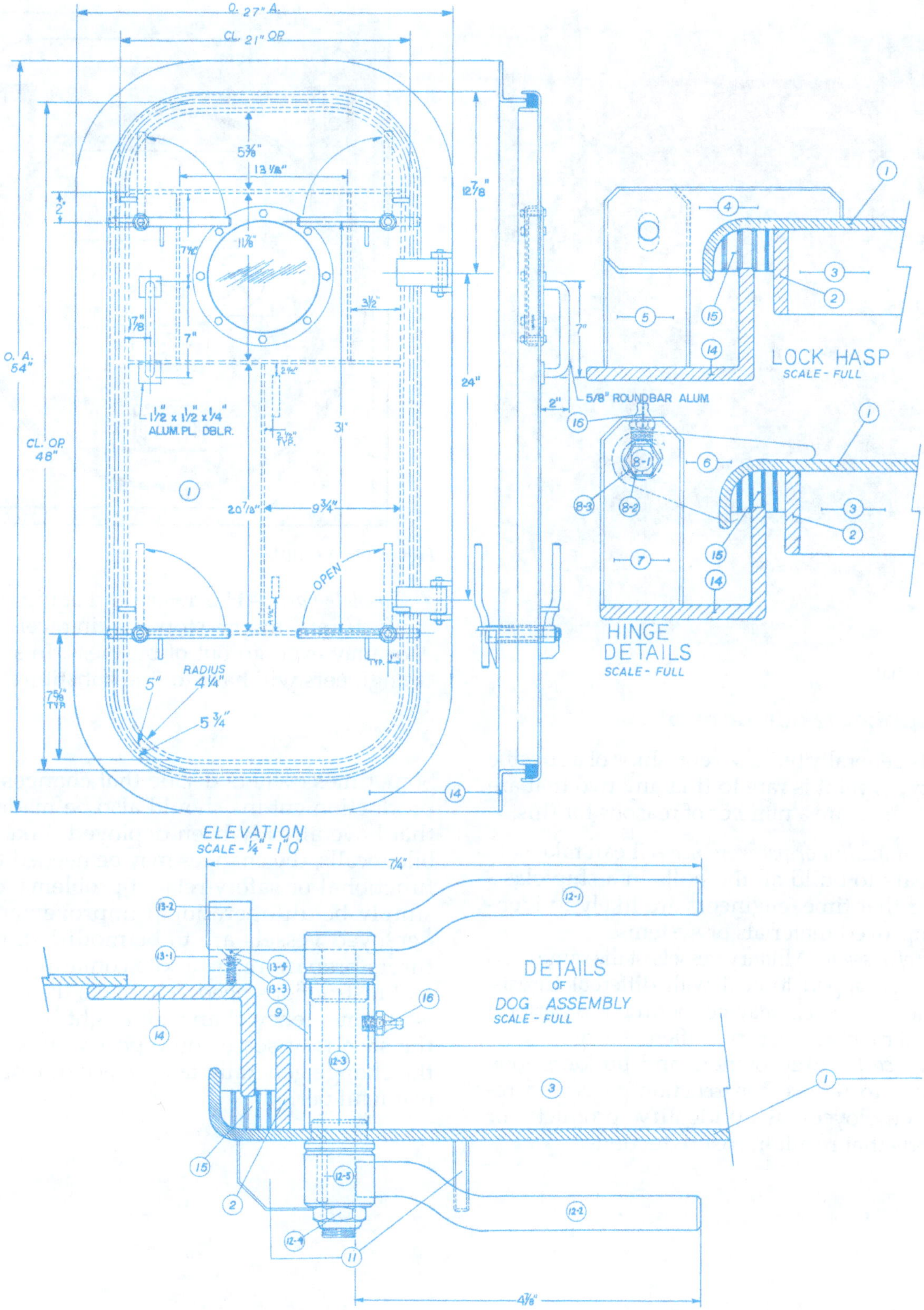

Figure 11 Example of a door drawing.

Figure 12 A hatch.

3.1.0 Configuration Control

Shipyards generally build several ships of a specific class or type, yet it is rare to find any two that are identical. There are a number of reasons for this:

- *Improved quality or performance* – It can take several years to build all the hulls in a ship class. During that time, engineers are likely to identify improved materials or systems.
- *Change of mission* – Military vessels of the same class may be equipped to deal with different threats. Commercial vessels may be modified to perform different functions or carry different cargos.
- *Reduced cost* – Ship owners and builders look for ways to reduce construction costs and reward employees who identify products or methods that result in cost reductions.

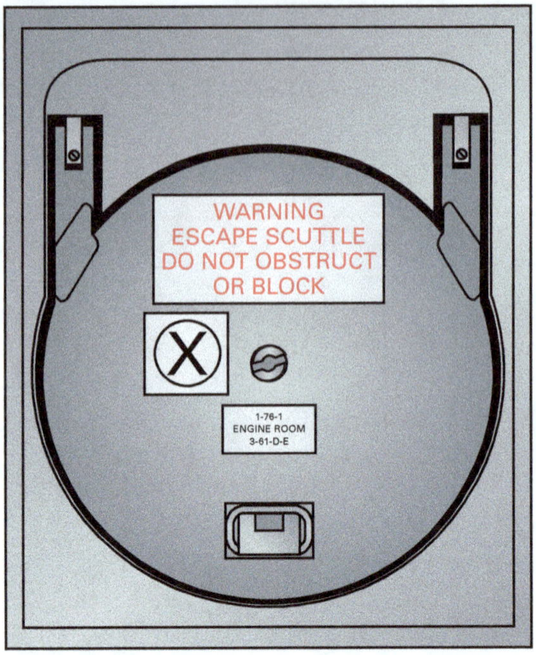

Figure 13 A scuttle.

- *Obsolete parts* – Hardware and equipment manufacturers might stop making certain parts or may even go out of business. In such cases, engineers will have to find substitutes.

3.2.0 Retrofit Kits

Sometimes owners decide that changes made as production cut-ins should also be made to hulls that have already been deployed. Like automobile recalls, the changes may be needed to correct functional or safety-related problems, or it may simply be an operational improvement. When deployed vessels are to be modified, manufacturers prepare **retrofit** kits containing the parts and instructions needed to make the change. If it is an emergency change, it might be done while the ship is at sea or on a port visit. Otherwise, the change generally takes place during the next overhaul period.

Bollinger Shipyards, Inc.

BOLLINGER SHIPYARDS LOCKPORT, LLC
New Construction Division NC/FRM023 Rev. (3) 12/06

ENGINEERING CHANGE NOTICE (ECN)

Hull Number(s)		ECN Number	
Project Number		Date	
Project Engineer		Revision Date	

Reason for ECN	
Authorization	

Document(s) Affected			
Contract Section		Page	
Specification Section		Page	
Functional Drawing Number Name		Rev Date	
Detailed Drawing Number Name		Rev Date	

Description of Revision

Department	Name	Qty	Department	Name	Qty

Figure 14 ECN form.

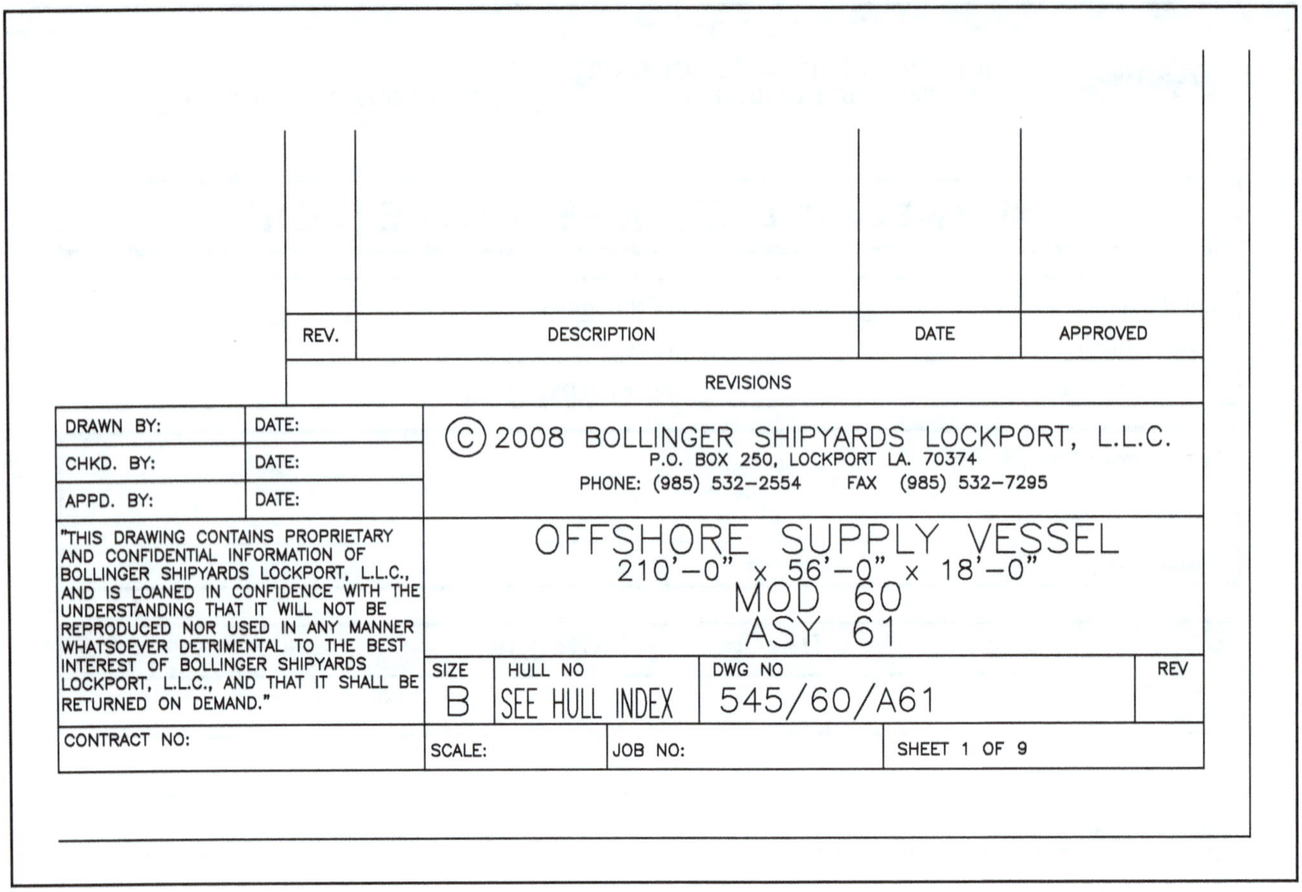

Figure 15 Example of a revision block.

Configuration Management

Configuration management is a science, and many companies employ people who specialize in this science. Why is it necessary?

Let's say that Acme Car Company built 9,000 of their Superba GT model in a given year. Although the transmissions for these vehicles were built to the same specifications, they were actually provided by three different manufacturers, and are therefore not identical. Acme learns from their dealer network that defective parts have been found in transmissions built by Company A, so they decide to do a recall to replace the defective parts. How do they know which of the 3,000 owners among the 9,000 have those transmissions? That's where configuration management comes in; this group maintains records of which vehicle serial numbers have which parts, so they know which vehicles have Company A transmissions. Furthermore, assuming that Company A also has a configuration management system, they might find that only 500 of the 3,000 transmissions they manufactured contain those particular parts. So, instead of having 9,000 owners bring their vehicles to the dealership, and then reworking 3,000 transmissions, only 500 vehicles will need to be recalled.

SUMMARY

On the job, a fitter will be given a package of drawings for a structure, and be directed to assemble that structure using the drawings. Drawings define how parts are assembled and provide details on where to locate parts and assemblies. Therefore, the ability to accurately read drawings is an essential skill for all structural fitters.

Drawings are arranged in a hierarchy, which can be viewed as a pyramid. At the top is the general arrangement (GA) drawing for the entire vessel, which identifies the major modules and decks of the vessel. The drawings for the vessel are commonly divided into systems, mechanical, structural, and habitability. Fitters are involved primarily with structural drawings, although they may also be required to use some propulsion system and anchor handling system drawings.

Each module of the vessel is further divided into units and assemblies. The assemblies will contain both subassemblies and individual piece parts such as castings. Subassemblies such as foundations consist of two or more parts and are generally fabricated in a shop, then sent to an assembly bay where they are incorporated into larger assemblies, or units. The drawing for the next higher assembly will show precisely where the foundation or other assembly is to be positioned.

Drawings are subject to change for a variety of reasons. Changes to drawings are identified on engineering change notices (ECNs) or configuration change requests (CCRs), which are then used to modify the drawings. All changes made to a drawing are recorded on the drawing's revision block.

Review Questions

1. Foundations are generally considered _____.
   a. blocks
   b. units
   c. assemblies
   d. subassemblies

2. Berthing drawings would be included under _____.
   a. system drawings
   b. habitability drawings
   c. mechanical drawings
   d. structural drawings

3. A GA drawing for a vessel is used to _____.
   a. identify all parts
   b. locate major assemblies
   c. describe assembly procedures
   d. allocate work to assembly bays and shops

4. A GA drawing is *not* likely to include _____.
   a. a plan view
   b. an elevation view
   c. a bill of materials
   d. frame numbering

5. Drawing practices are standardized across the industry.
   a. True
   b. False

6. An Engineering Change Proposal (ECP) is used to _____.
   a. inform fitters and CAD operators about drawing changes
   b. obtain management approval for a change
   c. cancel a change previously released
   d. identify a new project engineer

7. When a change is made to a vessel that is already deployed, it is known as a _____.
   a. renovation
   b. remodel
   c. retrofit
   d. renewal

Refer to *Figure A4* in the supplemental drawing package to answer Questions 8 through 10.

8. The approximate weight of assembly 61 is _____.
   a. 5,200 pounds
   b. 10,000 pounds
   c. 23,000 pounds
   d. 97,000 pounds

9. Assembly P15 is part of the _____.
   a. engine girder assembly
   b. tank top
   c. bottom shell
   d. tank top supports

10. How far is frame 76 located from the forward edge of assembly A61?
    a. 9'-11"
    b. 11'-11"
    c. 13'-11"
    d. 15'-11"

Trade Terms Introduced in This Module

Floor: The upright structures used to support the tank top.

Piece part: An individual part.

Retrofit: Incorporation of a change into a deployed system.

Stanchion: An upright member used to support plating.

Additional Resources

This module presents thorough resources for task training. The following resource material is suggested further study.

Ship Construction, D.J. Eyres, G. J. Bruce. Elsevier: Waltham, MA.

Construction of Marine and Offshore Structures, Ben C. Gerwick, Jr. CRC Press: New York, NY.

Figure Credits

Bollinger Shipyards, Figures 4, 5, 8, 11, 14, 15, and Supplemental Drawing Package

NCCER CURRICULA — USER UPDATE

NCCER makes every effort to keep its textbooks up-to-date and free of technical errors. We appreciate your help in this process. If you find an error, a typographical mistake, or an inaccuracy in NCCER's curricula, please fill out this form (or a photocopy), or complete the online form at **www.nccer.org/olf**. Be sure to include the exact module ID number, page number, a detailed description, and your recommended correction. Your input will be brought to the attention of the Authoring Team. Thank you for your assistance.

Instructors – If you have an idea for improving this textbook, or have found that additional materials were necessary to teach this module effectively, please let us know so that we may present your suggestions to the Authoring Team.

NCCER Product Development and Revision

13614 Progress Blvd., Alachua, FL 32615

Email: curriculum@nccer.org
Online: www.nccer.org/olf

❑ Trainee Guide ❑ Lesson Plans ❑ Exam ❑ PowerPoints Other ______________

Craft / Level: Copyright Date:

Module ID Number / Title:

Section Number(s):

Description:

Recommended Correction:

Your Name:

Address:

Email: Phone:

86302-15

Fitting III

Module Two

Trainees with successful module completions may be eligible for credentialing through the NCCER Registry. To learn more, go to **www.nccer.org** or contact us at 1.888.622.3720. Our website has information on the latest product releases and training, as well as online versions of our *Cornerstone* magazine and Pearson's product catalog.

Your feedback is welcome. You may email your comments to **curriculum@nccer.org**, send general comments and inquiries to **info@nccer.org**, or fill in the User Update form at the back of this module.

This information is general in nature and intended for training purposes only. Actual performance of activities described in this manual requires compliance with all applicable operating, service, maintenance, and safety procedures under the direction of qualified personnel. References in this manual to patented or proprietary devices do not constitute a recommendation of their use.

86302-15
Fitting III

Objectives

When you have completed this module, you will be able to do the following:

1. Describe tools and methods use to lay out mold lines, radius corners, and penetrations.
2. Describe proper fit procedures related to fitting tasks in all positions:
   - Longitudinals
   - Shell plating
   - Bulkheads/deck plates
   - Frames
   - Foundations
   - Miscellaneous structural items such as chocks, headers, collars, brackets, and clips
   - Watertight and non-watertight doors and hatches
3. Describe how to assemble, erect, and set a modular unit.

Performance Tasks

Under the supervision of your instructor, you should be able to do the following:

1. Lay out a door opening.
2. Lay out bulkhead and deck penetrations.
3. Lay out, build, and install a simple foundation.
4. Install doors and hatches.
5. Complete an inspection checklist using a structural drawing and common fitting practices.

Trade Terms

Camber
Mold line
Retrofit
Scribe
Set marks
Shell plates
Stiffeners
Stringers
Work line

Industry Recognized Credentials

If you are training through an NCCER-accredited sponsor, you may be eligible for credentials from NCCER's Registry. The ID number for this module is 86302-15. Note that this module may have been used in other NCCER curricula and may apply to other level completions. Contact NCCER's Registry at 888.622.3720 or go to **www.nccer.org** for more information.

Contents

Topics to be presented in this module include:

Figures and Tables

1.0.0 INTRODUCTION

In Levels One and Two of this program, trainees were introduced to the basic tasks performed by ship fitters, including tack welding, flame cutting, print reading, fitting tools, and basic fitting methods. This module covers layout, assembly, installation, and alignment of decks, shell plates, foundations, and other structural members using the skills learned in the preceding levels.

1.1.0 Safety Review

In order to ensure your own safety as well as that of others around you, it is essential that you remain constantly aware of safety hazards, and practice established safety procedures. Also, be aware that the environment in which you are working may be continually changing, so an environment that is safe at one moment can be hazardous in the next. The following is a review of basic safety practices:

- Wear required PPE, including work gloves, at all times.
- Wear face protection when chipping or grinding.
- Use fall protection when working at heights.
- Wear the required eye protection when cutting or welding, or when working near arc welding.
- Know the procedures to follow in case of fire or medical emergency.
- Know the locations of fire extinguishers and how to use them on incipient fires.
- Consult the applicable SDS for hazards and safety procedures before using chemicals.
- Use proper rigging equipment and lifting procedures.
- Be alert for moving cranes and for loads being moved by cranes.
- Do not enter a confined space unless it has been cleared and all required protections are in place.

2.0.0 LAYOUT

The ability to measure accurately and to lay out precise locations for structural elements and subassemblies are critical skills for structural fitters. Layout involves the use of tools and instruments to measure and mark lines, circles, arcs, bolt patterns, and equipment locations.

2.1.0 Layout Tools

Common tools used to perform basic layout are listed here. Each of these tools is described in the paragraphs that follow.

- Steel rules
- Tape measures
- Folding rules
- Steel squares
- Combination sets
- Dividers
- Punches
- Chalk lines

2.1.1 Steel Rules

A steel rule is a flat measuring tool made of tempered steel that has graduations down to 1⁄64 of an inch. Some steel rules are marked on both sides. One side may be marked in eighths and sixteenths of an inch, and the other side in thirty-seconds and sixty-fourths of an inch. They are available in lengths from 12 to 144 inches. The steel rule is used to measure distances and to mark lines. *Figure 1* shows two types of steel rules. The rule on the bottom has metric graduations.

2.1.2 Steel Tape Measures

A steel tape measure is an important tool for measuring distances. Tape measures range in length from 6 to 100 feet and have widths up to 1 inch. A 25-foot tape is the one carried by most fitters. Some tapes have locking mechanisms that allow them to remain drawn. The hook at the end of the tape measure should have a small amount of play so that it can be used to measure both inside and outside dimensions. *Figure 2* shows a tape measure being used to check the position of a bulkhead. It is common for tapes to show graduations in both inches and millimeters.

2.1.3 Folding Rules

Many fitters carry folding rules like the one shown in *Figure 3*. They serve as an easy-to-carry 6' rule with large, easy-to-read numbers. These rules are sometimes used as references for leveling instruments in place of leveling rods.

Figure 1 Steel rules.

Figure 2 Using a tape measure.

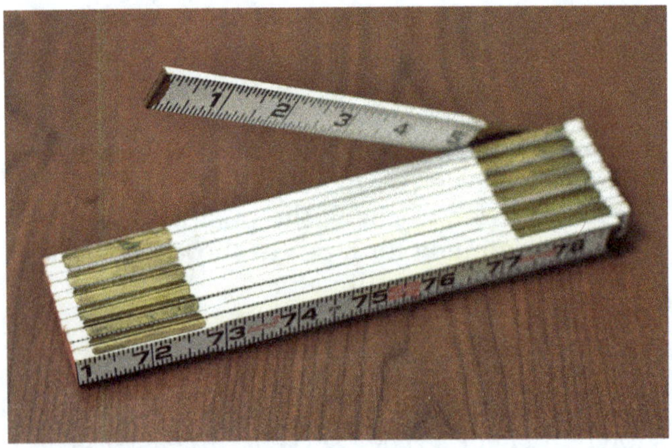

Figure 3 A folding rule.

2.1.4 Steel Squares

The steel square, also called a framing square, is used to square-up large stock and patterns and to lay out and check right angles. It is also used as a rule or straightedge. The steel square has a 24-inch blade and a 16-inch tongue. The blade and the tongue are marked in graduations down to one sixty-fourth of an inch. A machinist try square is similar, but has a 6-inch blade and 4-inch tongue. *Figure 4* shows a steel square and a machinist try square. In *Figure 5* (A), the fitters are using a framing square to square-up a corner. In *Figure 5* (B), it is being used to mark a piece of plate for trimming.

2.1.5 Combination Sets

The combination set (*Figure 6*) consists of a steel rule or blade, a square head, a center head, and a protractor head. The square head contains a spirit level, and the protractor head may also contain a spirit level. The combination set can be used as a combination square to lay out 45- and 90-degree angles, draw parallel lines, and measure heights. It

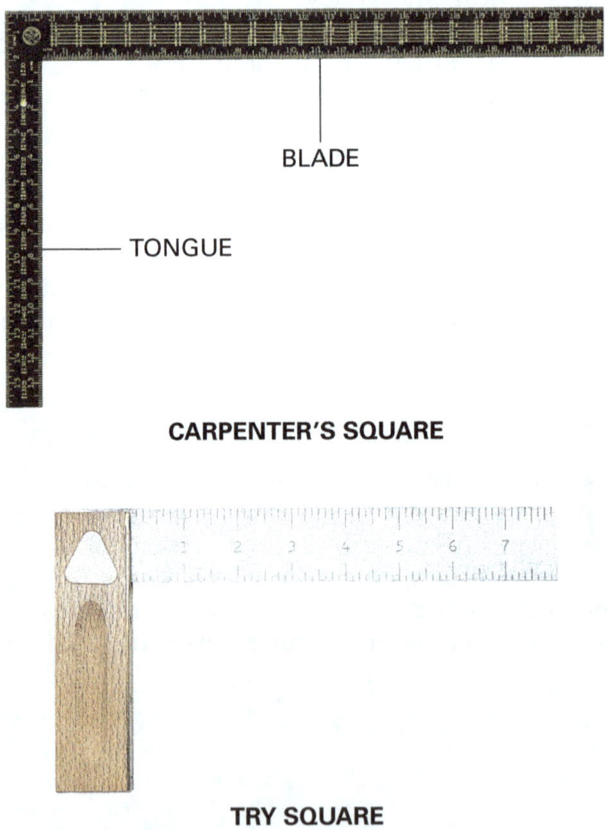

Figure 4 Steel squares.

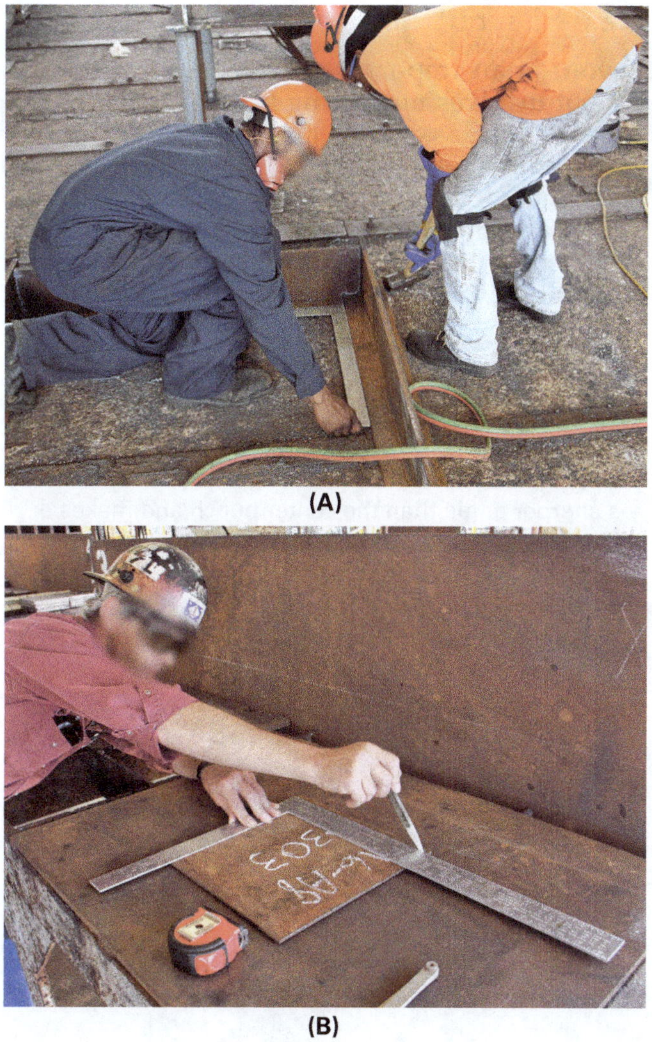

Figure 5 Framing squares in use.

can also be used as a protractor to measure and lay out angles to within 1 degree of accuracy.

The combination set can be used to locate the centers of round or square bars or serve as a steel rule, level, or marking gauge. *Figure 7* shows a fitter using a combination square with a square head to draw a straight line.

The protractor head is used to lay out and measure angles. The protractor is shaped like a half circle with the curved part divided into 180 degrees. The straight bottom edge has a center mark and is usually marked in sixteenths of an inch.

2.1.6 Dividers

Dividers are used to transfer measurements, to compare distances, and to scribe circles and arcs. Hardened steel points, which should be kept sharp, are attached to the two legs. The distance between the points is adjusted by a screw. Some dividers, such as the wing divider shown in *Figure 8*, have a holder for chalk, allowing them to be used to mark an arc or circle.

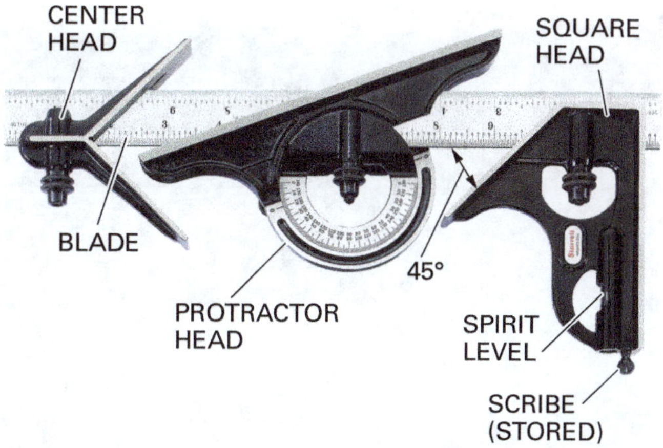

Figure 6 Combination set.

Figure 7 Combination square in use.

2.1.7 Punches

Punches (*Figure 9*) are used to mark points on steel for drilling and marking lines. The prick punch is made of steel and is about 4 to 6 inches long. The end of the punch is ground to a point with an angle of about 60 degrees. It is used with a small hammer to mark centers and to make small indentations along layout lines.

Figure 8 Using dividers to mark the circumference of a penetration.

Figure 9 Punches.

A center punch is similar to a prick punch except that the point is ground at a 90-degree angle. It is used to enlarge center marks made with a prick punch. This is done so that a drill can start easily when drilling holes.

A drift (alignment) punch is used to align holes before bolting or riveting.

2.1.8 Chalk Lines

A chalk line is used to transfer a straight line to a flat workpiece, such as a piece of plate. The chalk line case contains a reel of string, and is filled with colored chalk which coats the string. When the string is pulled taut and snapped onto the workpiece, it leaves a straight colored line that can be used for cutting or alignment (*Figure 10*). Blue chalk is commonly used; it is not water-resistant or permanent. The end of the string has a hook that can be used to attach it to one end of the workpiece. This allows a line to be snapped by one person. If there is no hook, a clamp can be used.

2.2.0 Layout Tasks

Layout is an important part of a fitter's work. Layout involves measuring from established references, such as a ship center line, and marking surfaces to show locations of structures. Layout tasks include locating **stiffeners** on decks, shell plates, and bulkheads; placing foundations; and establishing the locations for bulkhead and deck penetrations.

Punches

Although a prick punch and center punch may look similar, there is a difference. The prick punch forms a sharper angle than the center punch and makes a smaller mark. When a hole is to be drilled, the prick punch is first used to mark the location, then the center punch is used to enlarge the mark so that the drill or other tool will not walk away from the mark.

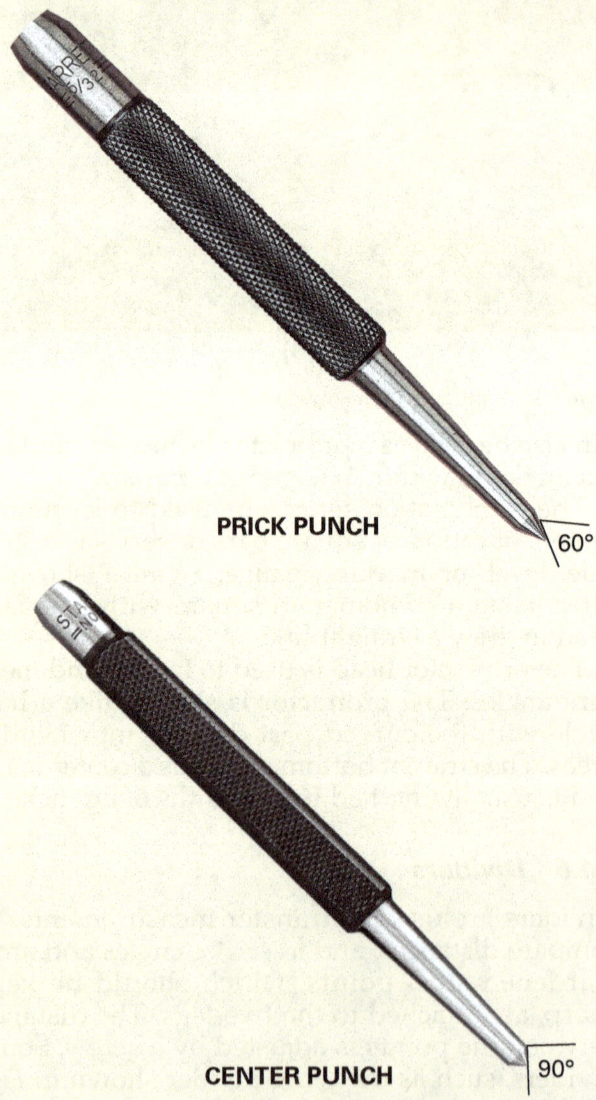

Figure 10 Snapping a line.

2.2.1 Laying Out Mold Lines

Mold lines are used to mark the exact locations of installed structures such as stiffeners and bulkheads. In some cases, mold lines are etched into plates when they are prepared for use. Other times, it will be necessary for the fitter to locate and lay out the mold lines.

When a structure such as a bulkhead is placed, its thickness must be taken into account, so the placement of the structure in relation to a mold line is critical. The following are basic rules that apply to placement of structures:

- For structures in a transverse position, the material "toes" aft when laying out from the bow to the center of the ship. That is, the material would be placed behind the line with its forward edge touching the line. Material toes forward when laying out from stern to center.
- For longitudinal members, all material toes inboard toward the ship's center line, so that the line is always visible when viewed from the center line regardless of whether it is on the port side or the starboard side of the center line.

To make sure that the structures are placed on the correct side of a line, a mark is placed on that side. If the mold lines are laid out by machine, the line takes the shape of a small triangle as shown in *Figure 11* (*A*). When the fitter lays out the lines, three hash marks are used, as shown in *Figure 11* (*B*).

A basic rule to remember is that when looking outboard from the center line in a longitudinally framed vessel, you will always see a mold line. The same applies when looking fore and aft from the midship line.

When laying out mold lines, it is a good idea to start with a work line that can then be used as a reference for placing the mold lines. For example, when laying out longitudinal stiffeners, the ship center line is generally used as the reference. However, if a structural member is installed over the center line, the line will be covered up. A work line, such as the one shown in *Figure 12* , is placed a specific distance, such as 4 inches, from the reference point. It then becomes the reference point for laying out other structures.

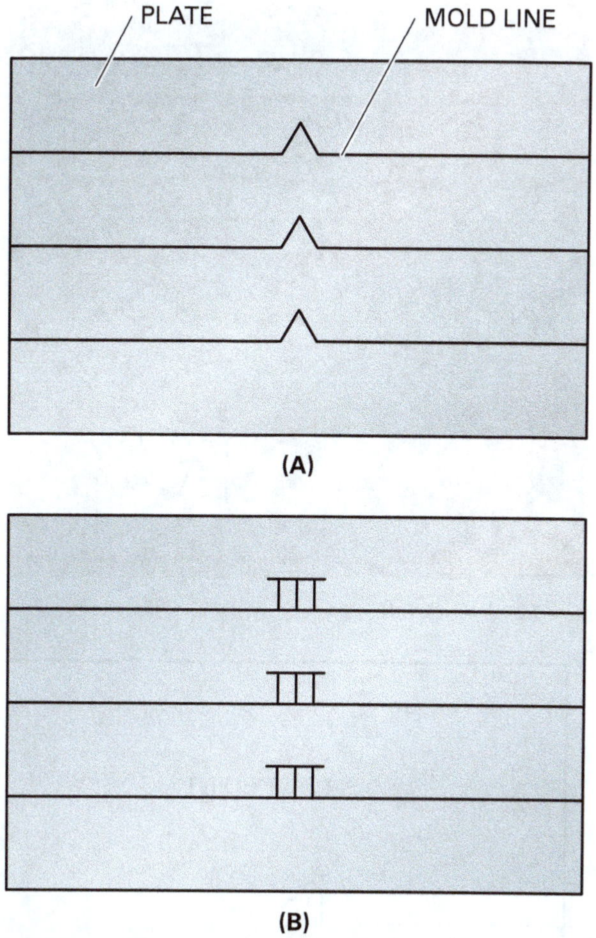

Figure 11 Markings for positioning stiffeners.

Unlike mold lines, the center line stiffener is placed directly on the center line. Therefore, if the stiffener is in place when laying out the work line in relation to the center line, half its thickness must be added to the measurement when placing the work line.

2.2.2 Laying Out Radius Corners

Radius (curved) corners are less likely to rip or be damaged due to ship stresses than square corners. Radius corners can also prevent personal injury in tight areas where people are likely to bump into them. *Figure 13* shows how to lay out a radius corner.

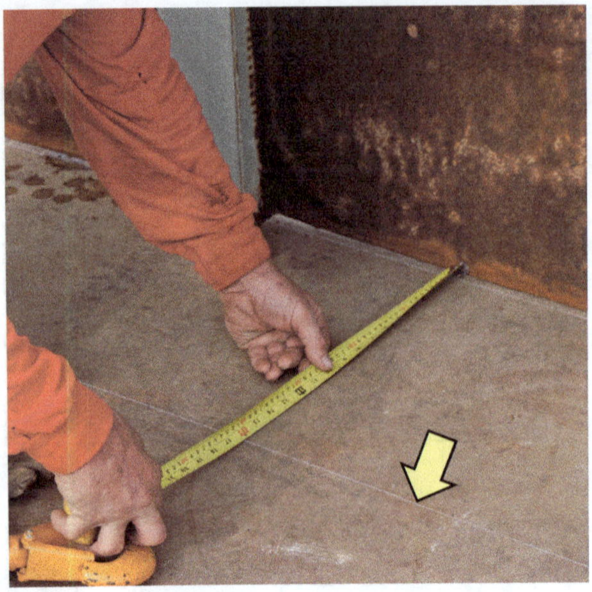

Figure 12 A work line.

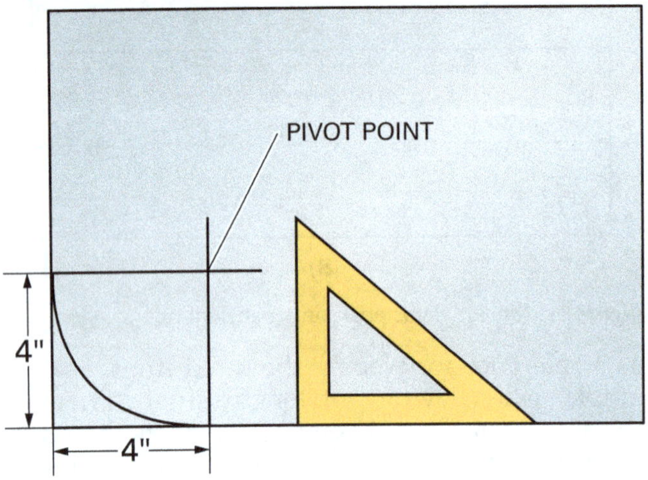

Figure 13 Laying out a radius corner.

The drawing will specify the radius. Start by marking off the required distance and drawing intersecting lines as shown. Center-punch the point where the two lines intersect. Then use dividers with a pen or soapstone to mark the radius.

2.2.3 Laying Out Bulkhead and Deck Penetrations

It is sometimes necessary to create openings in bulkheads and decks to run piping, conduit, and ductwork through bulkheads or from deck to deck (*Figure 14*). In most cases, the openings are precut during fabrication of the plate. In a retrofit situation where a new penetration is to be installed, it will be up to the fitter to make the penetration at the correct location and in the correct size and shape. In such cases, the applicable drawing will show the location and dimensions of the opening, which can be round, oval-shaped, or rectangular.

Figure 14 Examples of bulkhead penetrations.

These penetrations usually have sleeves, known as transits, that support the material passing through the opening and also provide protection from damage. The fitter must fit and install these sleeves so that an equal portion of the sleeve is on either side of the penetration. If the openings penetrate watertight compartments, they will be filled with a sealant once the cables, pipes, etc. are installed in them.

A typical method for laying out and installing penetrations is as follows:

- Locate the center of the new opening using the dimensions and reference points provided on the drawing. Use a prick punch followed by a center punch to mark the center of the opening.
- Locate and mark the outer edges of the opening at several points based on the dimensions provided on the drawing.

- For a circular opening, use a divider and soapstone to mark the circumference. Use a straightedge to mark rectangular openings.
- Use a cutting torch to cut out the opening. Clean the edges of the opening with a grinder.
- Insert the transit in the opening (*Figure 15*). If it does not fit, scribe the edge of the opening and adjust the opening.
- Measure and mark the center of the transit. Divide the plate thickness in half and make reference marks on the transit. If necessary, tack clips at the reference line, to keep the transit from falling through the opening.
- Insert the transit into the opening to the reference line, then tack-weld the transit to the bulkhead.

Figure 15 Transit in a penetration.

2.2.4 Laying Out Door Openings

Although door openings are generally cut out during bulkhead fabrication, it will sometimes be necessary for a fitter to create the opening. A drawing will show the location of the door. *Figure 16* is an example of a drawing for a watertight door.

The following is a general procedure for laying out an opening for a watertight door:

- Using the drawing provided, locate and mark the center point of the door opening, then locate and mark the sill height.
- Using the dimension on the door drawing, mark the edges of the opening with a punch. Snap lines on the top, bottom, and sides to mark the opening for cutting.
- Mark the radius corners of the door frame as required by the drawing.

3.0.0 Fitting Tasks

Ship fitters are involved in every aspect of ship assembly and erection, from preparing the lowest-level subassembly to erecting and joining the major ship modules in the drydock. This section covers some of the common tasks that are performed by fitters.

3.1.0 Installing Stiffening Members

Stiffeners are used to stiffen plating and bulkheads. Stiffeners can be made of angle, T-bar, or channel. In some vessels, stiffeners run fore and aft (longitudinal), while in other vessels, they run athwartship (transverse). The latter are referred to as transverse frames. *Figure 17* shows how longitudinal T-bar stiffeners are attached to the underside of deck plating.

When stiffeners are installed, there must be solid contact between the stiffener and the plate along its entire length. Minor distortions in the plate will sometimes leave gaps between the stiffener and the plate, so stiffeners may need to be forced down to a shell plate or deck plate. This is known as pulling down a stiffener. A wedge and saddle can be used for this purpose, as shown in *Figure 18*. If this method is not adequate, a porta-power and saddle can be used as shown in *Figure 19*. Check the plumb of the stiffener and adjust it, if necessary, using a wedge. Once the stiffener is flush and plumb, tack it on both sides and support it with diagonal braces (*Figure 20*) that will remain until the stiffener is welded.

Trammel Points

Trammel points are used to scribe large circles and arcs. They have two legs with hardened steel points. The legs can be installed on a blade of any length and can be adjusted along the length of the blade. The point on the trammel can be replaced with a marking tool such as a pen.

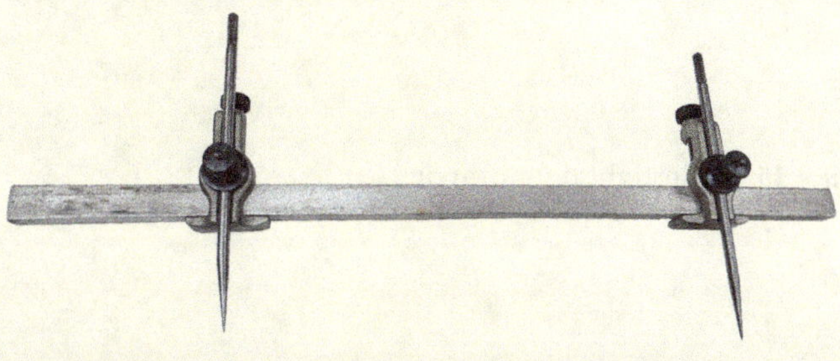

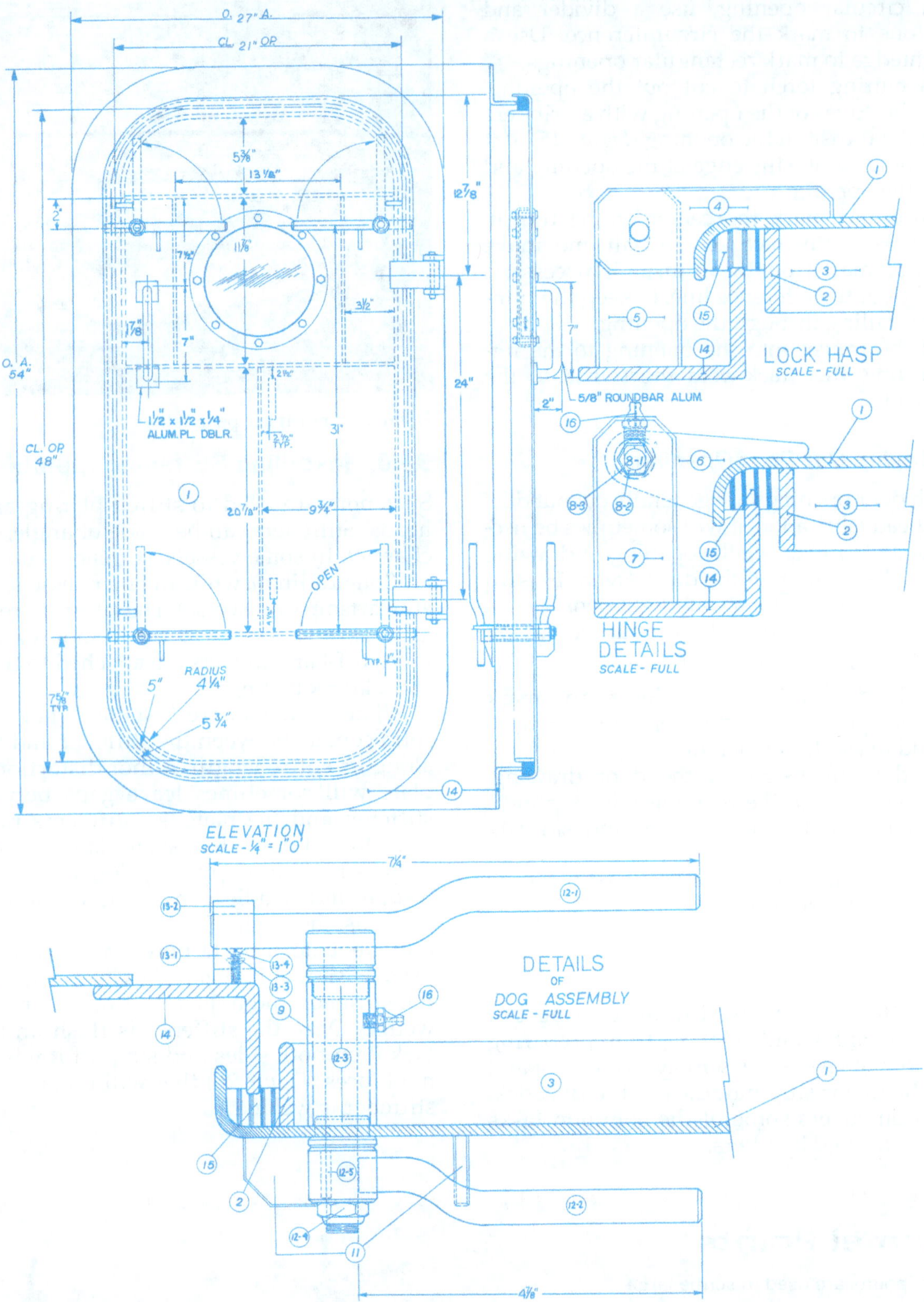

Figure 16 Watertight door drawing.

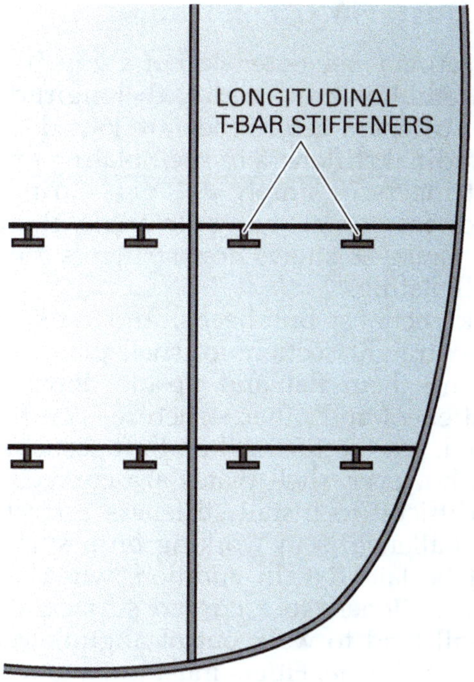

Figure 17 Longitudinal T-bar stiffeners.

Figure 18 Using a saddle and wedge to pull down a stiffener.

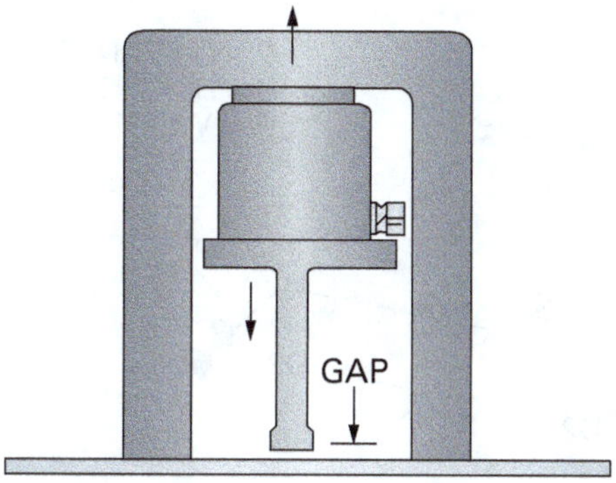

Figure 19 Using a saddle and porta-power to pull down a stiffener.

3.2.0 Fitting a Frame Over Stiffeners

Fitters are often required to lay out and burn cutouts on web plates so that frames can be installed over stiffeners. *Figure 21* shows an example of a frame installed in this manner. Collars are used to provide support in order to keep the web plates from racking (twisting) with the motion of the vessel. The drawing will specify whether watertight or non-watertight collars are to be used. A watertight collar is shown in *Figure 22* (A); a non-watertight collar is shown in *Figure 22* (B).

The following is a general procedure for installing a web plate over stiffeners:

- Measure the height, width, and distance for each longitudinal member, and transfer the measurements to the web plate (*Figure 23*).
- Check the drawing for proper cut-out size based on the type of stiffener (*Figure 24*), then lay out and center-punch cutouts for each of the stiffeners.
- Wearing appropriate PPE, use a torch to cut out each opening. Chip off any slag and clean the area around the cut.

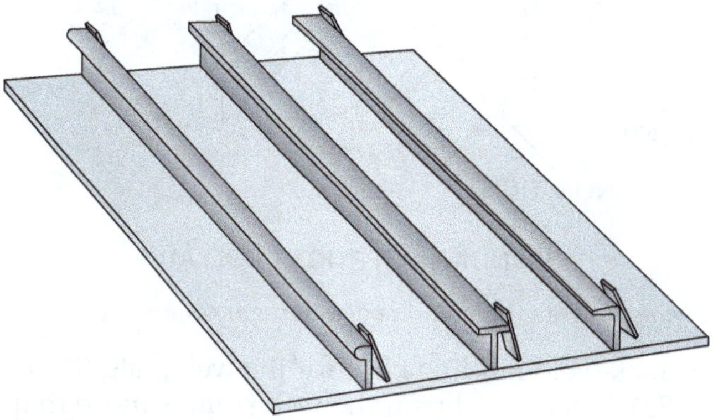

Figure 20 Braced stiffeners.

Figure 21 Frame installed over stiffeners.

(A) WATERTIGHT COLLAR

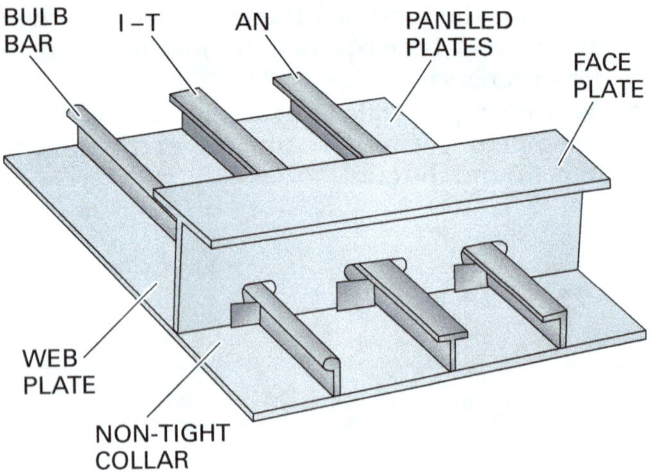

(B) NON-WATERTIGHT COLLAR

Figure 22 Watertight and non-watertight collars.

- Tack-weld landing clips for the web plate (*Figure 25*). Brace the clips if the web plate is more than 4' high.
- Lower the frame assembly into place and fit the web to the plate assembly using a saddle with a wedge or porta-power. Tack-weld the web plate to the deck plate.

3.3.0 Installing Shell Plating

Shell plating is the outer skin of a ship. Shell plating assemblies are fitted with longitudinal or lateral stiffeners before they are joined together. Longitudinal stiffeners in shell plating are called shell stringers, or simply **stringers**. Stringers can be made from angle bar, bulb angle, channel, or plates. *Figure 26* shows how stringers are placed on shell plating.

Deck sections, bulkheads, and tank tops, as well as straight sections of shell plate, are built by laying them flat and upside down so that the stiffeners and other structures can be easily installed. On the forward and aft sections of the vessel, however, shell plates are curved, so it is more difficult to install stiffeners correctly. The main challenge is in working on a surface that cannot be laid flat. In addition, when applying straight stiffeners to a curved surface, the stiffener will tend to walk out of alignment as it is faired to the plate. Fitters must learn to recognize how the stiffeners will react as they are faired to the shell plating and make necessary adjustments as they progress along the length of the stiffener.

Fairing-up curved shell plate panels can also be more difficult than fairing-up deck plate panels. Attachments such as eye lugs are attached to the shell plate panels so they can be pulled together with come-alongs or steamboat jacks. Then, the fitters must work along the lengths of the seams and plate butts using saddles and wedges or other means of fairing-up the shell plates (*Figure 27*) so they can be tack-welded.

3.4.0 Building and Setting a Foundation

Main engines, boilers, pumps, fans, generators, and other items of heavy equipment are mounted on foundations which carry and distribute the load. The equipment is bolted to the foundation, as shown in *Figure 28*, so that it can be removed if necessary. Heavy spare parts such as pistons,

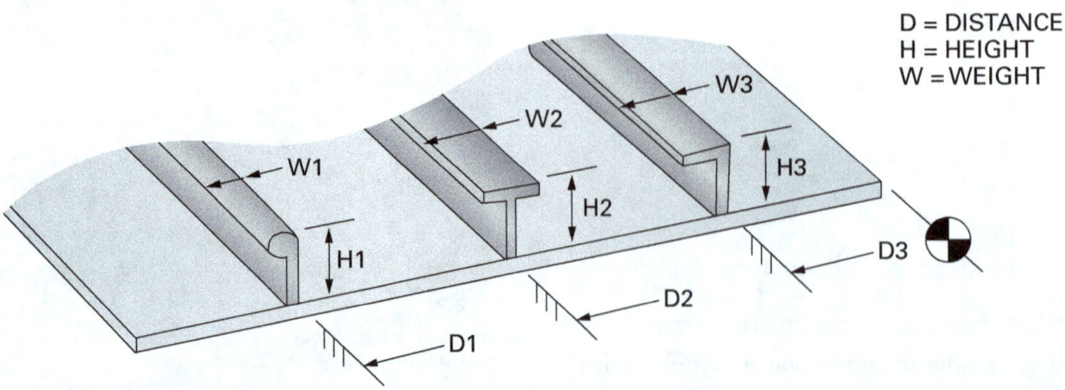

Figure 23 Web plate layout.

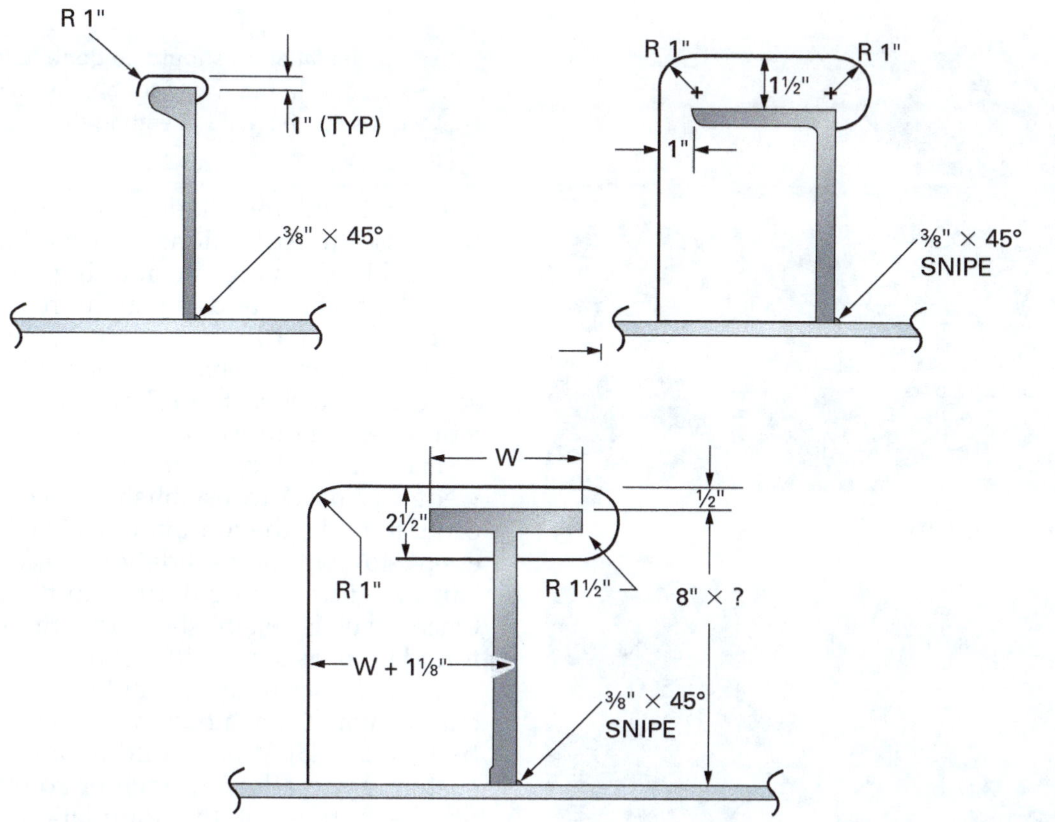

Figure 24 Cut-outs for stiffeners.

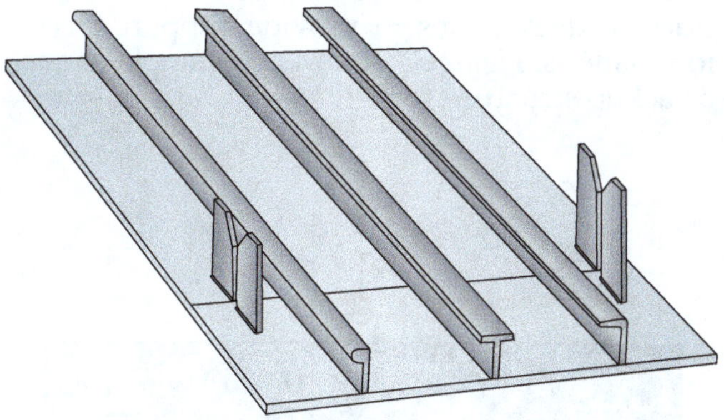

Figure 25 Landing clips in place.

propellers, and tailshafts are also fastened to foundations known as stowages. Foundations are typically made using steel plate, flat bar, and angle. Fitters have the responsibility to fabricate the foundations, lay out their locations, and install them.

Foundations are generally fabricated in the shop based on drawings such as the one shown in *Figure 29*, and are then sent to an assembly bay for installation.

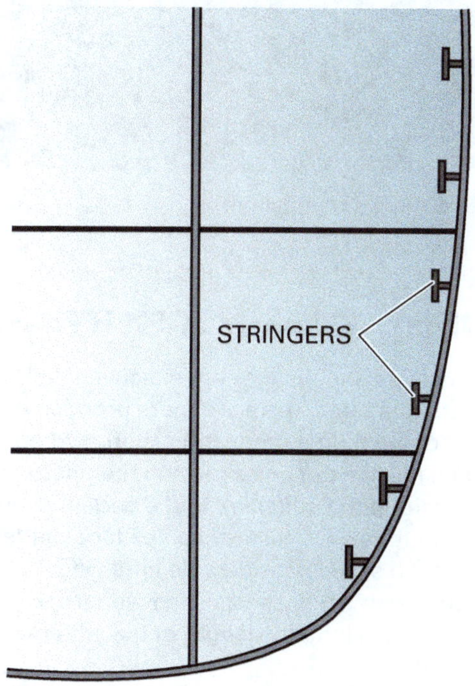

Figure 26 Shell plating with stringers.

Figure 27 Fairing-up shell plates.

Figure 28 Setting a foundation.

NOTE

The installation should be done before the vessel is in the water so that levels and plumb bobs can be used in setting them.

Layout and installation of the foundation on the vessel must be done in accordance with the applicable drawing. Positioning of foundations must be precise, and is usually referenced to the base line, center line, and a frame line of the vessel. Most foundations are mounted on decks, but some are attached to bulkheads. *Figure 30* shows a bulkhead-mounted foundation.

The foundation must be level. In *Figure 31*, a wedge is used to establish the correct level. In order to make the foundation level, it may be necessary to trim the foundation to account for deck camber and/or irregularities in the deck surface. Once level is established, the fitter scribes the foundation, as shown in *Figure 32*.

Once the scribing is complete, a torch is used to cut off material as necessary to level the foundation (*Figure 33*). When building a foundation, it is customary to allow an inch or so of extra length on the bottom of the foundation to allow for leveling.

Backup structures are attached to the undersides of deck plates to provide support for the foundations. *Figure 34* shows an example of such a backup structure.

Pulling Down a Stiffener

While saddles and wedges are effective in fairing stiffeners to a plate, there are more productive ways of doing it. In a production shop, hydraulic mechanisms like the one shown in the photo are used to fair-up the stiffeners to the deck plate and hold them in place until they can be tack-welded. This method saves considerable time when compared with using saddles and wedges at several points along the length of the stiffener.

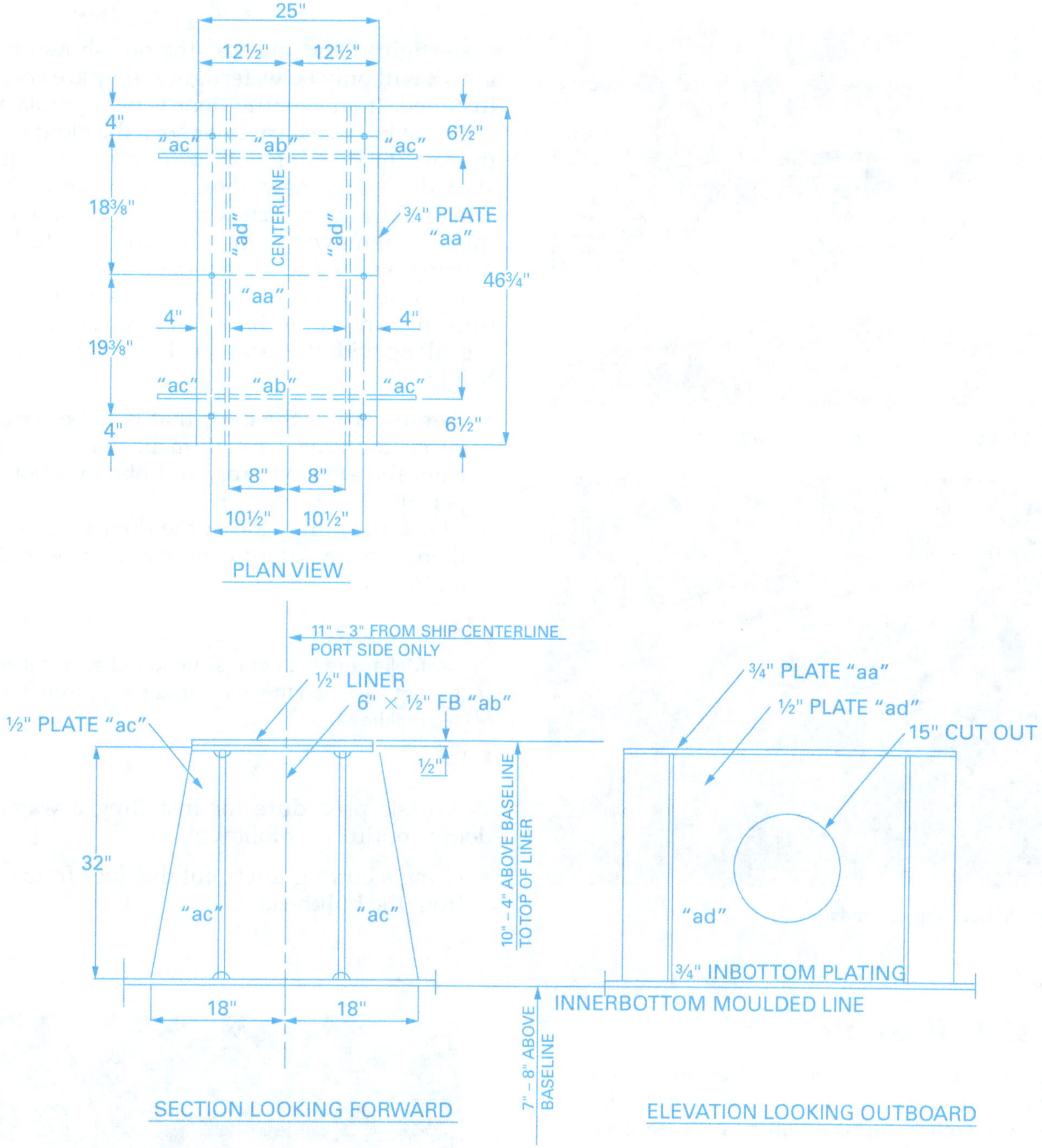

Figure 29 Foundation drawing.

Repetitive Tasks

When a number of identical foundations are needed, it is a good practice to make a template out of wood, strawboard, or red paper. This will make the process much faster and more efficient.

Figure 30 Bulkhead-mounted foundation.

Figure 31 Leveling a foundation.

3.5.0 Installing a Watertight Door

Watertight doors, such as the one shown in *Figure 35*, will only be watertight if they are correctly installed. If an existing door is being replaced, it will first be necessary to remove the old door and prepare the opening and surrounding area for installation of the new door. In such cases, the old door and frame, which is welded to the bulkhead, must be removed using a cutting torch. Before starting, verify the correct door and swing (right hand or left hand). For a new installation, such information should be written next to the opening, along with the location, door number, and sill height.

- Before cutting the door that is to be removed, check the area for flammables, remove paint from the affected area, and obtain a hot work permit.
- Check the blind side of the door to make sure there are no safety concerns that need to be addressed.

WARNING

Make sure the door is supported so that it will not fall when the frame is cut away from the bulkhead.

A basic procedure for installing a watertight door is outlined as follows:

- Using a cutting torch, cut the door frame away from the bulkhead.

Fishtail Jack

Many fitters prefer to use a fishtail jack like the one shown here instead of a saddle and wedge to align plates and other structural elements. A special tool is used to burn a stud onto one of the plates. One side of the fishtail jack fits over the stud and the other side has a screw mechanism that is rotated to align the two elements.

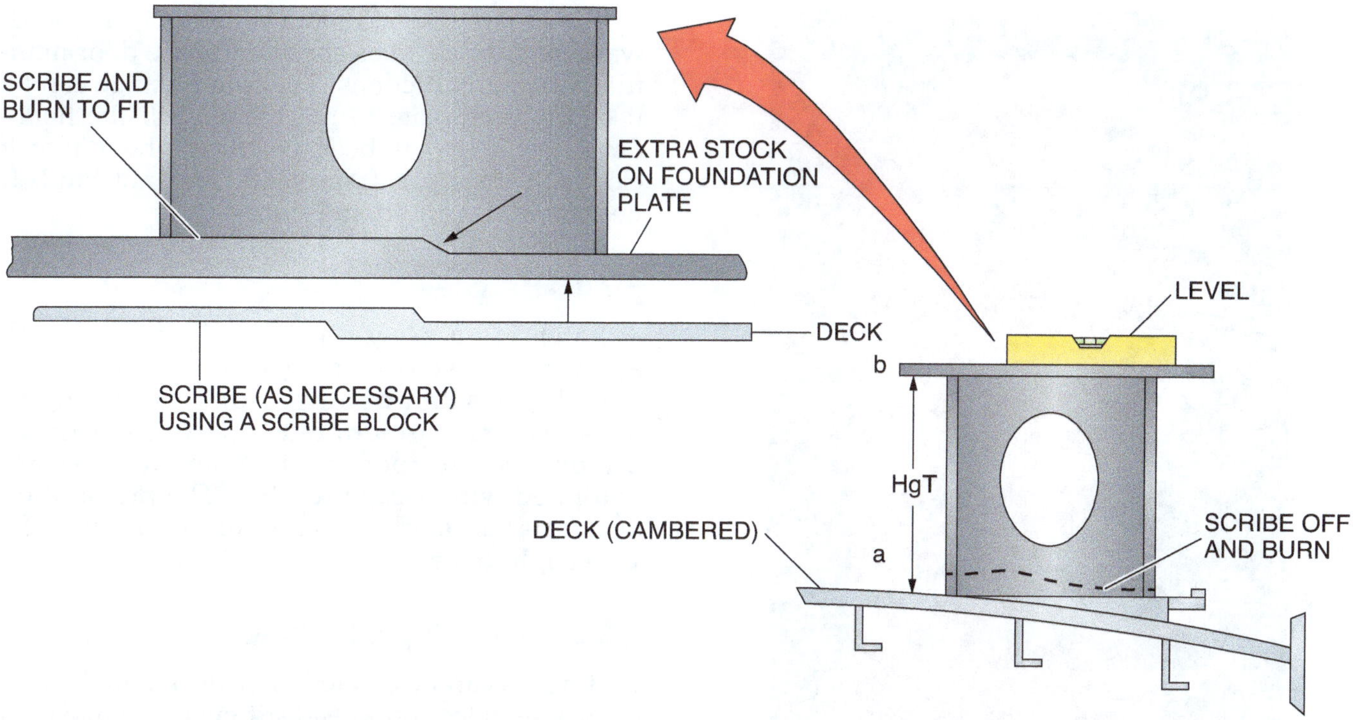

Figure 32 Scribing a foundation.

Figure 33 Trimming a foundation to make it level.

Figure 34 Foundation backup structure.

- Grind the bulkhead area around the frame to remove any residue from the previous weld.
- Inspect the bulkhead to make sure it is not twisted or warped. Applying intense heat can cause steel to warp and make it difficult to obtain a watertight seal. If necessary, install a strongback to straighten the bulkhead. Use shims to fill in any openings less than $\frac{1}{16}$" where the door frame is not flush to the bulkhead.
- If the bulkhead is bowed, make a relief cut, install a strongback, then re-weld the cut.
- When the opening and surrounding area have been prepared, use a come-along to align the door with the opening, ensuring that there is a $1\frac{1}{4}$" lap all around the frame.

Figure 35 Watertight door.

- Once the door is aligned in the opening, temporarily secure it to the bulkhead with C-clamps.
- Fair the frame to the bulkhead and tack-weld the frame using 1" tacks every 12 inches. Stagger the tacks side-to-side and corner-to-corner to prevent warping.
- Install reinforcing chocks as required by the drawing (*Figure 36*).

3.6.0 Joiner Doors

Joiner doors are non-watertight doors typically used as access and egress to compartments. *Figure 37* shows examples of joiner doors. Joiner doors are commonly installed using bolts or rivets. Before starting the installation, check the door frame to make sure the sides are vertical and the head and sill are completely horizontal. A string test can be used to check the trueness of the frame. This involves taping strings diagonally across the opening so that they cross at the center of the opening. If the frame is true, the strings will touch in the center. If they do not touch, the frame should be shimmed to compensate.

If bolts are used to mount the door, the hardware must be sized as specified in the door manufacturer's instructions. The frame flange can be used as a template to mark and drill the holes. Once the door has been installed, check that it closes flush to the frame and does not bind. If necessary, use additional shims.

3.7.0 Hatches

A hatch is an access opening in a deck with a spring-assisted closing mechanism (*Figure 38*). Hatches can be raised or flush with the deck. The flush-mounted type is used in areas exposed to foot traffic. Some hatches are equipped with escape scuttles. The preparation and installation of a hatch is similar to that of a watertight door.

3.8.0 Installing a Bulkhead

Bulkheads can be erected once deck panels have been assembled. Bulkhead stiffeners are installed before the bulkheads are erected. The bulkheads must be placed precisely using mold lines as a guide and must be faired-up to the deck plate. A

Figure 36 Chocks reinforcing a watertight door.

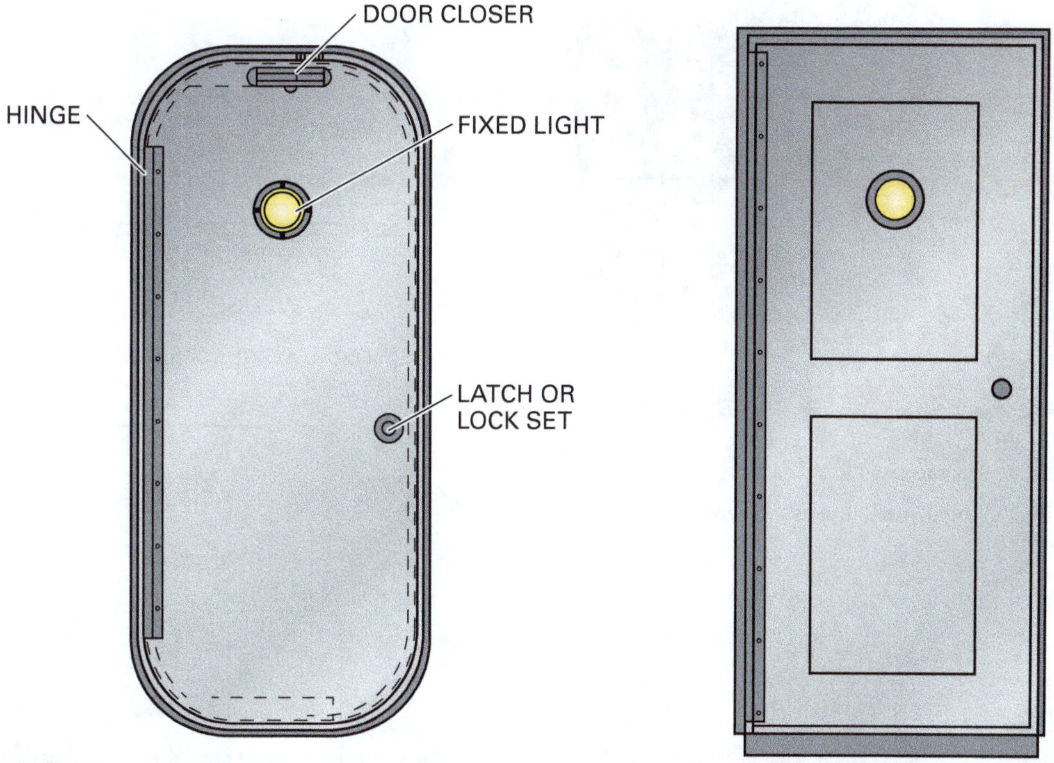

Figure 37 Joiner doors.

Figure 38 Example of a watertight hatch.

chain come-along can be used to close the gap between the deck and the bulkhead.

The following procedure is typical of a bulkhead installation:

- Before moving the bulkhead to the mold line, measure and snap a work line that can be used to verify correct placement of the bulkhead (*Figure 39*).
- Tack-weld landing clips to the deck plate to support the bulkhead during installation, then lower the bulkhead into position.
- Bring the bulkhead to plumb and tack-weld diagonal braces in place.
- Install a pad eye and tack-weld it, making sure to wrap the tack-weld around the far edge of the pad eye.
- Use the pad eye along with a plate dog and chain hoist to fair the bottom edge of the bulkhead to the deck. Once that is done, check that the bulkhead is plumb (*Figure 40*), verify the correct distance from the work line, and tack-weld the bulkhead to the deck.

Figure 39 A fitter using a work line as a reference.

Figure 40 Checking plumb.

3.9.0 Pulling a Sheerstrake to a Deck

A sheerstrake is the top shell plate, or the uppermost section of shell plating attached to the strength deck (main deck). When pulling a sheerstrake to the deck for welding, a special dog can be used (*Figure 41*).

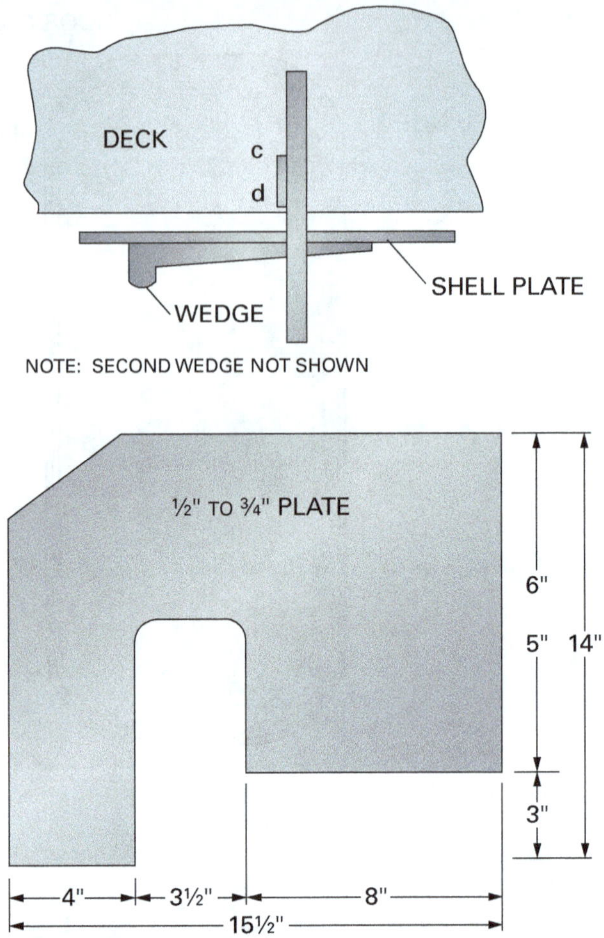

Figure 41 Dog used to pull a sheerstrake to the deck.

- The dog is welded to the deck plate from "c" to "d". In normal circumstances, a ¼" fillet weld about 2½" long will be sufficient. A stronger weld may be needed for very heavy plate.
- In *Figure 42*, the wedge placed at "a" is used to pull the shell plate in. This wedge must be level with the deck; otherwise, it will not be in line with the weld and the dog will tend to tear loose.
- Wedge "b" in *Figure 42* is used to raise the deck plate if required to put it on its **set marks**, which are marks made with a punch to show the final position of the deck plate in relation to the sheerstrake. Both wedges are driven from the side of the dog on which the weld is placed. If the deck must be pushed downward to meet its set marks, weld a small lug to the shell plate, as shown in *Figure 43*. The weld should be on the side of the lug from which the wedge is to be driven.

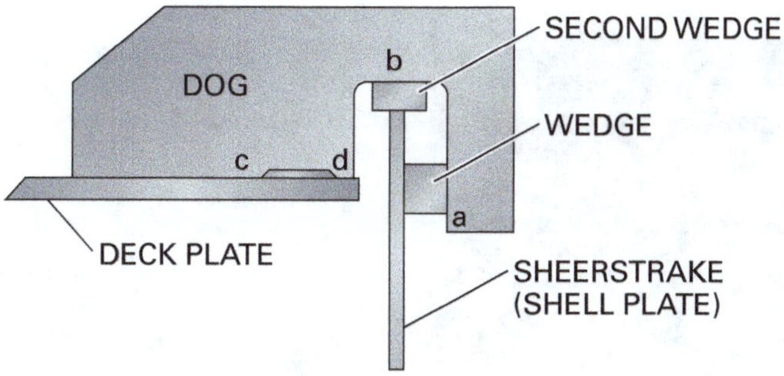

Figure 42 Positioning the wedges.

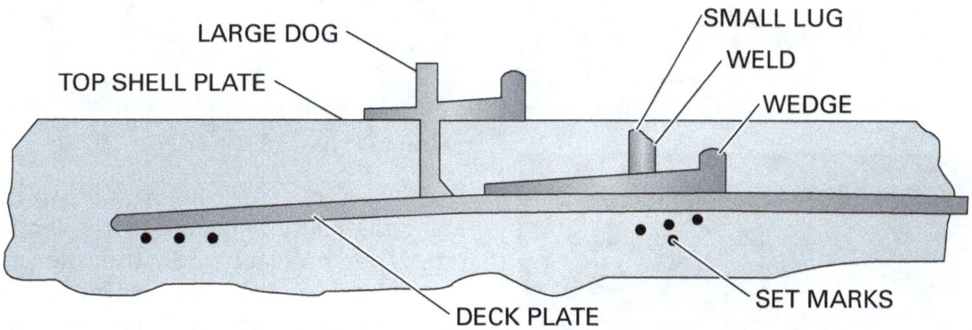

Figure 43 Wedging a deck down to its set marks.

4.0.0 Ship Construction

> **NOTE**
>
> Keep in mind that different shipyards use different terms to identify components of a vessel. For example, some yards use the term *modules* to identify the major sections of a ship that are erected in the drydock, while others refer to them as *blocks*. For this section, the following terms will be used (in order from smallest to largest):
>
> - Parts
> - Subassemblies
> - Assemblies (a collection of subassemblies)
> - Units (a collection of assemblies)
> - Modules

In a typical shipyard, raw materials such as steel plate, bar stock, and beams are staged at the rear of the yard (*Figure 44*). These materials are used in fabrication shops to fabricate deck, bulkhead, and shell plates, as well as other structural components. In the early stages of construction, a great deal of work involves installing and fairing-up stiffeners on deck plates, shell plates, and bulkheads (*Figure 45*). These subassemblies then become components of larger assemblies that are erected in assembly bays near the drydock (*Figure 46*).

Figure 44 Material staging area.

Figure 45 A deck subassembly under construction.

Figure 46 Assembly bays.

The large modules that are erected in the drydock are fitted with interior components such as machinery, partitions, ductwork, and piping before being moved to the drydock. The modules, or blocks, are then transferred to the drydock where the ship erection is being done (*Figure 47*). In the drydock, the modules are pulled together using steamboat jacks, chain come-alongs, and porta-jacks, and then are faired-up, secured in place, and tack-welded together in preparation for finish welding (*Figure 48*).

Tolerances are extremely tight in this type of construction because the sections need to fit together precisely, and they need to be precisely level and plumb. For example, piping, electrical conduit, and air conditioning duct runs have to match up perfectly because they must be connected from one section to the next. Modular sleeping compartments, complete with bathroom fixtures, closets, and stationary furniture, are commonly installed as a unit. It is the ship fitter's responsibility at every stage of construction to make sure that measurements are precise.

4.1.0 Quality Control

The only way a vessel will come together properly is if every subassembly, assembly, unit, and module is built with precision. If measurements are not precise, or if an assembly is not built level and plumb, it will not fit at the next level of assembly. This means costly rework and schedule slippages. Fitters have responsibility for their own work, and an important aspect of that responsibility is checking their own work. Additional inspections are also performed, but the fitter doing the work has the ultimate responsibility.

Figure 47 Transferring a module to the drydock.

Modular Construction

The breakdown of major modules depends on several factors, including the type of vessel and the facilities available at the erection site. The images below show the installation of a double-bottom module as well as a superstructure module.

Figure 48 Fairing-up modules prior to welding.

Because inspection is so important, the process is documented on inspection checklists. Formal, documented inspections are performed by the fitter following each completed task.

During the entire fitting process, the fitter should be referencing the drawing at every step to ensure the following:

- The materials being used are the right size and thickness.
- Spacing of the materials is correct.
- The direction of the material toe is correct.
- Material thickness is on the correct side of the layout.
- Tacking sequence is correct.
- Material is installed straight and square to the work surface.
- All brackets and chocks are installed.
- All snipes and weep holes are cut.
- Grinding of slag and burrs has been completed.
- Grinding of sharp edges for coating adhesion has been completed when required by specification.

These steps must be performed during the fitting process in strict compliance with the detail drawings. At the completion of the job, the fitter is required to inspect the work and make any required corrections. After that, the supervisor will do an inspection and generate an inspection form to document any defects that were found. This report will describe what type of defect occurred, where it was found, and how long it took to correct. Then the supervisor will release this work to the welders.

After the welding has been completed, the welding supervisor will do an inspection and generate an inspection form, documenting any defects that were found. This report will describe the type of defect, where it was found, and how long it took to correct. The supervisor will then release the work to the quality control department. They will then generate their own inspection sheet and list any defects that were found. These defects will need to be corrected before the customer makes the final inspection. The inspection reports are all controlled documents and are kept on file by the quality assurance department.

Figure 49 shows a typical defects list. It includes the location of the defect, the nature of the defect, and the date. Other general notes can be added to the list to provide additional information.

Quality Inspection

Quality Control Inspector is a career path open to structural fitters, as well as other shipyard crafts. QC inspectors check and verify that assembly and installation have been performed in accordance with the applicable drawings and specifications. These inspectors also check purchased items to verify that they meet the specifications supplied to the manufacturer. Some QC inspectors specialize in welding inspection. They must be able to recognize and identify various types of welding defects and perform a variety of tests to ensure that welds are properly made.

BSI TASK CHECK-OFF LIST

STRUCTURAL FIT-UP (BLN/FRM011) REVISION "G" 7/12

Task Final:	[ ]	**Task Partial:**	[ ]
Originator		**Date**	
Job Number		**Task Number**	
Location		**Controlling Document**	
Item/Task Description			

Section 1

ENSURE THAT THE FIT-UP IS IN ACCORDANCE WITH THE REFERENCED DRAWING.

PASS/FAIL:	**SAT**	**UNSAT**
1) DEFECT FREE MATERIALS		
2) PROPER MATERIALS		
3) PREPARED EDGES (NO SHARP OR JAGGED)		
4) PROPER FIT-UP		
5) DIMENSIONS WITHIN TOLERANCE		
6) PROPER TACK SIZE AND SEQUENCE		
7) GOUGE REPAIR		

TEST EQUIP. NO. ____________

NOUN NAME ____________

CAL. EXP. DATE ____________

EXPLAIN UNSAT BLOCKS:

CORRECTIVE ACTION:

Total Rework Man-Hours: ____________

Section 2

ITEM RELEASED:

RELEASED BY:		**Date**	
CRAFT / SUPERVISOR			
RELEASED TO:		**Date**	
CRAFT / SUPERVISOR			

Section 3

Figure 49 Inspection defects list.

SUMMARY

Fitters are involved in every aspect of ship construction, from the fabrication and placement of deck plates and bulkheads to final joining of the ship modules in the drydock. A fitter must be able to make precise measurements because even a tiny error in a subassembly can have a large impact on downstream assembly work. Every structure that a fitter builds must be properly laid out, level, and plumb before it is welded.

Every deck plate, shell plate, and bulkhead has longitudinal or lateral stiffeners, and sometimes both. Among the fitter's most important tasks is the layout and installation of these stiffeners using various tools and methods to make them flush to their mounting surface. Spirit levels and plumb bobs can be used to check smaller assemblies, but surveying instruments are needed for larger assemblies. The fitter must therefore be able use such leveling instruments to establish elevations and check the level and plumb.

Among the other tasks performed by fitters is the layout of watertight doors, joiner doors, hatches, and scuttles. Fitters must be able to read prints, make accurate measurements, and install components and structures as specified by the drawings. Each fitter is responsible for the quality of his or her work.

Review Questions

1. Information regarding hazards and safety procedures related to chemicals are found in the _____.
   a. employee handbook
   b. SDS
   c. drawings
   d. engineering specs

Review Question Figure 1

2. The tool shown in *Review Question Figure 1* is a _____.
   a. try square
   b. framing square
   c. protractor
   d. combination square

Review Question Figure 2

3. The tool shown in *Review Question Figure 2* is a(n) _____.
   a. center punch
   b. prick punch
   c. alignment punch
   d. trammel point

4. When looking outboard from the center line in a vessel laid out for longitudinal framing, you should see mold lines.
   a. True
   b. False

5. When laying out an opening for a watertight door, the fitter should start by locating and marking the _____.
   a. locations of the hinges
   b. center of the opening and the height of the sill
   c. top of the opening
   d. locations of the dogs

6. Stiffeners that run athwartship are called _____.
   a. longitudinal stiffeners
   b. L-brackets
   c. transverse frames
   d. web plates

7. Stringers are the _____.
   a. longitudinal stiffeners in shell plating
   b. tools used to mark straight lines
   c. transverse frames used to stiffen shell plating
   d. the uppermost section of shell plating

8. Joiner doors are installed using _____.
   a. C-clamps
   b. welding
   c. bolts or rivets
   d. contact cement

9. Of subassemblies, modules, assemblies, and units, the largest is generally the _____.
   a. subassembly
   b. assembly
   c. unit
   d. module

10. The responsibility for the initial inspection of a fitter's work lies with the _____.
    a. supervisor
    b. quality control inspector
    c. welding inspector
    d. fitter

Trade Terms Introduced in This Module

Camber: The rise of the crown of the main deck from side to side (athwartship).

Mold line: Lines showing the exact locations of installed structures.

Retrofit: Modification of an existing device or structure.

Scribe: To trace the contour of a surface with a pointed instrument.

Set marks: Groups of center-punch marks used to show exactly where the edge of a plate should be set.

Shell plates: The outer hull plating of a ship.

Stiffeners: An angle bar, T-bar, or channel used to stiffen the plating of a bulkhead, deck, or shell plate.

Stringers: A fore and aft (front to back) member used to provide longitudinal strength.

Work line: A line created by a fitter to serve as a reference line when laying out structures.

Additional Resources

This module presents thorough resources for task training. The following reference material is suggested for further study.

Ship Construction. D.J. Eyres, G. J. Bruce. Elsevier: Waltham, MA.

Construction of Marine and Offshore Structures. Ben C. Gerwick, Jr. CRC Press: New York, NY.

Figure Credits

Courtesy of Huntington Ingalls Industries Communications, Module opener, Figure 47

© Stanley Black & Decker, Figures 4, 9, Review Questions Figures 1, 2

The LS Starrett Company, Figure 6, SA01

Bollinger Shipyards, Figures 16, 29

Topaz Publications, Inc., SA02

Slingmax®, Inc., SA05, SA06

Courtesy of Tecnico Corporation, Figure 49

NCCER CURRICULA — USER UPDATE

NCCER makes every effort to keep its textbooks up-to-date and free of technical errors. We appreciate your help in this process. If you find an error, a typographical mistake, or an inaccuracy in NCCER's curricula, please fill out this form (or a photocopy), or complete the online form at **www.nccer.org/olf**. Be sure to include the exact module ID number, page number, a detailed description, and your recommended correction. Your input will be brought to the attention of the Authoring Team. Thank you for your assistance.

Instructors – If you have an idea for improving this textbook, or have found that additional materials were necessary to teach this module effectively, please let us know so that we may present your suggestions to the Authoring Team.

NCCER Product Development and Revision

13614 Progress Blvd., Alachua, FL 32615

Email: curriculum@nccer.org
Online: www.nccer.org/olf

❑ Trainee Guide ❑ Lesson Plans ❑ Exam ❑ PowerPoints Other ____________

Craft / Level: Copyright Date:

Module ID Number / Title:

Section Number(s):

Description:

Recommended Correction:

Your Name:

Address:

Email: Phone:

GMAW and FCAW: Equipment and Filler Metals

29205-09

Trainees with successful module completions may be eligible for credentialing through the NCCER Registry. To learn more, go to **www.nccer.org** or contact us at **1.888.622.3720.** Our website has information on the latest product releases and training, as well as online versions of our *Cornerstone* magazine and Pearson's product catalog.

Your feedback is welcome. You may email your comments to **curriculum@nccer.org**, send general comments and inquiries to **info@nccer.org**, or fill in the User Update form at the back of this module.

This information is general in nature and intended for training purposes only. Actual performance of activities described in this manual requires compliance with all applicable operating, service, maintenance, and safety procedures under the direction of qualified personnel. References in this manual to patented or proprietary devices do not constitute a recommendation of their use.

29205-09
GMAW and FCAW: Equipment and Filler Metals

Objectives

When you have completed this module, you will be able to do the following:

1. Explain gas metal arc welding (GMAW) and flux-cored arc welding (FCAW) safety.
2. Explain the characteristics of welding current and power sources.
3. Identify and explain the use of GMAW and FCAW equipment:
   - Spray transfer
   - Globular transfer
   - Short circuiting
   - Pulse
4. Identify and explain the use of GMAW and FCAW shielding gases and filler metals.
5. Set up GMAW and FCAW equipment and identify tools for weld cleaning.

Trade Terms

Axially
Globular transfer
Pulsed transfer
Short-circuiting transfer
Spray transfer
Stickout

Required Trainee Materials

1. Pencil and paper
2. Appropriate personal protective equipment

Prerequisites

Before you begin this module, it is recommended that you successfully complete *Core Curriculum; Maritime Structural Fitter Level One;* and *Maritime Structural Fitter Level Two.*

Industry Recognized Credentials

If you are training through an NCCER-accredited sponsor, you may be eligible for credentials from NCCER's Registry. The ID number for this module is 29205-09. Note that this module may have been used in other NCCER curricula and may apply to other level completions. Contact NCCER's Registry at 888.622.3720 or go to **www.nccer.org** for more information.

Contents

Topics to be presented in this module include:

Figures and Tables

Figures and Tables (continued)

1.0.0 INTRODUCTION

Gas metal arc welding (GMAW), shown in *Figure 1*, uses an electric arc to melt a consumable wire electrode and fuse it with the base metal. The arc generates an intense heat of 6,000°F to 10,000°F. The arc is stabilized and the molten filler and base metals are protected from oxidation by a shielding gas dispensed from a GMAW gun nozzle. The wire is automatically fed into the weld puddle as it melts. The process usually operates with direct current electrode positive (DCEP).

Flux-cored arc welding (FCAW), shown in *Figure 2*, uses an electric arc to melt a flux-cored (tubular) wire electrode and fuse it with the base metal to form a weld. A wire feeder automatically feeds the flux-cored wire as it is being consumed in the arc. The equipment used for FCAW is basically the same as that used for GMAW. The main difference between the two processes is that GMAW uses solid wire or composite core wire electrodes with a shielding gas, while FCAW uses a tubular flux-cored wire electrode with or without shielding gas.

In the two basic FCAW processes, one uses an external shielding gas while the other is self-shielding. Whether a shielding gas is used is determined by the type of flux-cored wire that is used. When the flux-cored wire does not require an external shielding gas, the process is called self-shielding FCAW or FCAW-S. When an external shielding gas is required, the process is called gas-shielded FCAW or FCAW-G. Equipment specifically designed for self-shielding FCAW does not have provisions for shielding gas (gas connections, solenoids, and gas preflow and postflow timers), and the guns do not have gas nozzles. Equipment used for gas-shielded FCAW welding is the same as that used for GMAW except that the wire feed drive assemblies have to be able to handle the core wire, which is softer and generally larger in diameter than solid wire.

This module provides an overview of the different items of equipment required for gas metal arc

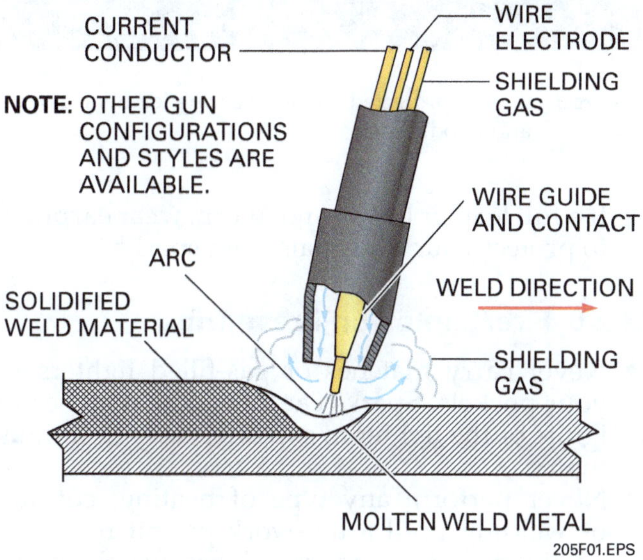

Figure 1 Basic GMAW process.

> **Hot Tip**
>
> **FCAW**
>
> FCAW is presently limited to the welding of ferrous metals. These include low- and medium-carbon steels, some low-alloy steels, cast irons, and some stainless steels. When critical welds are required, the gas-shielded FCAW process is often used because the combination of flux and separate shielding gas produces welds of very high quality.

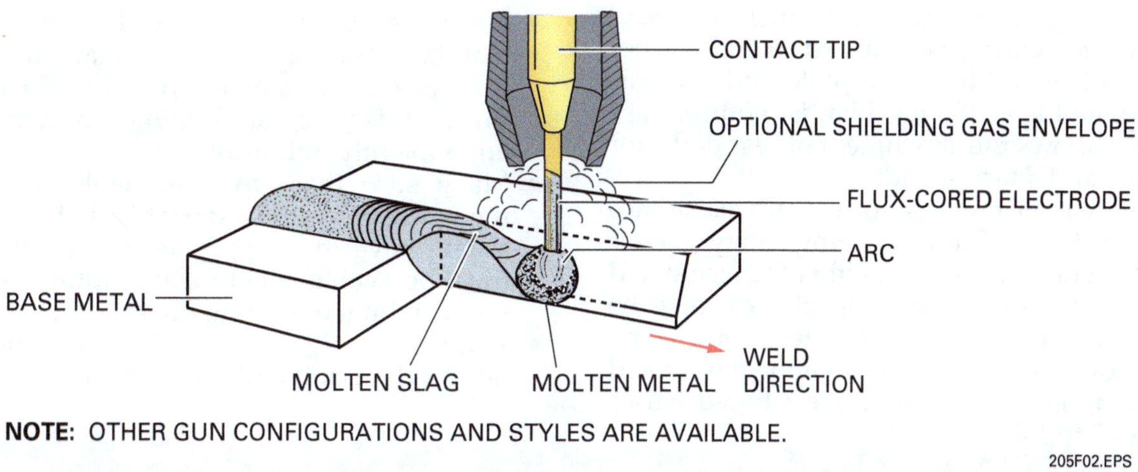

Figure 2 Basic FCAW process.

welding. Topics include safety and equipment, power sources, wire feeders, guns, equipment setup, and filler metals. Upon completion of this module, you will be aware of the safety concerns associated with both GMAW and FCAW, and you will be able to select and set up both GMAW and FCAW equipment safely and efficiently.

2.0.0 Safety Practices

The following is a summary of safety procedures and practices that must be observed while cutting or welding. Keep in mind that this is only a summary. Complete safety coverage is provided in the *Welding Level One* module, *Welding Safety*. If you have not completed that module, do so before continuing. Above all, be sure to wear appropriate protective clothing and equipment when welding or cutting.

2.1.0 Protective Clothing and Equipment

- Always use safety glasses with a full face shield or a helmet. The glasses, face shield, or helmet lens must have the proper light-reducing shade for the type of welding or cutting to be performed. Never directly or indirectly view an electric arc without using a properly tinted lens.
- Wear proper protective leather and/or flame-retardant clothing along with welding gloves that will protect you from flying sparks, molten metal, and heat (*Figure 3*).
- Wear 8-inch or taller high-top safety shoes or boots. Make sure that the tongue and lace area of the footwear will be covered by a pant leg. If the tongue and lace area is exposed or the footwear must be protected from burn marks, wear leather spats under the pants or chaps over the front and top of the footwear.
- Wear a solid material (non-mesh) hat with a bill pointing to the rear or, if much overhead cutting or welding is required, a full leather hood with a welding face plate and the correctly tinted lens. If a hard hat is required, use one that allows the attachment of rear deflector material and a face shield.
- If a full leather hood is not worn, wear a face shield and snug-fitting welding safety glasses for gas welding or cutting. Either the face shield or the lenses of the welding glasses must be an approved shade 5 or 6 filter. For electric arc welding or cutting, wear safety glasses and a welding hood with the correct tinted lenses (shades 5 to 14).

205F03.EPS

Figure 3 A fully protected welder wearing leathers, gloves, and hood.

- If a full leather hood is not worn, wear earplugs to protect your ear canals from sparks.

2.2.0 Fire/Explosion Prevention

- Never carry matches or gas-filled lighters in your pockets. Sparks can cause the matches to ignite or the lighter to explode, causing serious injury.
- Never perform any type of heating, cutting, or welding until a hot-work permit has been obtained and an approved fire watch established. Most work-site fires caused by these types of operations are started by cutting torches.
- Never use oxygen to blow dust or dirt from clothing. The oxygen can remain trapped in the fabric for a time. If a spark hits the clothing during this time, the clothing can burn rapidly and violently out of control.
- Make sure that any flammable material in the work area is moved or shielded by a fire-resistant covering. Approved fire extinguishers must be available before attempting any heating, welding, or cutting operations.
- Always comply with any site requirement for a hot-work permit and/or fire watch.

- Never release a large amount of oxygen or use oxygen as compressed air. Its presence around flammable materials or sparks can cause rapid and uncontrolled combustion. Keep oxygen away from oil, grease, and other petroleum products.
- Never release a large amount of fuel gas, especially acetylene. Methane and propane tend to concentrate in and along low areas and can ignite at a considerable distance from the release point. Acetylene is lighter than air, but is even more dangerous than methane. When mixed with air or oxygen, acetylene will explode at much lower concentrations than any other fuel.
- To prevent fires, maintain a neat and clean work area, and make sure that any metal scrap or slag is cold before disposal.
- Before cutting or welding containers such as tanks or barrels, check to see if they have contained any explosive, hazardous, or flammable materials, including petroleum products, citrus products, or chemicals that decompose into toxic fumes when heated. As a standard practice, always clean and then fill any tanks or barrels with water or purge them with an appropriate purging gas to displace any oxygen.

2.3.0 Work Area Ventilation

- Make sure confined space procedures are followed before conducting any welding or cutting in the confined space.
- Never use oxygen for ventilation in confined spaces.
- Always perform cutting or welding operations in a well-ventilated area. Cutting or welding operations involving zinc or cadmium materials or coatings result in toxic fumes. For long-term cutting or welding of such materials, always wear an approved full-face, supplied-air respirator (SAR) that uses breathing air supplied from outside the work area. For occasional, very short-term exposure, a high-efficiency, particulate arresting (HEPA)-rated or metal-fume filter may be used on a standard respirator.
- Make sure that confined spaces are properly ventilated for cutting or welding purposes.
- Use powered extraction systems when available. *Figure 4* shows a welder working with a portable extraction (vacuum) system designed to remove and filter welding fumes and vapors.

205F04.EPS

Figure 4 Welder working with a portable extraction (vacuum) system.

3.0.0 Characteristics of Welding Current

The current produced by a welding machine has different characteristics from the current flowing through utility power lines. Welding current has low voltage and high amperage, while the power line current has high voltage and low amperage.

3.1.0 Voltage

Voltage is the measure of the electromotive force or pressure that causes current to flow in a circuit. There are two types of voltage associated with welding current: open-circuit voltage and operating voltage. Open-circuit voltage is the voltage present when the machine is on but no welding is being done. There are usually ranges of open-circuit voltages that can be selected, up to about 40 volts. Operating voltage, or arc voltage, is the voltage after the arc has been struck.

Arc voltage is generally slightly lower than the open-circuit voltage. The arc voltage is typically 2 to 3 volts lower than the open-circuit voltage for each 100 amperes of current, but it depends on the range selected.

3.2.0 Amperage

Amperage is the electric current flow in a circuit. The unit of measure for amperage is the ampere (amp).

In welding, the current flows in a closed loop through two welding cables; the ground lead connects the power source to the base metal, and the other cable connects the power source to the wire electrode. During welding, an arc is established between the end of the wire electrode and the work. The arc generates intense heat at 6,000°F to 10,000°F, melting the base metal and the wire electrode to form the weld. The amount of amperage produced by the welding machine determines the intensity of the arc and the amount of heat available to melt the work and the electrode.

4.0.0 Types of Welding Power Sources

Direct current (DC) power sources are usually designed to produce either constant current (CC) or constant voltage (CV) DC. A constant current machine produces a constant current output over a wide voltage range; this is the typical of a SMAW welding machine. A constant voltage machine maintains a constant voltage as the output current varies; this is the typical power source used with GMAW or FCAW machines. The output voltages and output currents of a power source can be plotted on a graph to form a curve. These curves show how the output voltage relates to the output current as either changes (*Figure 5*).

> **Hot Tip**
>
> **Constant Current/ Constant Voltage Power Sources**
>
> Transformer rectifiers can be designed to produce constant current or constant voltage DC. Some will produce either variable or constant voltage by setting a switch. These machines are sometimes referred to as multiprocess power sources, since they can be used for any welding process, including GMAW and FCAW.

Different types of welding machines (power sources) are available for GMAW, including the following:

- Transformer-rectifier power sources
- Engine-driven generator power sources
- Inverter power sources

4.1.0 Transformer-Rectifier Power Sources

The transformer-rectifier power source uses a transformer to convert the primary current to welding current and a rectifier to change the current from AC to DC. Transformer-rectifier power sources can be designed to produce either AC and DC or DC only. Transformer rectifiers that produce both AC and DC are usually lighter duty than those that produce DC only. Transform-

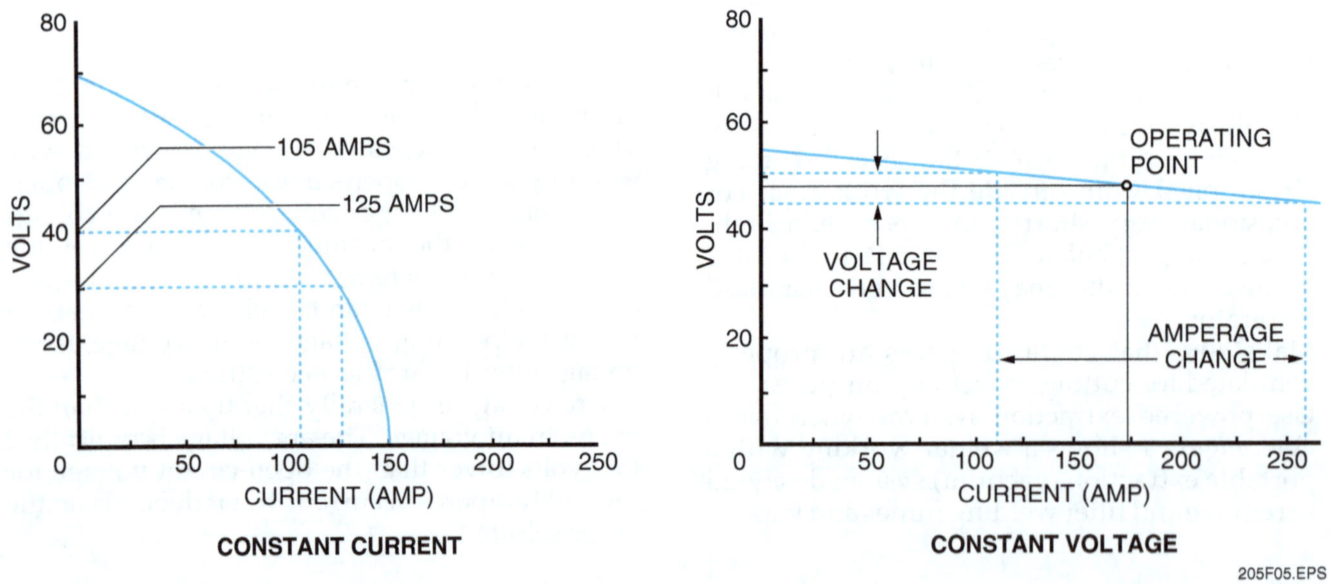

Figure 5 Constant current and constant voltage output curves.

er-rectifier power sources that produce DC only are sometimes called rectifier power sources or just rectifiers. Depending on the size, transformer-rectifier power sources may require 120/240-volt single-phase input power, 208-volt three-phase input power, or 480-volt three-phase input power.

> **WARNING!**
> Coming into contact with the primary voltage of a welding machine can cause shock or electrocution. Make sure that all welding machines are properly grounded to prevent injury.

Transformer rectifiers used for GMAW and FCAW have an ON/OFF switch and a voltage control. If the machine has a CC/CV or mode switch, it also has an amperage control for SMAW or GTAW. When the switch is set to CV for either GMAW or FCAW, the amperage control is disabled or becomes the voltage control.

The welding cables (electrode cable and workpiece cable) are connected to terminals marked ELECTRODE and GROUND, or POSITIVE (+) and NEGATIVE (–). Cables often have selector switches to select direct current electrode negative (DCEN) or direct current electrode positive (DCEP). If there is no selector switch, the cables must be manually changed on the machine terminals to select the type of current desired.

Figure 6 shows a typical industrial transformer-rectifier CC/CV power source with a wire feeder. These units usually have a 100 percent duty cycle at their rated output and a 60 percent duty cycle at their maximum output. They require a wire feeder and a GMAW/FCAW gun or spool gun. These devices are described in a later section.

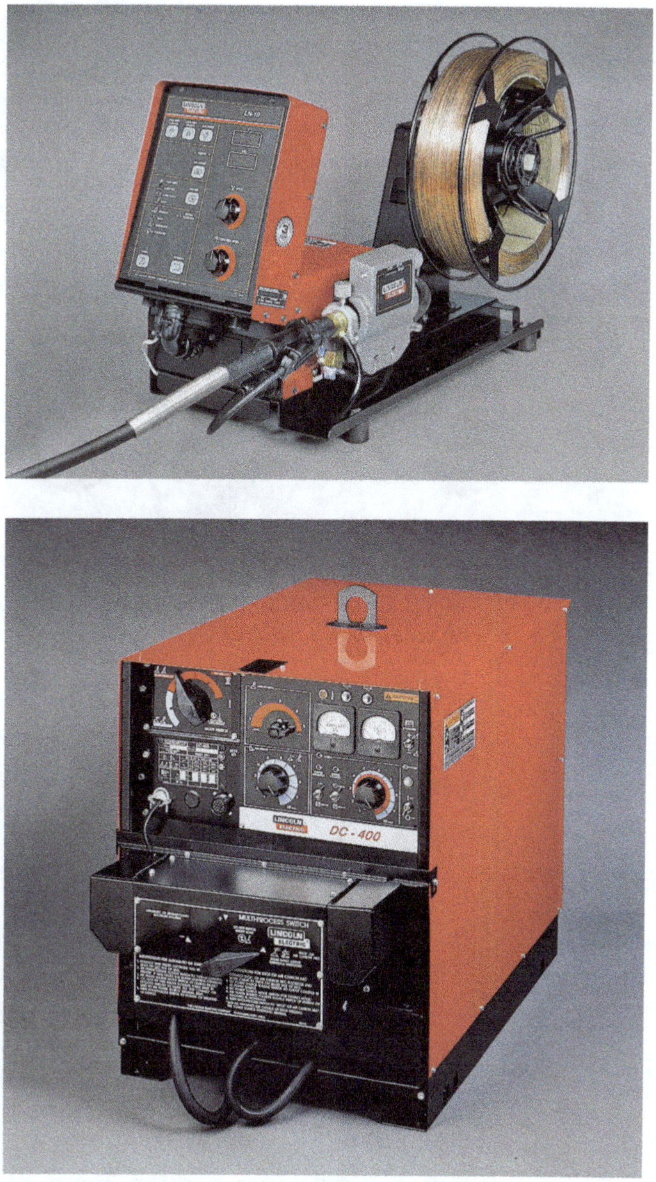

205F06.EPS

Figure 6 Industrial transformer-rectifier CC/CV power source with wire feeder.

4.2.0 Engine-Driven Generator Power Sources

Power sources can also be powered by gasoline or diesel engines. The engine can be connected to a generator that will produce the DC power needed for welding. The size and type of engine used depend on the size of the power source.

To produce welding current, the generator must turn at a required number of revolutions per minute (rpm). The engines of power generators have governors to control the engine speed. Most governors have a welding speed switch. The switch can be set to idle the engine when no welding is taking place. When the electrode is touched to the base metal, the governor automatically increases the speed of the engine to the required rpm for welding. If no welding takes place for about 15 seconds, the engine will automatically return to idle. The switch can also be set for the engine to run continuously at the welding speed.

Figure 7 shows a diesel engine-driven power source. This particular unit can produce CC/AC or DC and CV/DC outputs up to 400 amps at a 100 percent duty cycle for SMAW, GTAW, GMAW, and FCAW. For GMAW and FCAW, it is available with a portable wire feeder that provides constant wire feed speed for use with CV power sources. When used with CV power sources, the wire feeder provides wire as needed when it senses that an arc is struck. The wire feeder is shown with an FCAW-S gun, but other guns can also be used.

4-WHEEL TRAILER WITH
MULTI-PROCESS WELDER

PORTABLE INDUSTRIAL
WIRE FEEDER

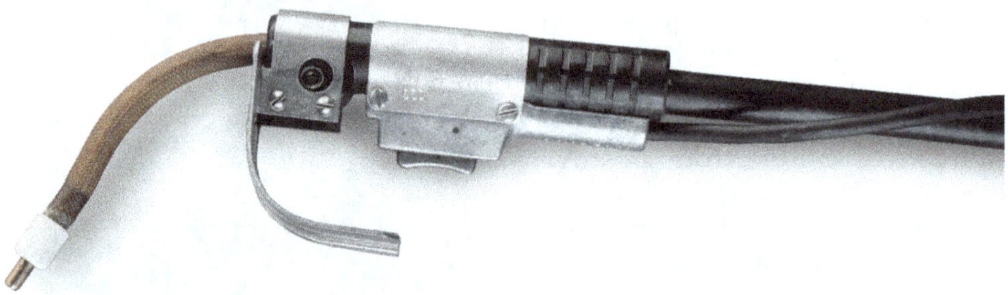

SEMIAUTOMATIC FLUX-CORED,
SELF-SHIELDING (FCAW-S) GUN

205F07.EPS

Figure 7 Diesel engine-driven power source with wire feeder and FCAW-S gun.

Engine-driven power sources often have an auxiliary power unit that produces 120 volts alternating current (VAC) for lighting, power tools, and other electrical equipment. When 120 VAC is required, the engine-driven generator or alternator must run continuously at the welding speed.

Engine-driven generators have engine controls and welding current controls. The engine controls vary with the type and size of the unit but normally include the following:

- Starter
- Voltage meter
- Temperature gauge
- Fuel gauge
- Hour meter

Engine-driven generators have a voltage/amperage control. They may have a polarity or mode selector switch, or you may have to manually change the welding cables at the welding current terminals to change the polarity. If the machine has a CC/CV or mode selector switch, it also has an amperage control. When the switch is set to CV or GMAW/FCAW mode, the amperage control is disabled and becomes a voltage control.

Hot Tip

GMAW/FCAW Feeders for Engine-Driven Power Sources

If possible, select wire feeders or spool guns that operate from 24VDC battery power and arc voltage. If AC-powered feeder or spool gun controls are used, the power source will have to operate in the continuous run mode to supply AC power to operate the feeder or spool gun.

The advantage of engine-driven generators is that they are portable and can be used in the field where electricity is not available for other types of welding machines. The disadvantage is that engine-driven generators are costly to purchase, operate, and maintain.

4.3.0 Inverter Power Sources

Inverter power sources (*Figure 8*) increase the frequency of the incoming primary power. This provides a smaller, lighter power source with a faster response time and more control for pulse welding. Inverter welding machines are used where space is limited and portability is important. The controls on these welding machines vary according to size and application. These smaller inverter power sources are often mounted on lightweight carts for greater portability.

Typical industrial advanced inverter power sources designed specifically for GMAW or FCAW operations usually have controls similar to those shown in *Figure 9*. For specific control functions for the power supply you are using, refer to the operator's manual.

A typical program sequence for one of the internal welding programs available in an advanced inverter power source is shown in *Figure 10*.

4.4.0 Power Source Ratings

The rating (size) of a welding machine is determined by the amperage output of the machine at a given duty cycle. The duty cycle of a welding machine is based on a 10-minute period. It is the percentage of 10 minutes during which the machine can continuously produce its rated amperage without overheating. For example, a machine with a rated output of 300 amps at 60 percent duty cycle can deliver 300 amps of welding current for 6 minutes out of every 10 without overheating.

The duty cycle of a welding machine will be 10, 20, 30, 40, 60, or 100 percent. A welding machine with a duty cycle of 10 to 40 percent is considered a light- to medium-duty machine. Most industrial, heavy-duty machines for manual welding have 60 to 100 percent duty cycle ratings. Machines designed for automatic welding operations have 100 percent duty cycle ratings.

COMPACT CV POWER SOURCE WITH TOP-MOUNTED WIRE FEEDER

CART-MOUNTED CV POWER SUPPLY WITH WIRE FEEDER AND GMAW/FCAW-GS (gas-shielded) GUN ASSEMBLY

205F08.EPS

Figure 8 Inverter power sources.

KEY CONTROLS

A. Large, bright, easy to read meters
B. Local/remote output control
C. Mode selection switch
D. Negative and Positive Twist-Mate® or Tweco® Output Terminals
E. Wire Feeder Voltmeter Polarity Switch
F. Circuit Breaker Reset Buttons
G. On/Off Switch
H. Hot Start Control
I. Arc control knob
J. Weld Terminals on or remotely controlled
K. Preset Output Amps and Voltage Control
L. Thermal Overload Indicator Light

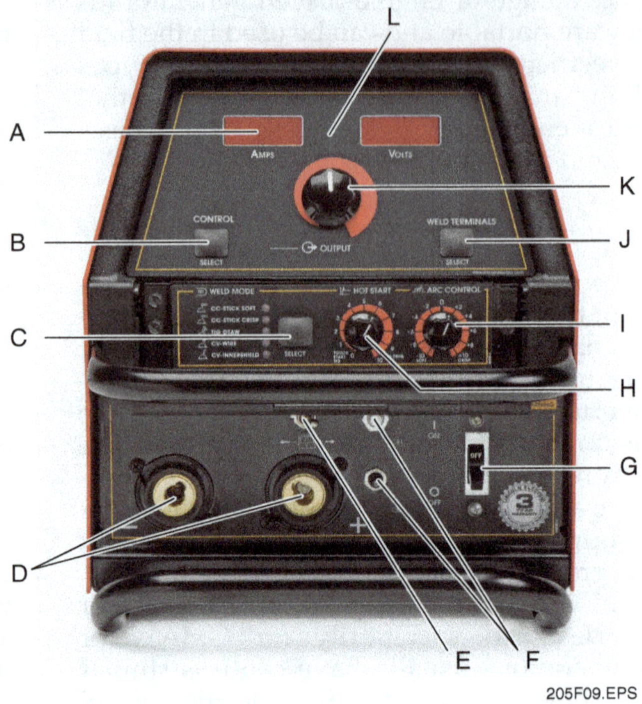

Figure 9 Typical industrial inverter power source controls.

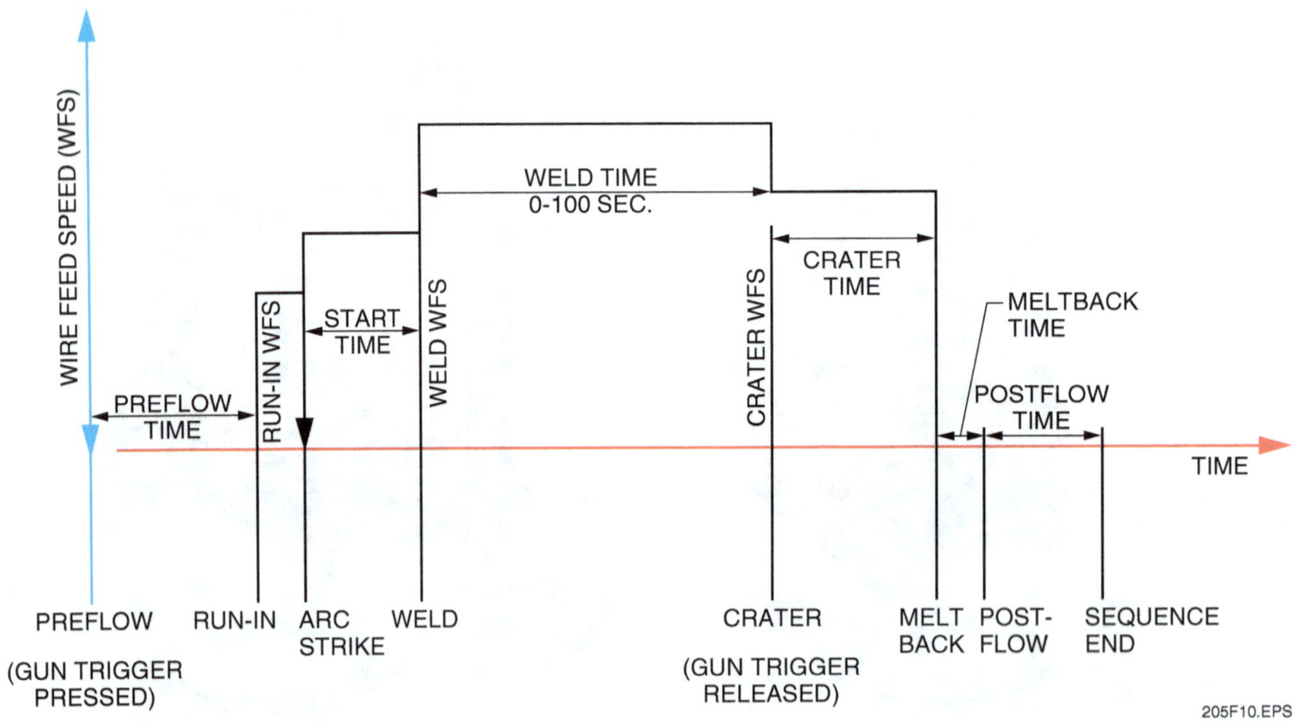

Figure 10 Typical welding program sequence.

Hot Tip

All-in-One Welding Machines

Portable, light industrial GMAW/FCAW welding machines are also available. The MIG unit shown is rated at 90 amps with a 20 percent duty cycle. It is equipped with an internal wire feeder. The TIG unit is rated at 90 amps at 100 percent duty cycle.

205SA01.EPS

With the exception of 100 percent duty cycle machines, the maximum amperage that a welding machine can produce is always higher than its rated capacity. A welding machine rated 300 amps at a 60 percent duty cycle will generally put out a maximum of 375 to 400 amps. But since the duty cycle is a function of its rated capacity, the duty cycle will decrease as the amperage is raised over 300 amps. Welding at 375 amps with a welding machine rated 300 amps at 60 percent duty cycle will lower the duty cycle to about 30 percent. If welding continues for more than 3 out of 10 minutes, the machine will overheat.

> **NOTE**
> Most welding machines have a heat-activated circuit breaker that shuts off the machine automatically when it overheats. The machine cannot be turned back on until it has cooled.

If the amperage is set below the rated amperage, the duty cycle increases. Setting the amperage at 200 amps for a welding machine rated 300 amps at 60 percent duty cycle will increase the duty cycle to 100 percent. *Figure 11* shows the relationship between amperage and duty cycle.

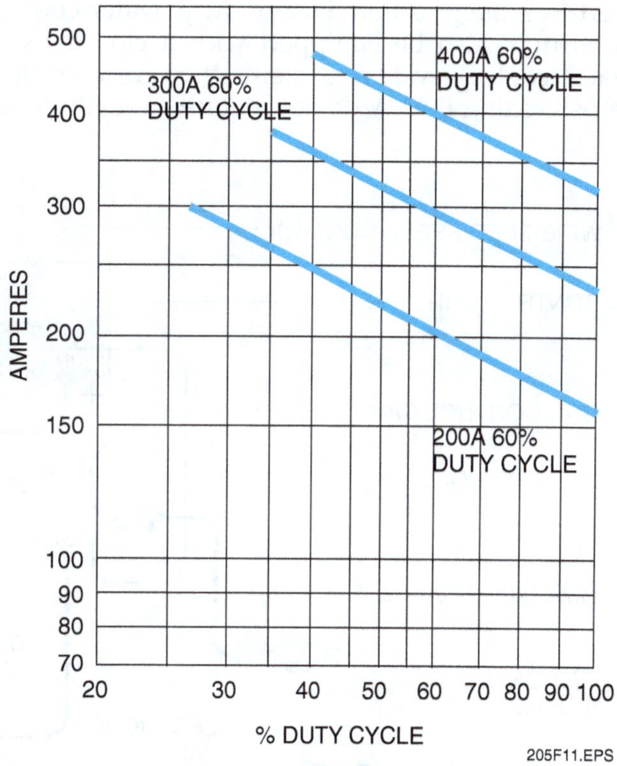

205F11.EPS

Figure 11 Relationship between amperage and duty cycle.

5.0.0 GMAW and FCAW Equipment

Typical GMAW and FCAW systems (*Figure 12*) are made of four interconnected pieces of equipment, including the following:

- DC power source (welding machine)
- Wire feeder (if required)
- Welding gun
- Shielding gas supply (if required)

In some systems, equipment functions may be integrated or combined. For example, the wire feeder can be built into the same cabinet as the power source, or it may be incorporated into the welding gun. Welding guns are shown in a later section of this module.

Most systems have a separate wire feeder (push system) that pushes the wire from the feeder through the gun cable and gun. Other systems may incorporate the wire feeder into the welding gun (pull system) or do both (push-pull system). There are some systems that have synchronized wire feeders located at intervals along very long feeder conduits. Various types of wire feeders are covered in a later section. Heavy duty, water-cooled guns may also be equipped with a closed loop water cooling system. Systems that are used for self-shielding FCAW do not require gas cylinders.

5.1.0 GMAW Metal Transfer Modes

GMAW can use any of several modes (types of metal transfer) to conduct the filler metal to the molten weld pool. The principal modes include the following:

- Spray transfer
- Globular transfer
- Short-circuiting transfer
- Pulsed transfer

5.1.1 Spray Transfer

In the spray transfer mode (*Figure 13*), the welding voltage is set in the higher ranges to keep the end of the wire electrode away from the work and to maintain a continuous arc. The high heat of the arc melts the end of the wire electrode, and fine droplets are carried by the arc axially to the molten weld puddle on the base metal. The spray transfer deposits filler metal at a high rate and produces very little spatter. However, the weld puddle is relatively large, and the mode works best in the flat or horizontal fillet positions. Also, only certain shielding gases or gas combinations can be used with spray transfer. Shielding gases are covered later in this module.

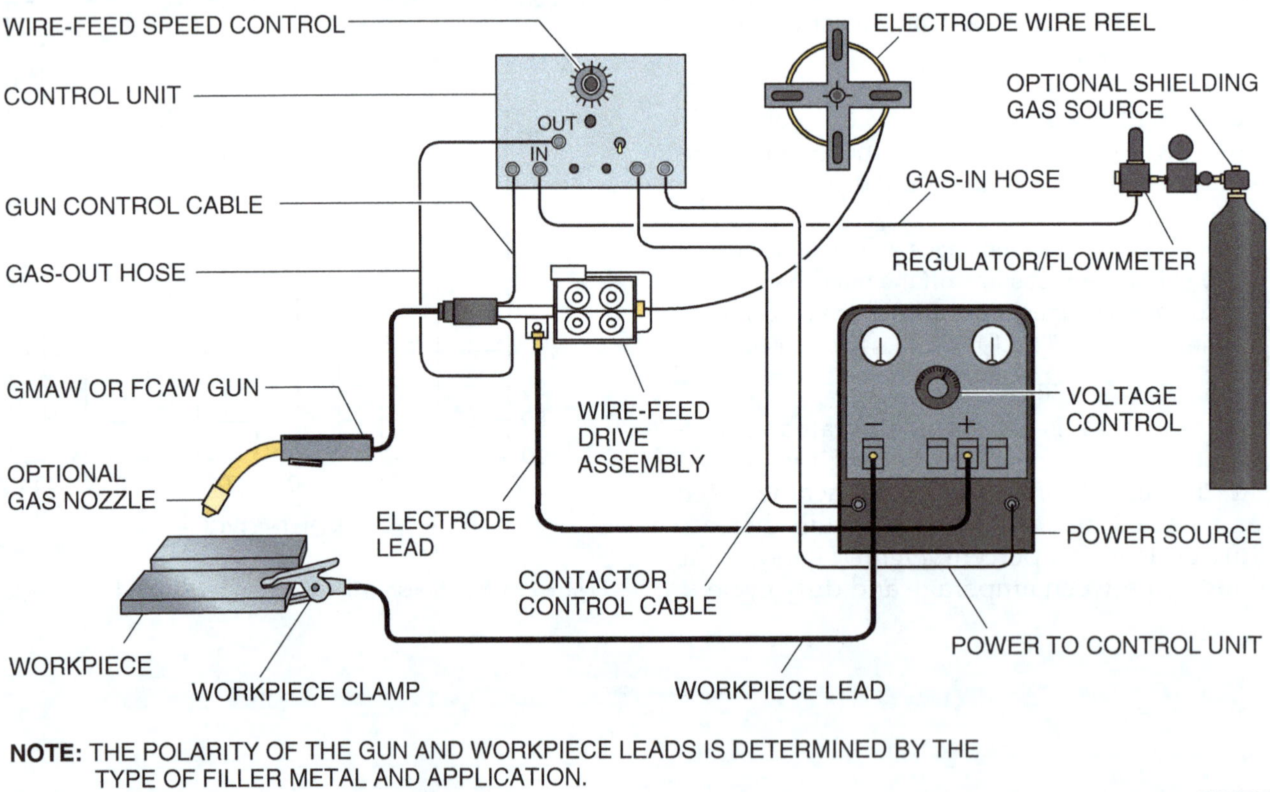

Figure 12 Simplified GMAW/FCAW system diagram.

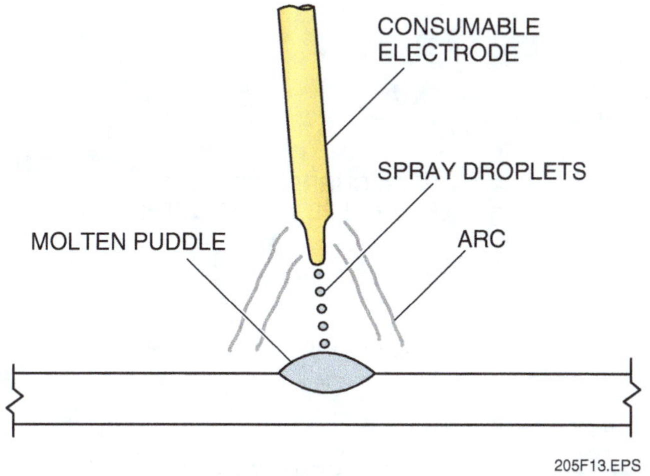

Figure 13 Spray transfer.

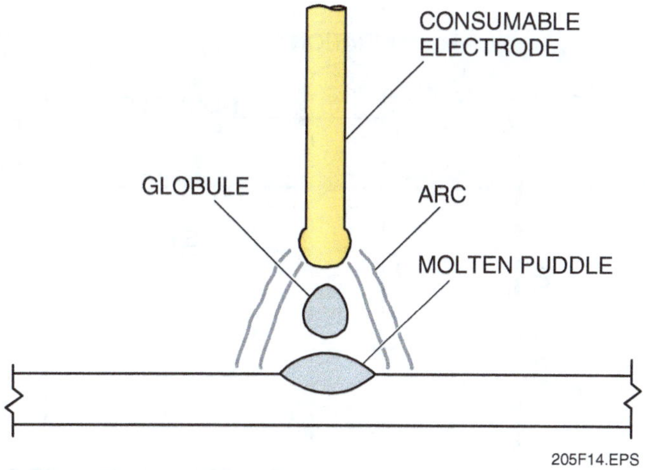

Figure 14 Globular transfer.

5.1.2 Globular Transfer

In the globular transfer mode (*Figure 14*), the welding voltage is set lower than it is for spray transfer. The end of the wire electrode remains a short distance from the work to maintain a continuous arc. Because of its weight, the electrode wire tip periodically melts off in a large glob, which the arc then carries to the workpiece, where it mixes with the weld puddle. Globular transfer can be used to deposit filler metal at high rates, and it is capable of producing high quality welds in all positions. However, bead appearance can be rough, and spatter may be heavy in some applications.

5.1.3 Short-Circuiting Transfer

Short-circuiting transfer (GMAW-S) uses the lowest voltage settings of the three modes (*Figure 15*). It is the best mode for welding light-gauge sheet metal. The wire electrode is fed against the workpiece. When contact is made, current flow heats the wire electrode until it melts enough for a droplet to separate and mix with the weld puddle. When the droplet separates, a gap is formed, and an arc is initiated. The continuously feeding wire closes the gap and contacts the workpiece, shorting out the arc. The current again heats the wire until another droplet separates, and the arc is again established. This process repeats many times each second. The short-circuiting process can be used to weld thin materials, and it is suitable for all positions. However, it does not penetrate deeply and may produce incomplete fusion on thicker materials because of the low heat input.

5.1.4 Pulsed Transfer

Pulsed transfer (GMAW-P) welding is a modified combination of spray transfer welding and short-circuit transfer arc welding (*Figure 16*).

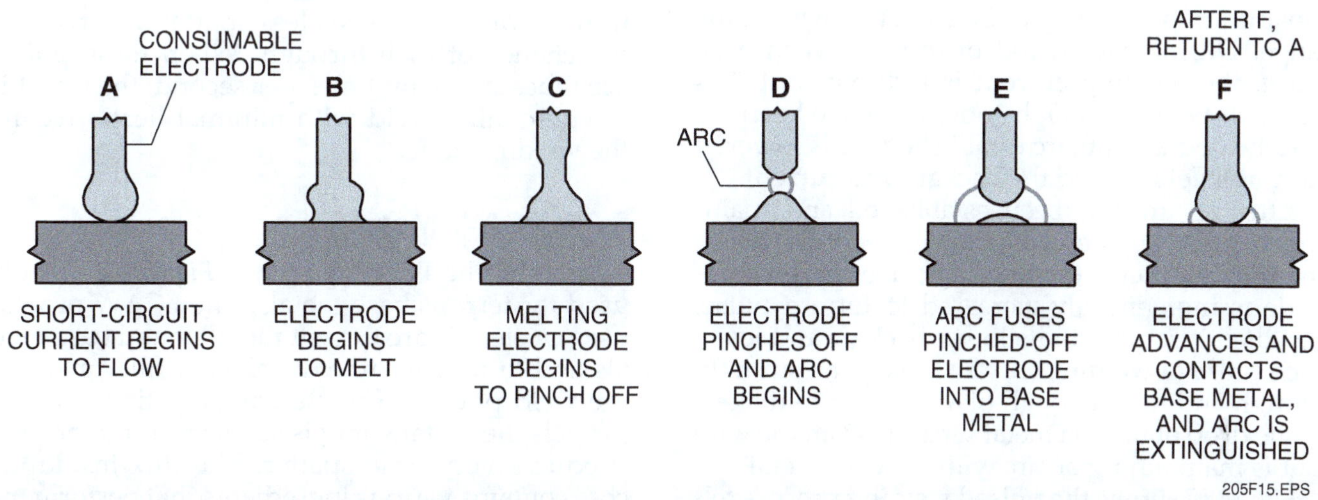

Figure 15 Short-circuiting transfer.

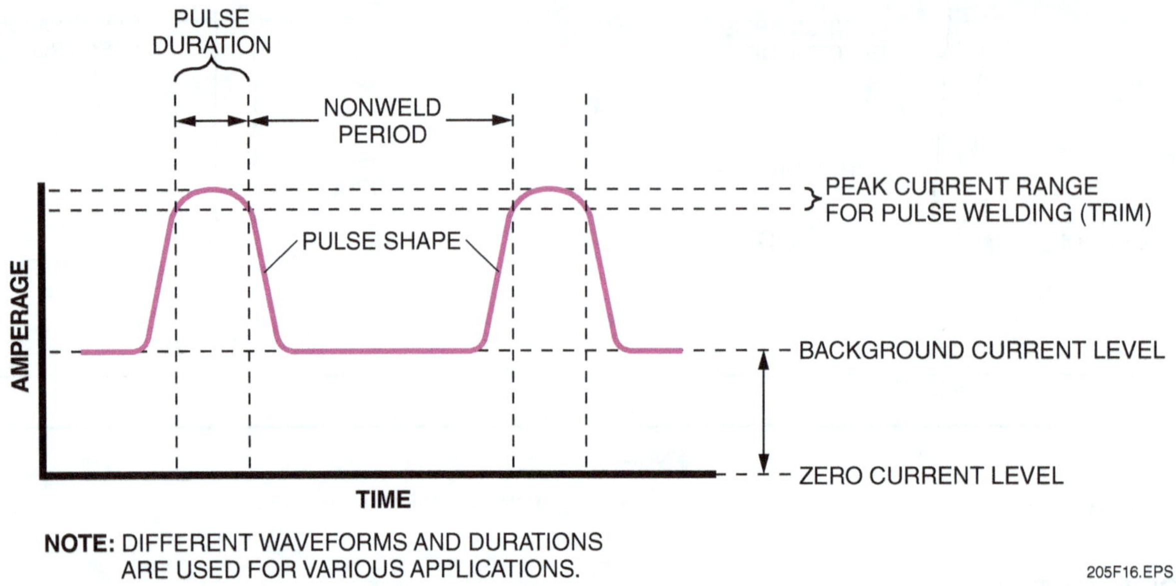

Figure 16 Output pulses of high welding current occurring over time.

The process transfers molten metal in droplets while maintaining an arc with the base metal. The pulsed GMAW process also allows the transfer action to occur while lower current levels are used. The advantages of the lower current levels are less spatter and better penetration without melting through the base metal. Pulsed transfer is especially good for aluminum welding because it results in a smaller heat affected zone.

In pulsed transfer welding, the welding power is electronically controlled by a special power source that pulses the output power to the consumable electrode (wire). Although different GMAW machines may have different types of special power sources, the results are the same for each. The welding current is generated in pulses of high current at specified intervals of time. The frequency of these pulses varies from one power source to the next, but some power sources are capable of producing pulses over a wide time range. During the period of time between each pulse, the welding current is reduced, but it is still maintained at a high enough level to keep the wire heated and an arc established. This lowered current level is called the background current.

After an arc has been established and maintained between the consumable electrode (wire) and the base metal, the welding current to the electrode is momentarily increased to form a pulse, and then it returns to its lower background level. Each pulse of welding current must peak strongly enough and have enough of a duration to make a single drop of molten metal separate from the wire that is maintaining an arc with the base metal.

Figure 17 shows the pulsed welding process relative to one of the pulse shapes. Points A through E on the diagram relate to the electrode arc illustrations, labeled A through E, above the diagram.

When the single drop of molten metal separates from the electrode, it is transferred across the arc to form the molten puddle on the base metal. This transfer process creates less spatter and reduces the chance of melt-through. When these pulses are repeated many times per second, the result is a high-quality weld with minimal heat affecting the welding zone.

Hot Tip

Adaptive Control Welding

Some newer inverter power sources are capable of using adaptive control welding during pulse transfer. These units sense and alter the frequency of the pulse, its width, the peak current, and wire speed automatically to maintain optimum welding characteristics during welding.

5.2.0 FCAW Metal Transfer Process

Currently, the FCAW process (*Figure 18*) is only used for ferrous base metals. The FCAW process uses an electric arc to melt the filler and base metals and to transfer fine droplets of filler metal to the weld puddle. The electromagnetic force that propels the metal spray is strong, but the process produces very little spatter. The flux inside the core contains various ingredients that perform the following functions:

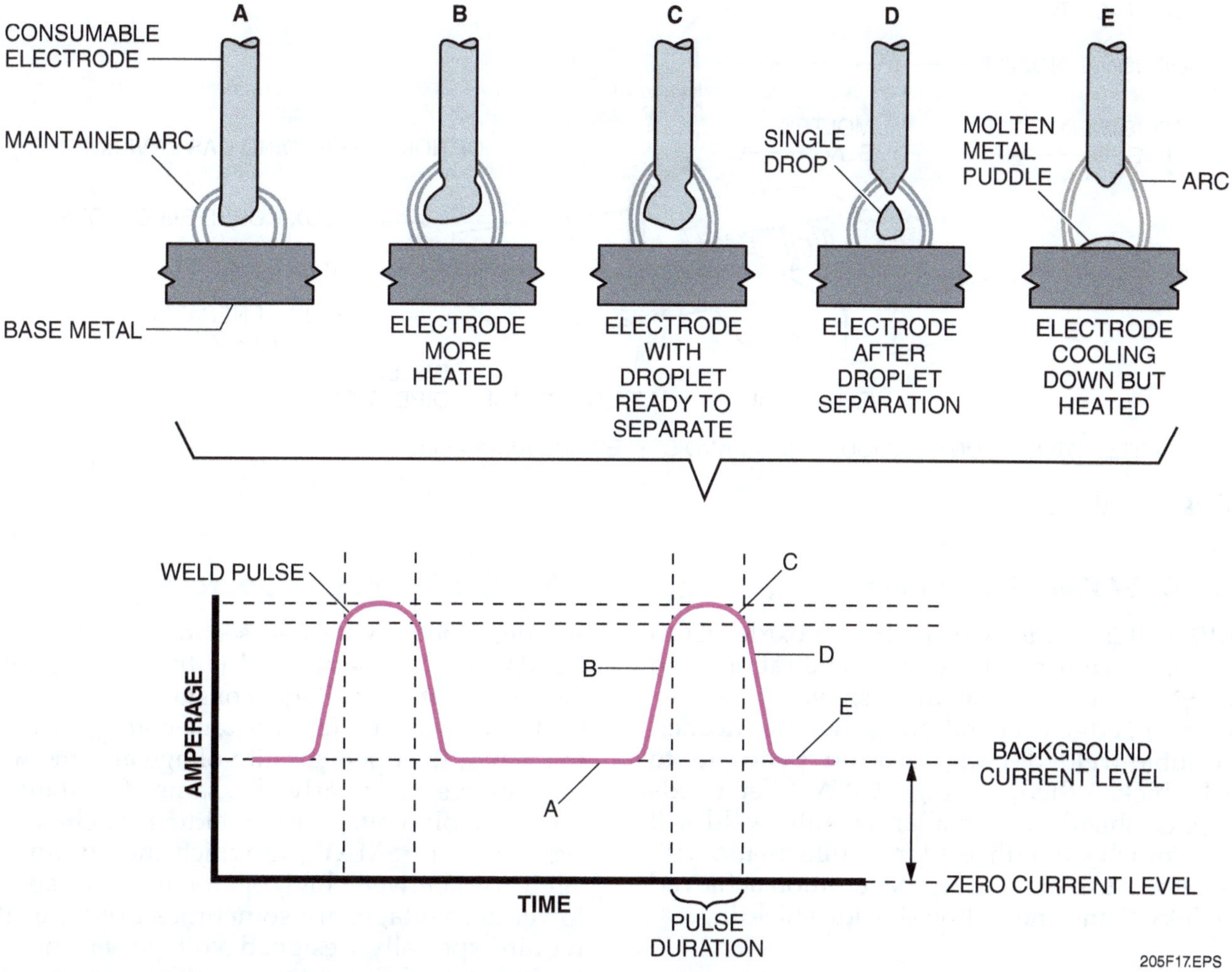

Figure 17 Typical pulsed welding process associated with one pulse shape.

- Ionizers to stabilize the arc
- Deoxidizers to purge the weld of gases and slag
- Additional metals and elements to enhance the quality of the weld metal
- Ingredients to generate shielding gas
- Ingredients to form a protective slag over the weld bead

The gas shield generated by flux in the wire core, as well as that of the external gas shield, protects the arc and molten weld metal. The slag cover produced by the flux helps to support the weld metal and allows the process to be used for making high-quality welds in all positions. With larger diameter flux-cored wires, deposition rates of 25 pounds per hour can be achieved.

Hot Tip

Pulsed Transfer Advantages

The following are advantages of pulsed transfer welding over welding processes using continuous currents:

- Rivals GTAW for weld quality
- Permits all-position welding
- Allows smooth welds to be made on thin materials without melt-through
- Controls thick plate weld pool more easily
- Reduces cost because larger diameter electrode wire can be used
- Can be used on base metals such as aluminum, stainless steel, copper alloy, and nickel steels, as well as carbon steels
- Reduces weld contamination from surface oxides because less surface area is exposed during the weld
- Creates smaller heat-affected zone (HAZ)

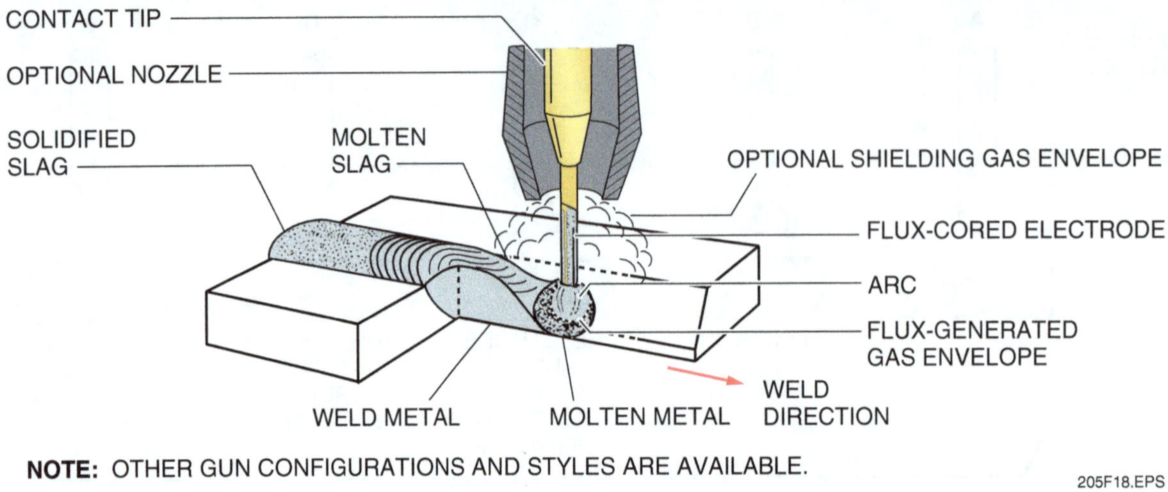

Figure 18 FCAW process.

5.2.1 FCAW Weld Penetration

FCAW welds made with carbon dioxide (CO_2) shielding have a much deeper penetration than either SMAW or FCAW without shielding gas. For this reason, edge preparation is often not needed for double-welded butt joints on plate up to ⅜-inch thick. When making FCAW fillet welds using CO_2 shielding, a smaller size fillet weld will have as much strength as a larger fillet made with SMAW because of the deep penetration achieved with FCAW and the carbon dioxide shielding.

5.2.2 FCAW Joint Design

FCAW uses the same basic joint designs (*Figure 19*) as SMAW. However, when making groove welds with FCAW and CO_2 shielding gas, the square groove (square butt) joint can be used for plate thicknesses up to ⅜ of an inch. Above this thickness, beveled joints should be used.

Because FCAW electrode wire is much thinner than comparable SMAW electrodes and FCAW penetration with carbon dioxide is superior to SMAW, the included angles of V-joints can be reduced to half the angle required for SMAW. For example, an SMAW included angle of 60 degrees can be reduced to 30 degrees when using FCAW with CO_2 shielding. This can save approximately 50 percent on filler metal and considerable welding time when compared with using SMAW.

CAUTION

Always refer to your WPS or site quality standards for specific information on joint requirements. Information in this module is provided as a general guideline only.

5.3.0 GMAW Power Sources

As described in earlier sections, GMAW and FCAW are best performed with a constant-voltage or constant voltage/constant current (CV/CC) DC power. In constant-voltage welding machines, the open-circuit voltage and the welding voltage are nearly the same. Constant-current variable voltage DC welding machines, the type used for SMAW, in which the output voltage varies between high open-circuit voltage and lower arc voltage, are sometimes used, but they require specially designed voltage-sensing wire feeders. CV/CC welding machines are required for GMAW-P operations.

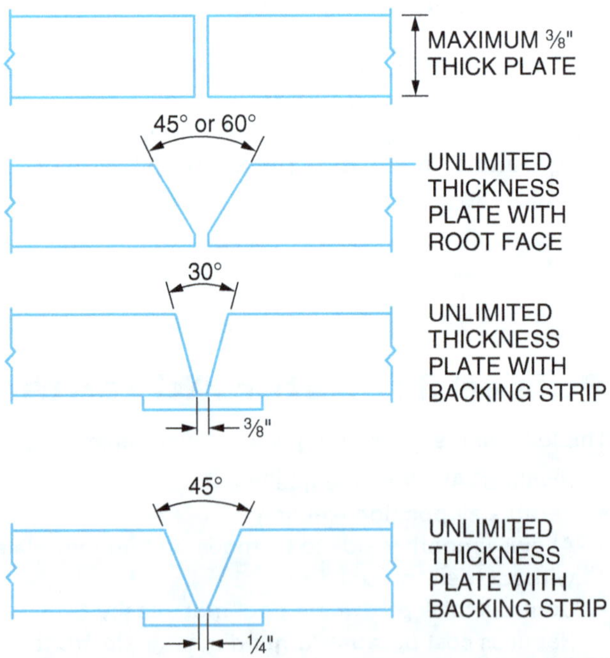

Figure 19 Joint designs for FCAW with CO_2 shielding.

Power sources usually contain controls for adjusting arc voltage, slope, and inductance, and they have meters to monitor amperage and voltage. The simplest machines are power sources with built-in wire feeders. A simple combination welder and wire feeder may have only a wire-feed speed control and a voltage control.

5.3.1 Slope

The constant voltage power sources used for GMAW and FCAW are not truly constant in their voltage output, because the voltage always drops a little as the current (amperage) increases. This voltage-to-amperage relationship forms a slight curve when plotted as a graph. The general angle of the volt-ampere curve is known as the slope. This slope is adjustable on some power sources. A flatter (more horizontal) slope is better for spray transfer welding, while a steeper slope is better for short-circuiting welding (*Figure 20*).

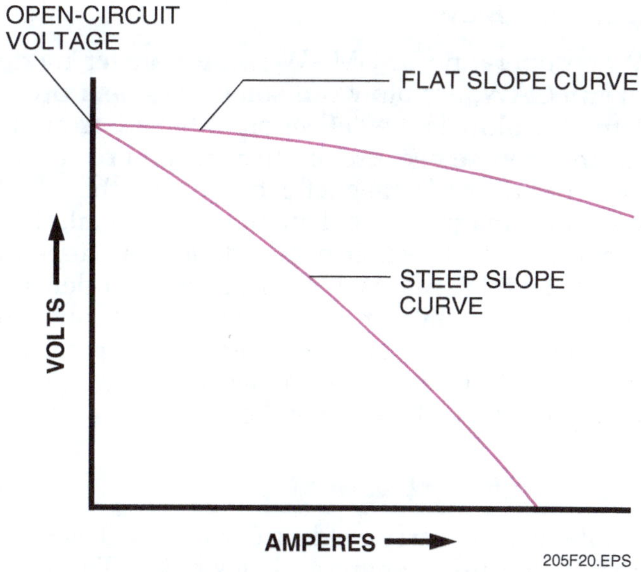

Figure 20 Flat slope and steep slope volt-ampere curves for GMAW.

5.3.2 Pinch Effect and Inductance

When using short-circuiting welding in GMAW, the current passing through the wire electrode heats and then melts the wire. This is sometimes called the pinch effect because the molten wire appears to be pinched off, forming a droplet of molten metal (*Figure 21*).

Low inductance produces high pinch effect, and high inductance produces low pinch effect. With low circuit inductance, the current rise is very fast, and the high pinch effect can cause the wire droplet to explode or spatter. When the circuit inductance is high, the current rises more slowly, decreasing the number of short circuits per second and extending the arc duration. The pinch effect forms the droplet more gently, which results in a more fluid puddle, smoother weld, and less spatter.

Constant voltage power sources are self-regulating and automatically produce current as it is required. Inductance controls the rate at which the current output rises. Some inductance is desirable in FCAW because it prevents explosive arc starts by slowing down the rate of current rise. Many FCAW power sources have a control to adjust the inductance.

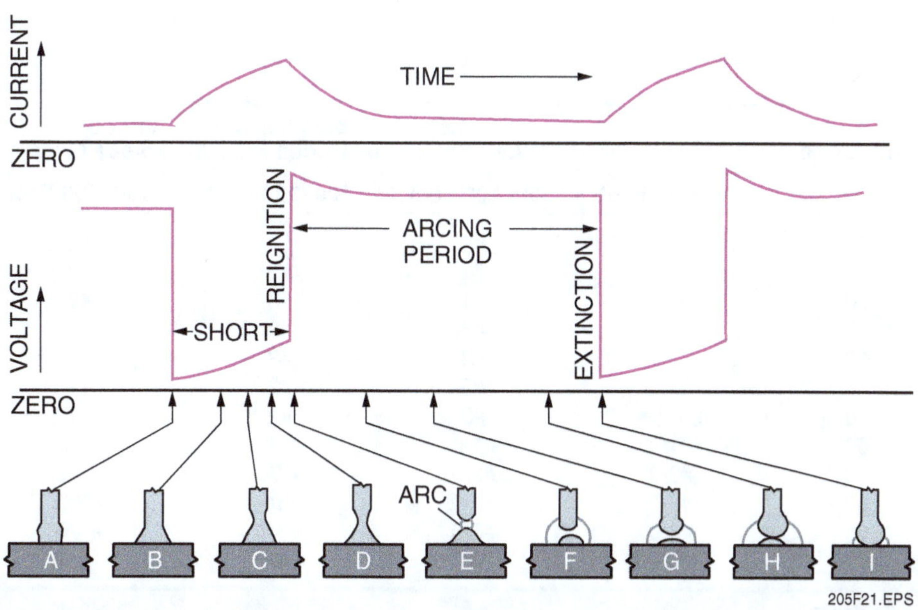

Figure 21 Pinch effect.

5.3.3 Arc Blow

When operating in GMAW spray transfer mode or in FCAW, arc blow can sometimes be a problem. Arc blow is the deflection of the arc from its normal course because of the attraction or repulsion of the arc's magnetic field with the weld current's magnetic field in the base metal. The amount and direction of arc deflection depend upon the relative position, direction, and density of the base metal's magnetic field. Arc blow can result in excess spatter and weld defects. It can be minimized by relocating the workpiece clamp or changing the weld angle of the gun.

5.4.0 Welding Cable

Cables used to carry welding current are designed for maximum strength and flexibility. The conductors inside the cable are made of fine strands of copper wire. The copper strands are covered with layers of rubber reinforced with nylon or Dacron cord (*Figure* 22).

Figure 22 Cut-away section of welding cable.

The size of a welding cable is based on the number of copper strands it contains. Large-diameter cable has more copper strands and can carry more welding current. Typical welding cable sizes range from #4 to #3/0 (3 aught), as shown in *Table 1*.

When selecting the welding cable size, the amperage load as well as the distance the current will travel must be considered. The longer the distance the current has to travel, the larger the cable must be to reduce voltage drop and heating caused by the electrical resistance in the welding cable. When selecting welding cable, use the rated capacity of the welding machine for the cable amperage requirement. For the distance, measure both the electrode and the workpiece leads, and add the two lengths together. To identify the welding cable size required, refer to a recommended welding cable size table, furnished by most welding cable manufacturers.

5.4.1 Welding Cable End Connections

For efficient use, welding cables must be equipped with the proper end connections or terminals.

CAUTION

If the end connection is not tightly secured to the cable, the connection will overheat and oxidize. An overheated connection will cause variations in the welding current and permanent damage to the connector and/or cable. Check connections for tightness and repair loose or overheated connections.

Lugs are used at the end of the welding cable to connect the cable to the welding machine current

Table 1 Welding Cable Sizes

Machine Size in Amperes	Duty Cycle (%)	Copper Cable Sizes for Combined Lengths of Electrodes Plus Workpiece Lead				
		Up to 50 Feet	50–100 Feet	100–150 Feet	150–200 Feet	200–250 Feet
100	20	#8	#4	#3	#2	#1
180	20	#5	#4	#3	#2	#1
180	30	#4	#4	#3	#2	#1
200	50	#3	#3	#2	#1	#1/0
200	60	#2	#2	#2	#1	#1/0
225	20	#4	#3	#2	#1	#1/0
250	30	#3	#3	#2	#1	#1/0
300	60	#1/0	#1/0	#1/0	#2/0	#3/0
400	60	#2/0	#2/0	#2/0	#3/0	#4/0
500	60	#2/0	#2/0	#3/0	#3/0	#4/0
600	60	#3/0	#3/0	#3/0	#4/0	* * *
650	60	#3/0	#3/0	#4/0	* *	* * *

* * Use double strand of #2/0
* * * Use double strand of #3/0

205T01.EPS

terminals. The lugs come in various sizes to match the welding cable size, and are mechanically connected onto the welding cable.

Quick disconnects are also mechanically connected to the cable ends. They are insulated and serve as cable extensions for splicing two lengths of cable together. Quick disconnects are connected or disconnected with a half twist. When using quick disconnects, care must be taken to ensure that they are tightly connected to prevent overheating or arcing in the connector. *Figure 23* shows typical lugs and quick disconnects used as welding cable connectors.

The workpiece clamp (*Figure 24*) provides the connection between the end of the workpiece lead and the workpiece. Workpiece clamps are mechanically connected to the welding cable and come in a variety of shapes and sizes. The size of a workpiece clamp is the rated amperage that it can carry without overheating. Workpiece clamps must at least have the same rating as the rated capacity of the power source on which they will be used.

5.5.0 External Wire Feeders

An external wire feeder pulls the electrode wire from a spool and pushes it through the gun cable and the gun. It consists of a wire spool holder with a drag brake, an electric motor that drives either one or two sets of opposing slotted rollers, and various controls. The wire feeder also usually contains the shielding gas connections, the gas control solenoid, and (optionally) the cooling-water flow-control solenoid. Some of the controls that may be located on a wire feeder include the following:

- *Spool drag*– This is a manual adjustment used to adjust the drag on the wire spool to prevent the uncontrolled unwinding of the wire.
- *Wire feed speed control*– This is a variable control used to set the wire feed speed. Since GMAW power sources are self-regulating (they produce the required current on demand), increasing or decreasing the wire feed speed will cause the welding machine to automatically increase or decrease its current output.
- *Voltage control*– This control is used to set the welding voltage.
- *Wire jog*– The manual wire jog switch is used to start and stop the wire feeder without energizing the welding current.
- *Wire meltback timer (burnback)*– This is an adjustable time delay that causes the welding current to continue after the trigger has been released. This consumes the length of wire that continues to feed after the trigger has been released so that the proper **stickout** is maintained to start the next weld.
- *Gas purge*– The manual gas purge switch is used to operate the gas solenoid in order to purge the shielding gas system prior to actual welding. It is also used to adjust the flowmeter.
- *Trigger hold*– This switch enables or disables a trigger hold function activated by the gun trigger. It enables welding without holding the trigger.

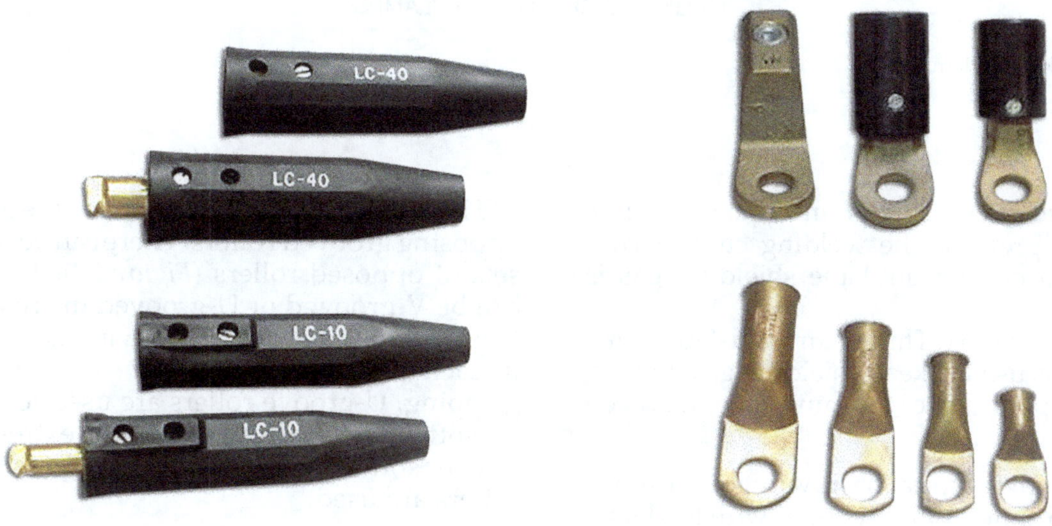

205F23.EPS

Figure 23 Quick disconnects and lugs.

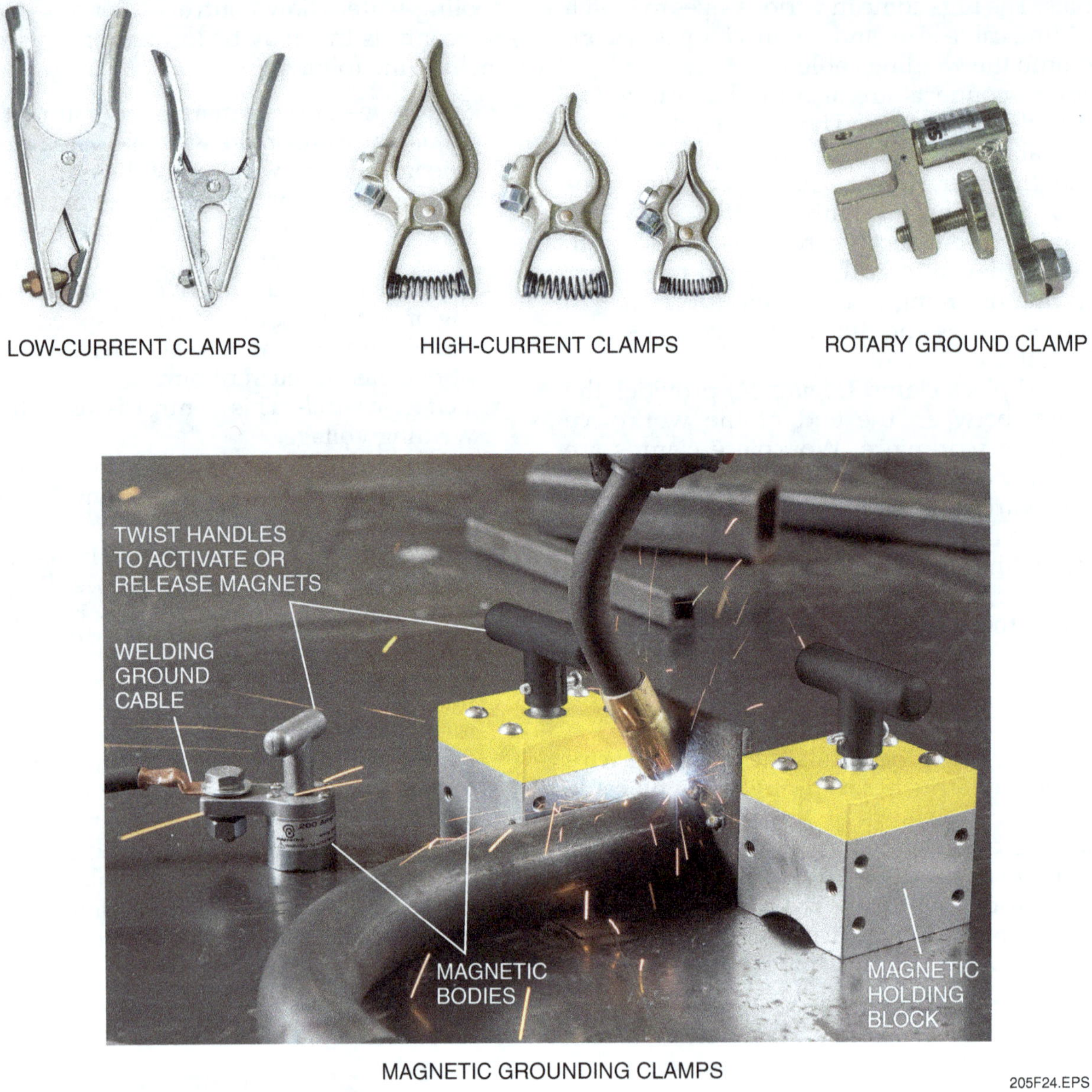

Figure 24 Workpiece clamps.

- *Gas preflow timer–* This is an adjustable time delay that prevents the welding current contactor from closing until the shielding gas is flowing.
- *Gas postflow timer–* This is an adjustable time delay that is used to keep shielding gas flowing for a short period after the gun trigger has been released.

Figure 25 shows typical basic wire feeder units, and *Figure 26* shows other types of wire feeders.

The wire feeder grips the wire in the grooves of opposing grooved rollers. There can be one or two sets of opposed rollers (*Figure 27*). The grooves can be V-grooved or U-grooved in cross section. V-grooves are used for hard wire. One roller slot in each V-groove set may be serrated for better gripping. U-groove rollers are used for soft wire or soft-shelled cored wire. For extremely soft wire or soft-shelled cored wire, cogged U-groove rollers are used.

Figure 25 Basic GMAW/FCAW wire feeder units.

DOUBLE REEL BENCH
GMAW/FCAW WIRE FEEDER

SINGLE REEL PROGRAMMABLE BENCH
WIRE FEEDER FOR GMAW-P

WEATHERPROOF PORTABLE
GMAW/FCAW WIRE FEEDER WITH
FCAW-S GUN

PROGRAMMABLE PORTABLE PUSH-PULL
WIRE FEEDER AND PUSH-PULL GUN

205F26.EPS

Figure 26 Various types of wire feeders.

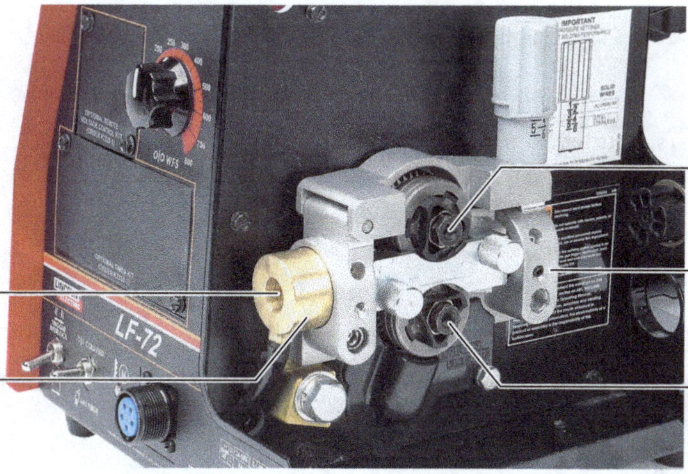

205F27.EPS

Figure 27 Wire feeder drive roll systems.

Some rollers are made in two pieces with different-size half-grooves formed on the outside edges of each half-roller. The half-rollers are assembled to fit two different ranges of wire size. *Figure 28* shows several different groove styles and combination rollers.

When dust or dirt is a problem, a felt wiper can be added to the wire between the spool and the inlet guide to the feed rollers. The felt wipers are treated with a silicone lubricant that cleans and lubricates the wire. Special liquid silicone cleaner/lubricants can occasionally be added to the felt wiper.

CAUTION

Use only special cleaner/lubricants made specifically for wire feed wipers. Use of other cleaners or lubricants may cause porosity in the weld.

Hot Tip

Knurled Roller

In addition to the U-groove rollers, knurled V-groove rollers are used for hard-shelled cored wire. Knurled rollers have small serrations formed into the edge of the groove to help the roller grab the wire and push it out to the gun without crushing it.

205SA02.EPS

Hot Tip

Wire Straighteners

A wire straightener can be added to some wire feeders. This device can be adjusted to remove the spool cast (curl) imparted to solid wire or hard-shelled cored wire to prevent wire jams and wear in the gun cable.

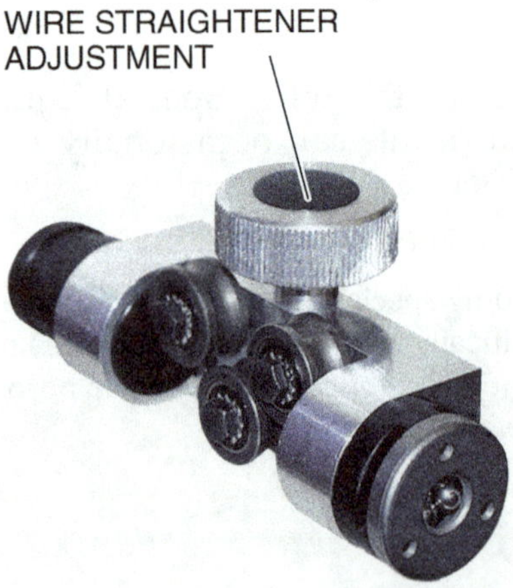

205SA03.EPS

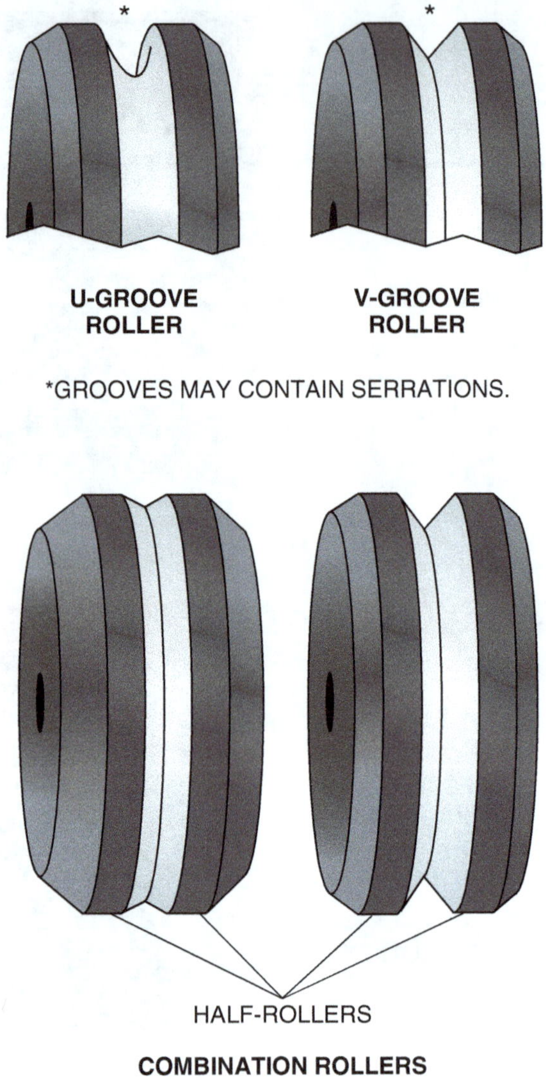

205F28.EPS

Figure 28 Groove styles and combination rollers.

5.6.0 GMAW/FCAW-G Guns

The GMAW/FCAW-G gun (*Figure 29*) supports and guides the wire electrode and provides the electrical contact between it and the welding machine electrode lead. The gun also contains a remote switch to start and stop the welding current and the shielding gas flow. The gun tip contains a nozzle to direct the shielding gas around the arc. The tip of the electrode wire extends through the contact tube, located within the center of the nozzle. The contact tube is replaceable and must be matched to the electrode wire size. Because the contact tube is the welding current's electrical connection with the electrode wire, it is also subject to wear from friction with the wire and to electrical erosion. The contact tube must be replaced periodically. The duty cycle of GMAW

Hot Tip

Programmable Wire Feeders

Programmable wire feeders are normally used with compatible CC/CV power sources for GMAW-P applications. Custom welding programs can be created, or preset welding programs can be stored in the feeder or in the power source. The program can be activated by the various controls and displays on the feeder.

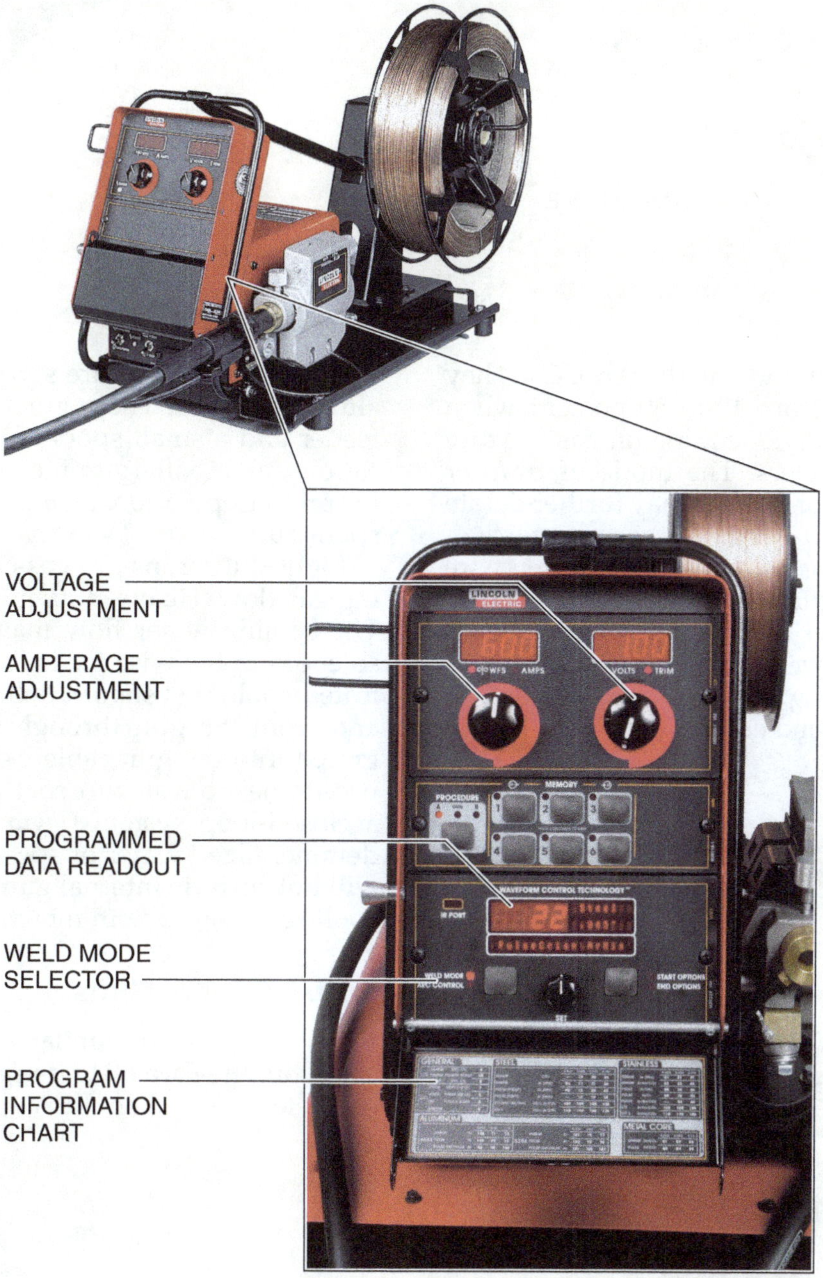

205SA04.EPS

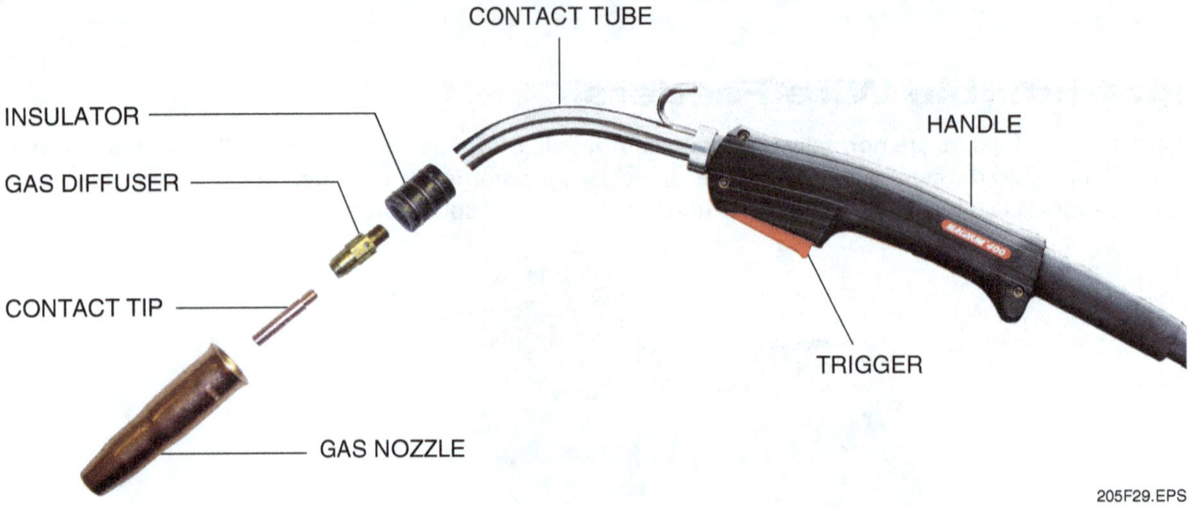

Figure 29 Construction of a typical GMAW/FCAW-G gun.

guns are determined and rated with CO_2. They are further derated from 10 to 50 percent when mixed gases are used, depending on gas mixture and welding parameters. The mode of transfer (short circuit, spray, or pulsed) may further derate their duty cycle.

Some types of guns contain motor-driven wire pullers within the gun handle (*Figure 30*). These work in conjunction with a push wire feeder located with the power source. This system was developed for feeding soft wire, such as aluminum, that tends to bend easily.

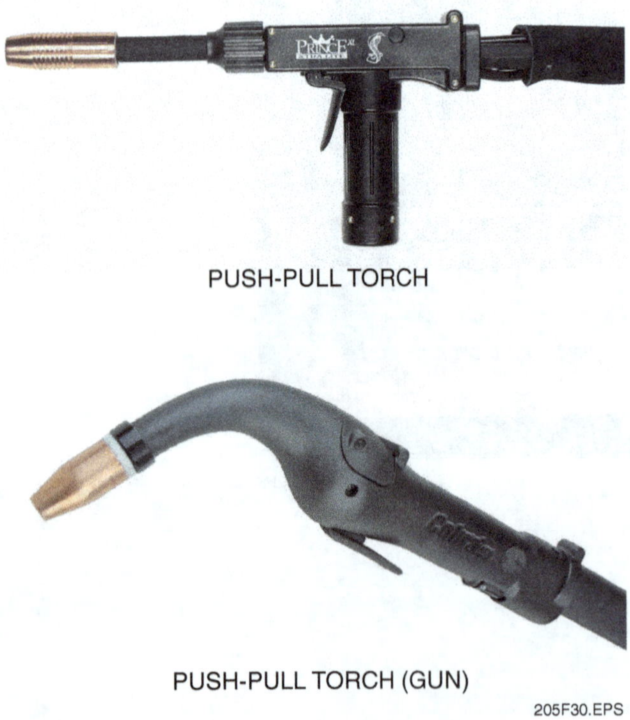

Figure 30 Pull guns for push-pull wire feeders.

In addition, there are spool guns designed for aluminum wire. These guns contain both a wire feeder and a small spool of wire. In some applications, an auxiliary gun control unit is required for proper operation. *Figure 31* shows one style of spool gun.

Light-duty guns are gas-cooled by the shielding gas flow. Heavy-duty guns are water cooled. The cooling water flow may be controlled by a solenoid in the wire feeder or by a separate dedicated cooling system. Cooling water is piped to and from the gun through flexible tubing integrated into the gun cable. Some cooling systems use domestic water to cool the gun. Others use a closed-loop system (*Figure 32*) that circulates demineralized water or special cooling fluid that will not corrode internal gun surfaces or plug the cooling passages with mineral scale.

5.7.0 FCAW-S Guns

FCAW-S guns are similar to GMAW/FCAW-G guns, but they are often designed to handle larger diameter wire. The FCAW gun supports and

Figure 31 Spool gun.

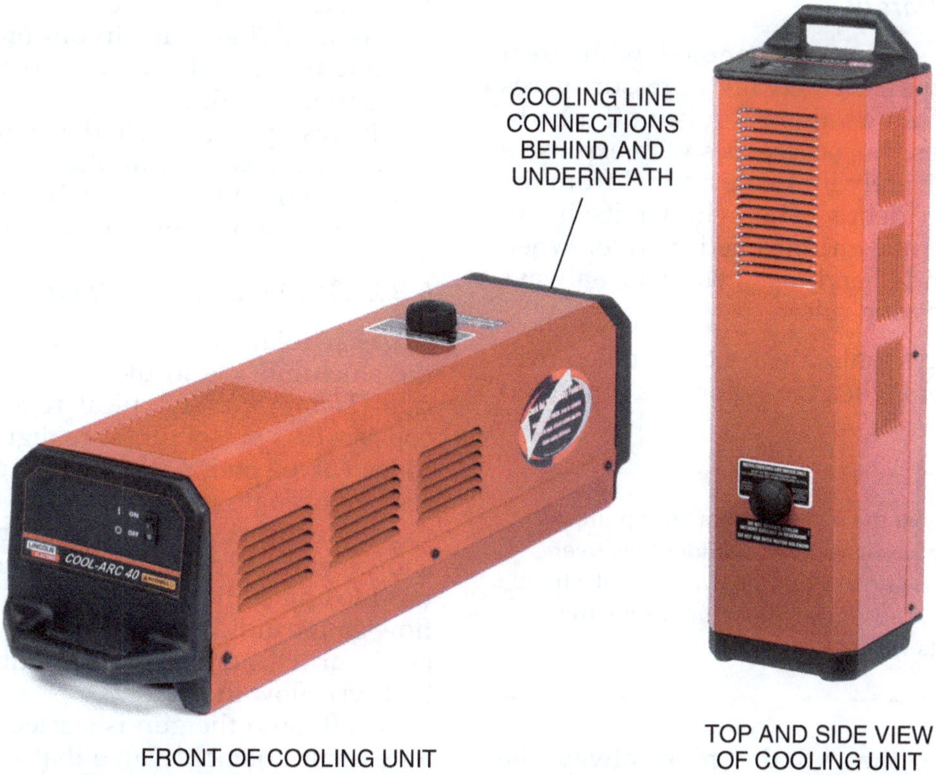

Figure 32 Typical closed-loop cooling unit.

guides the wire electrode. It also contains a remote switch for starting and stopping the welding current. The electrode wire extends through the contact tube located at the end of the gun. The contact tube provides the electrical connection from the welding machine electrode lead to the electrode wire. The contact tube is replaceable and must always be matched to the electrode wire size. If it is too small, the wire will jam; if it is too large, poor contact will cause damage from arcing between the tip and the wire. The contact tube is subject to wear from sliding friction and electrical erosion caused by the wire, and it must be replaced periodically.

Guns designed for self-shielded FCAW do not have provisions for delivering shielding gas. This makes the gun nozzles smaller in diameter. To protect the contact tube, these guns have a fiber insulator that threads onto the end of the gun nozzle.

FCAW-S generates smoke, which can be removed at the gun tip with a special smoke extraction nozzle. The nozzle is connected to a vacuum pump or blower by a flexible tube that draws off smoke, which improves visibility.

5.8.0 Shielding Gas Supply

The GMAW/FCAW-G process always uses shielding gas to protect the welding electrode (wire) and the weld from contamination. The shielding gas is directed around the end of the electrode and the weld zone by the nozzle of the gun. The most common shielding gases are argon (Ar), helium (He), and carbon dioxide (CO_2). These gases may be used alone or mixed. A small amount of oxygen (O_2) is often added to argon to stabilize the arc, increase the filler transfer rate, and improve wetting and bead shape.

Some FCAW core wires require the use of a shielding gas. The shielding gas is directed around the electrode end and the weld zone by the nozzle of the gun. The shielding gas used depends on the filler wire and base metal being welded. The most common shielding gases for FCAW carbon and low-alloy steels are carbon dioxide or an argon-carbon dioxide mixture. For FCAW high-alloy or stainless steels, argon or an argon-carbon dioxide mixture is generally used. These gases are available in pressurized cylinders of various sizes and in bulk liquid.

CAUTION

Check your WPS or site quality standards for the type of shielding gas to use. If these are not available, refer to the filler metal supplier's specifications.

5.8.1 Cylinder Safety

Cylinders must always be handled with great care because of the extremely high-pressure gases they contain. When a valve is broken off a high-pressure cylinder, the gas escapes with explosive force and can severely injure personnel from the blast and blown debris. The cylinder itself can become a powerful, uncontrolled missile. When transporting and handling cylinders, always observe the following rules:

- The safety cap should always be installed over the valve except when the cylinder is connected for use.

> **WARNING!**
> Do not remove the protective valve cap unless the cylinder is secured. If a cylinder falls over and the nozzle breaks off, the cylinder will shoot like a rocket, causing severe injury or death to anyone in its way.

- When in use, the cylinder must always be secured to prevent it from falling. It should be chained or clamped to the welding machine or to a post, beam, or pipe.
- A cylinder cart should be used to transport a cylinder.
- Cylinders should never be hoisted by slings or magnets. They can slip out of the sling or fall from the magnets. Always use a hoisting basket or similar device.
- Always open the cylinder valve slowly. Once pressure has been applied to the system, the valve should be completely opened to prevent leakage from around the valve stem.

5.8.2 Gas Regulators/Flowmeters

A gas regulator/flowmeter is required to meter the shielding gas to the gun at the proper flow rate (*Figure 33*). A typical regulator/flowmeter consists of a preset pressure regulator with a cylinder valve spud, a flowmetering (needle) valve, and a flow-rate gauge. The pressure regulator is usually equipped with a tank pressure gauge to show cylinder pressure. The metering valve is used to adjust the gas flow to the gun nozzle. The flow gauge indicates the gas flow rate in cubic feet per hour (cfh) or liters per minute as shown by a ball-type flow indicator.

Gas flow to the gun is started and stopped by an electric solenoid valve that is usually located inside the control unit or the wire feeder unit. The solenoid is controlled by the operator with the trigger switch on the gun. The same trigger switch simultaneously starts and stops the wire feeder and welding current.

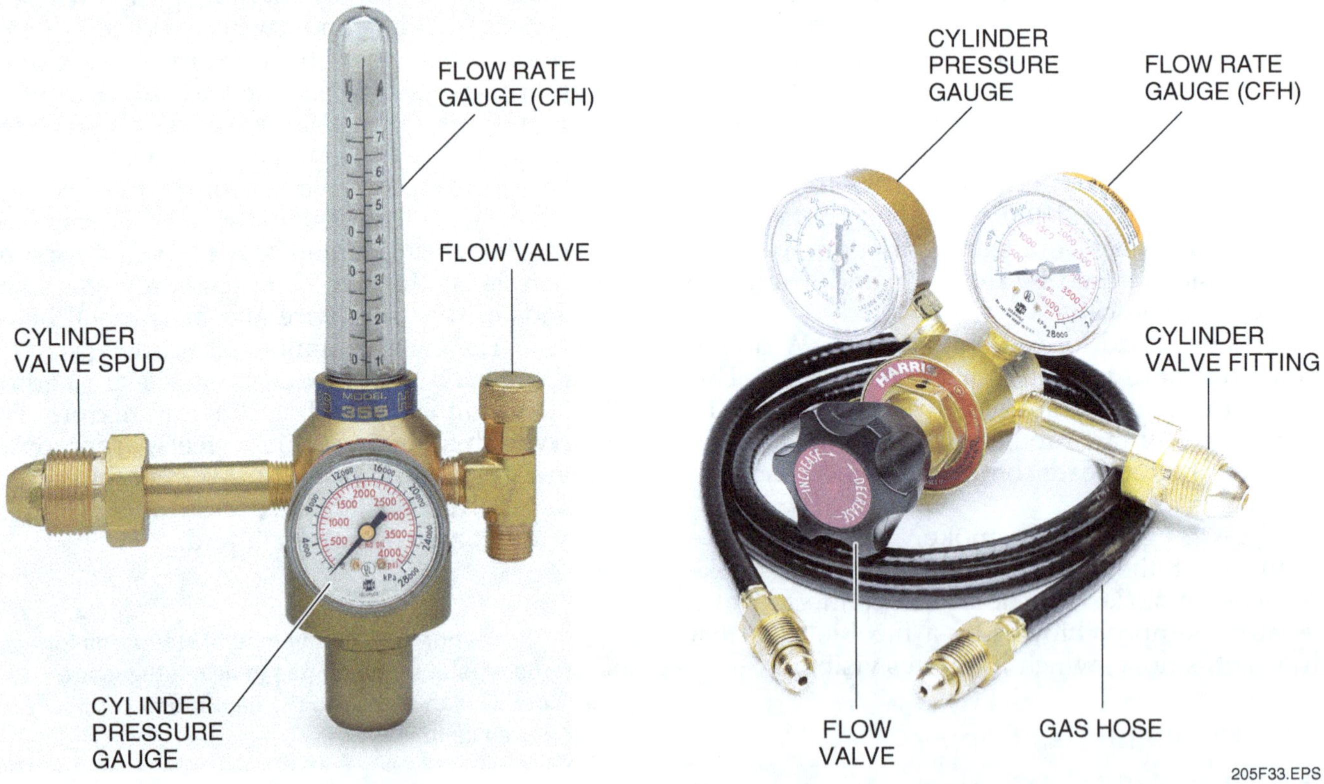

Figure 33 Shielding gas regulator/flowmeter.

6.0.0 Shielding Gases

Shielding gases are used to displace atmosphere from the weld zone and to prevent contamination of the weld puddle by oxygen, nitrogen, and moisture. The three principal shielding gases are argon, helium, and carbon dioxide. These gases are often mixed in various proportions for specific applications.

6.1.0 Shielding Gas Characteristics

Each of the shielding gases has distinctive performance characteristics. Each welds and works differently and has different arc modes and base metals. Mixtures of gases often have the best features of the individual constituents. The following sections explain some of the features of the principal gases and their primary uses.

6.1.1 Argon

Argon (Ar), an inexpensive inert gas, is the most commonly used shielding gas. Because of its current transfer and ionization properties, argon can be used alone or in combination with other shielding gases. It forms a tight arc column with high current density, which concentrates the arc in a small area and produces deep penetration.

Pure argon is used with aluminum, nickel-based alloys, copper alloys, and the reactive metals zirconium, titanium, and tantalum. It provides excellent arc stability, penetration, and bead shape. For ferrous metals, argon is usually mixed with oxygen, helium, hydrogen, carbon dioxide, and/or nitrogen.

6.1.2 Helium

Because of thermal conductivity (heat transfer), helium is used when deeper penetration and higher travel speed are required.

Arc stability with helium is not as good as it is with argon. The helium arc column is wider than the argon arc column, resulting in reduced current density. Helium is better for welding aluminum, magnesium, and copper alloys because its use results in better puddle fluidity and bead wetting. Some mixtures of helium and argon have the beneficial characteristics of both gases. Helium mixtures of 90 percent helium, 7.5 percent argon, and 2.5 percent carbon dioxide are widely used for short-circuit GMAW of stainless steel in all positions.

> **GOING GREEN**
>
> **Shielding Gases**
>
> Shielding gases, such as argon, helium, and nitrogen may be inert gases, but an excessive release of them, especially in confined spaces, could cause breathing problems. These gases are not defined as hazardous, but they are asphyxiates. Also, the fumes from the welds involving shielding gases may be hazardous to the environment if released in excessive amounts.

6.1.3 Carbon Dioxide

Carbon dioxide is commonly used to weld mild steel because of its weld performance and low cost. In the welding arc, it is not an inert gas because it breaks down into carbon monoxide and free oxygen. The oxygen combines with elements in the weld puddle to form slag. Carbon dioxide does not support spray transfer, so short circuiting transfer and globular transfer must be used. Carbon dioxide provides good penetration. However, the weld surface is heavily oxidized and has high weld spatter. Normally, argon is mixed with no more than 25 percent carbon dioxide to improve the performance characteristics. *Figure 34* shows the effects of shield gas on the welding arc in GMAW and FCAW.

6.1.4 Gas Mixtures

The principal shielding gases are often mixed to improve their overall welding characteristics. For steel, small amounts of oxygen added to argon stabilize the arc, increase filler metal deposition rate, and improve wetting and bead shape. Small percentages of oxygen added to carbon dioxide improve arc characteristics. Helium is added to argon to increase the arc voltage and heat for welding nonferrous metals. Small amounts of nitrogen and oxygen are added to argon for welding stainless steel when the austenitic characteristics should be preserved. Helium and carbon dioxide are added to argon to improve arc stability, wetting, and bead profile. Some examples and uses of gas mixtures are shown in *Table 2*. Many other typical gas mixtures are available.

6.2.0 Shielding Gas Selection

Selecting the best shielding gas for a particular welding task involves considering many factors. The shielding gas selected will affect arc shape, arc density, arc temperature, arc stability, rate of filler metal transfer, degree of spatter, weld penetration, weld bead shape, weld bead appearance, weld porosity, weld chemistry, and weld quality.

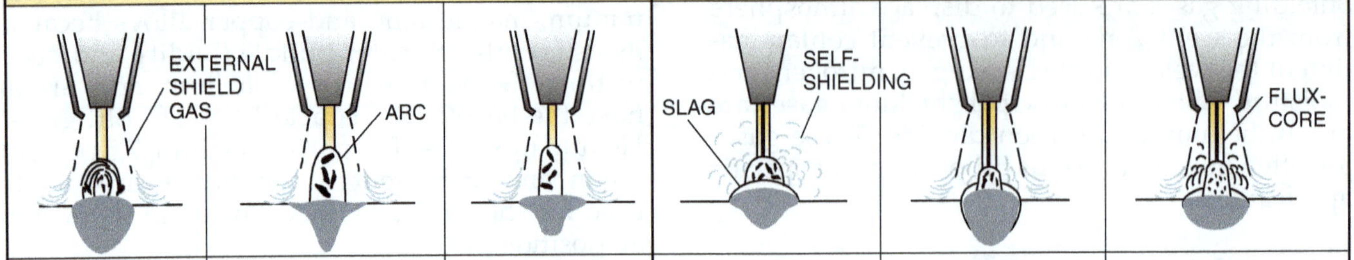

GMAW			FCAW		
CO_2	Ar-O_2	Ar-CO_2	SELF-SHIELDING	DUAL-SHIELDING with CO_2	DUAL-SHIELDING with Ar-CO_2
Shorter, Colder Arc	Hotter Arc Narrow Penetration	Eliminates Under Cutting	Better Arc Ionization Provides Covering (Slag)	Increases Penetration	Increases Arc Length, Improves Spray, and Decreases Penetration

205F34.EPS

Figure 34 Effects of shield gas on the welding arc.

Table 2 Examples and Uses of Gas Mixtures

Gas Mixture	Use
Argon with 1% oxygen	Spray transfer on stainless steels
Argon with 2% oxygen	Spray transfer on carbon steels, low-alloy steels, and stainless steels
Argon with 5% oxygen	General carbon steel welding using highly deoxidized wires
Argon with 8% to 12% oxygen	Single-pass spray transfer welds with high puddle fluidity
Argon with 3% to 10% carbon dioxide	Spray transfer and short-circuiting transfer on carbon steels
Argon with 11% to 20% carbon dioxide	High-speed and out-of-position welds on carbon steels and low-alloy steels
Argon with 21% to 25% carbon dioxide	Short-circuiting transfer on mild steels
Argon with 25% helium	Sometimes used for aluminum where bead appearance is important
Argon with 75% helium	Welding heavy aluminum
Argon with 90% helium	Welding aluminum over 3" thick and copper over 1" thick
Argon with 5% to 10% carbon dioxide and 1% to 3% oxygen	Welding carbon steel, low-alloy steel, and stainless steel with all transfer modes
Argon with 10% to 30% helium and 5% to 15% carbon dioxide	Pulse spray transfer on heavy out-of-position carbon and low-alloy steel
60% to 70% helium with 20% to 35% argon and 4% to 5% carbon dioxide	Short-circuiting transfer of high-strength steels in all positions
90% helium with 7.5% argon and 2.5% carbon dioxide	Short-circuiting transfer of stainless steels in all positions
Argon, helium, carbon dioxide, and oxygen	High-rate deposition of low-alloy, high-tensile metals, and mild steels

205T02.EPS

Shielding gases affect the welding process in three ways:

- Gas thermal conductivity affects arc voltage and heat delivered to the weld. The thermal conductivity of both helium and carbon dioxide is much higher than that of argon; both helium and carbon dioxide supply more heat to the weld. Because of this, they also require more voltage and power to maintain a stable arc.
- Some gases can react with the filler and base metals and can also affect arc stability. Carbon dioxide and most oxygen-bearing gases cannot be used with aluminum because they form oxides. However, they provide better fusion and arc stability when used with steels.
- Gases affect the mode of metal transfer and penetration depth. Mixtures containing more than 15 percent carbon dioxide do not allow true spray transfer.

Common short-circuiting shielding gases and their applications in short-circuiting transfer welding are shown in *Table 3*. Common shielding gases and their applications in spray transfer welding are shown in *Table 4*.

6.3.0 Shielding Gas Flow Rate

Gas flow rate from the gun tip is important because it affects the quality and the cost of a weld. Too low a flow rate will not shield the weld zone adequately, which will result in a poor-quality weld. Excessive gas flow wastes expensive gases and can cause turbulence at the weld. The turbulence can pull atmosphere into the weld zone, causing porosity in the weld. Welding specifications specify shielding gas flow rates.

Table 3 Short-Circuiting Shielding Gases and Their Applications

Metal	Shielding Gas	Thickness	Advantages
Carbon steel	75% Argon + 25% Carbon Dioxide	Less than ⅛" (3.2 mm)	High welding speeds without burn-through; minimum distortion and spatter
	75% Argon + 25% Carbon Dioxide	More than ⅛" (3.2 mm)	Minimum spatter; clean weld appearance; good puddle control in vertical and overhead positions
	50% Argon + 50% Carbon Dioxide	—	Deeper penetration; faster welding speeds
Stainless steel	90% Helium + 7.5% Argon + 2.5% Carbon Dioxide	—	No effect on corrosion resistance; small heat-affected zone; no undercutting; minimum distortion
Low-alloy steel	60 to 70% Helium + 25 to 35% Argon + 4.5% Carbon Dioxide	—	Minimum reactivity; excellent toughness; excellent arc stability, wetting characteristics, and bead contour; little spatter
	75% Argon + 25% Carbon Dioxide	—	Fair toughness; excellent arc stability, wetting characteristics, and bead contour; little spatter
Aluminum, copper, magnesium, nickel, and their alloys	Argon and Argon + Helium mixtures	Over ⅛" (3.2 mm)	Argon satisfactory on sheet metal; argon-helium preferred base material

205T03.EPS

Table 4 Spray Transfer Shielding Gases and Their Applications

Metal	Shielding Gas	Thickness	Advantages
Aluminum	100% Argon	0 to 1" (0 to 25 mm)	Best metal transfer and arc stability, least splatter
	65% Argon + 35% Helium	1 to 3" (27 to 76 mm)	Higher heat input than straight argon; improved fusion characteristics with 5XXX series Al-Mg alloys
	75% Helium + 25% Argon	Over 3" (76 mm)	Highest heat input; minimizes porosity
Magnesium	100% Argon	—	Excellent cleaning action
Carbon steel	Argon + 3 to 5% Oxygen	—	Improves arc stability; produces a more fluid and controllable weld puddle; good coalescence and bead contour; minimizes undercutting; permits higher speeds than pure argon
	Argon + 8 to 10% Carbon Dioxide	—	High-speed mechanized welding; low-cost manual welding
Low-alloy steel	98% Argon + 2% Oxygen	—	Minimizes undercutting; provides good toughness
Stainless steel	99% Argon + 1% Oxygen	—	Improves arc stability; produces a more fluid and controllable weld puddle; good coalescence and bead contour; minimizes undercutting on heavier stainless steels
	98% Argon + 2% Oxygen	—	Provides better arc stability, coalescence, and welding speed than 1% oxygen mixture for thinner stainless steel materials
Nickel, copper, and their alloys	100% Argon	Up to ⅛" (3.2 mm)	Provides good wetting; decreases fluidity of weld material
	Argon + Helium mixtures		Higher heat inputs of 50 and 75% helium mixtures offset high heat dissipation of heavier gauges
Titanium	100% Argon		Good arc stability; minimum weld contamination; inert gas backing is required to prevent air contamination on back of weld area

205T04.EPS

7.0.0 Filler Metals

The WPS specifies the filler metals to be used. The improper selection of filler metal can cause the weld to fail. GMAW and FCAW filler metals are explained in two sections that follow.

7.1.0 GMAW Filler Metals

GMAW uses a continuous solid wire electrode for filler material. This filler wire is drawn from high-grade pure alloys compounded for different applications. During manufacture, the wire is drawn, cleaned and inspected for defects several times. Standard wire diameters are 0.023", 0.030", 0.035", 0.045", 0.052", and 0.062". The finished wire is wound on spools 4", 8", 12", or 14" in diameter. The most common spool is 12" in diameter and weighs 30 pounds (mild steel wire). Reels up to 30" in diameter and weighing up to 1,000 pounds are also available for stationary production welding.

Several industry organizations and the U.S. government publish specification standards for welding wire. The most common are the Amer-

Hot Tip

Filler Wire and Filler Metal

The terms *filler wire* and *filler metal* are both used to describe the filler materials used in GMAW. Filler metal is also referred to as filler rod when used for GTAW.

ican Welding Society (AWS) specifications. The purpose of the AWS specifications is to set standards that all manufacturers must follow when manufacturing welding consumables. This ensures consistency for the user regardless of who manufactured the product. The specifications set standards for the following:

- Classification system, identification, and marking
- Chemical composition of the deposited weld metal
- Mechanical properties of the deposited weld metal

The following AWS specifications refer to gas metal arc welding wires and rods:

- *A5.7:1984, Specification for Copper and Copper Alloy Bare Welding Rods and Electrodes*
- *A5.9/A5.9M:2006, Specification for Bare Stainless Steel Welding Electrodes and Rods*
- *A5.10/A.510M:R2007, Specification for Bare Aluminum and Aluminum Alloy Welding Electrodes and Rods*
- *A5.14/A5.14M:2005, Specification for Nickel and Nickel Alloy Bare Welding Electrodes and Rods*
- *A5.16/A5.16M:2007, Specification for Titanium and Titanium Alloy Welding Electrodes and Rods*
- *A5.18/A5.18M:2005, Specification for Carbon Steel Electrodes and Rods for Gas Shielded Arc Welding*
- *A5.19:2006, Specification for Magnesium Alloy Welding Electrodes and Rods*
- *A5.28/A5.28M:2005, Specification for Low-Alloy Steel Electrodes and Rods for Gas Shielded Metal Arc Welding*

NOTE

The specification number is generally followed by a colon and a four-digit number, such as A5.18:2005 or A5.28:2005. The colon and the number that follows indicate the year that the specification was last revised.

Filler wire is graded for three major areas of use:

- *General use* – Wire meets specifications. No record of chemical composition, strength, or similar factors is supplied to the user with the wire purchase.
- *Rigid control fabrication* – A Certificate of Conformance is supplied with the wire at purchase. The stock is identified by code numbers located on the roll package.
- *Critical use* – A Certified Chemical Analysis report is supplied, and records are kept by the fabricator of welds and processes for later reference on aircraft, nuclear reactors, and pressure vessels.

Hot Tip

Rod and Electrode Designations

The letter R that follows the E in an AWS classification number indicates that the classification is used as a metal filler rod as well as an electrode. GMAW and GTAW filler metals are both labeled the same way when used for gas tungsten arc welding (GTAW).

The following factors affect the selection of a GMAW filler wire:

- Base metal chemical composition
- Base metal mechanical properties
- Weld joint design
- Service or specification requirements
- Shielding gas used

Commonly used filler wire types include the following:

- Carbon steel
- Stainless steel
- Aluminum and aluminum alloy
- Copper and copper alloy
- Low-alloy steel

7.1.1 Carbon Steel Filler Metals

Carbon steel filler metals are identified by specification *AWS A5.18*. The filler metal is available in wire reels and rod form. The wire classification (number) is located on a label on the side of the reel or container (*Figure 35*).

All carbon steel filler metals contain alloys such as silicon, manganese, aluminum, and carbon. Other alloys such as nickel, chromium, and molybdenum are also often added. The purpose of the alloys is as follows:

- *Silicon (Si)* – Concentrations of 0.40 to 1.00 percent are employed to deoxidize the puddle and to strengthen the weld. Silicon above 1 percent may make the welds crack-sensitive. Silicon also lowers the melting temperature of the wire and promotes wetting.
- *Manganese (Mn)* – Concentrations of 1 to 2 percent are also employed as a deoxidizer and to strengthen the weld. Manganese also decreases hot-crack sensitivity.
- *Aluminum (Al), titanium (Ti) and zirconium (Zr)* – One or more of these may be added in very small amounts for deoxidizing. Strength may also be increased.

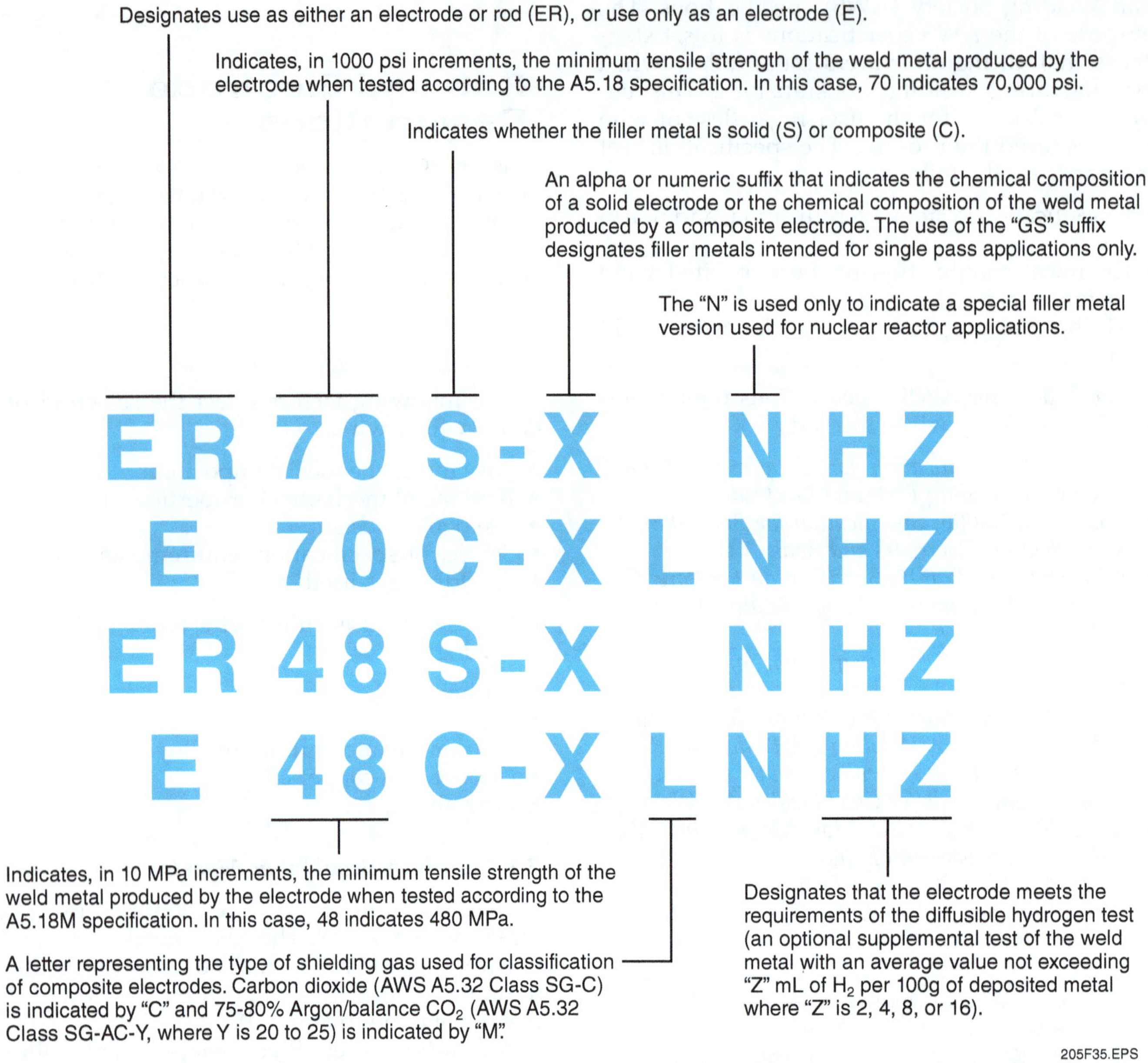

Figure 35 AWS classification for carbon steel filler metal.

- *Carbon (C)* – Concentrations of 0.05 to 0.12 percent are used to add strength without adversely affecting ductility, porosity, or toughness.
- *Nickel (Ni), chromium (Cr), and molybdenum (Mo)* – May be added in small amounts to improve corrosion resistance, strength, and toughness.

In many cases, carbon steel filler metal wire is copper coated. The copper coating protects the filler metal wire from corrosion and makes electrical contact easier as the wire passes through the copper contact tube at the end of a gun.

AWS carbon steel filler metal wire classifications and their uses are listed in *Table 5*.

7.1.2 Low-Alloy Steel Filler Metals

Low-alloy steel filler metals are identified by *AWS Specification A5.28*. The filler metal is available in wire reel and rod form. The wire classification (number) is located on a label on the side of the reel or container. The classification is similar to the carbon steel filler metal classification except that there is a letter in the suffix at the end of the classification (*Figure 36*). The letter suffix indicates that the filler metal contains alloys (nickel, manganese, molybdenum, etc.) that give the deposited weld metal special properties. Also, the tensile strength of low-alloy filler metals is often higher than that of carbon steel filler metals. For specific

Table 5 AWS Carbon Steel Wire Classifications and Uses

Wire Classification	Intended Uses
ER70S-2 ER48S-2	Welds all grades of carbon steel, including killed (highly deoxidized by adding aluminum powder or ferrosilicon), semikilled, and rimmed (slightly deoxidized by adding a small amount of aluminum powder). Also welds carbon steel with rusted surfaces. Acceptable shielding gases include Ar-O_2, Ar-CO_2, and CO_2. The viscous weld puddle is well suited to out-of-position welding. Wetting is helped by keeping the oxygen or CO_2 percentage relatively high.
ER70S-3 ER48S-3	Welds killed and semikilled steels (used in appliances, automobiles, and farm equipment) with CO_2, Ar-O_2, or Ar-CO_2 shielding gases.
ER70S-4 ER48S-4	Welds semikilled and rimmed steels (used in boilers and pressure vessels, piping, ship steels, structural steels, and A515 grades 55-70 steels) with CO_2, Ar-O_2, or Ar-CO_2 shielding gases.
ER70S-6 ER48S-6	Welds many carbon steels with CO_2 and high welding current. High-speed welding can be done with Ar-O_2 with 5% or more O_2. Either spray transfer or short-circuiting transfer can be used.
ER70S-7 ER48S-7	Welds a wide variety of steels with superior weld quality and appearance. Porosity and weld defects caused by mill scale or rust are minimized.
ER70S-G ER48S-G E70C-G E48C-G	Filler metals not included in the preceding classes that only have certain mechanical properties specified. Absence of a C or M suffix means that the manufacturer should be consulted for the recommended shielding gas and intended applications.
E70C-GS E48C-GS	Composite stranded or metal cored electrodes intended for only single pass applications. These electrodes may have higher alloy content and the manufacturer should be consulted for specific applications.
E70C-3 E48C-3 E70C-6 E48C-6	Composite stranded or metal cored electrodes intended for both single and multipass applications on steel. They are characterized by a spray arc and excellent bead wash characteristics.

205T05.EPS

information on low-alloy filler metal composition and intended use, refer to the manufacturer's specifications.

7.1.3 Stainless Steel Filler Metals

Stainless steel filler metals are identified by specification *AWS A5.9*. The filler metal is available in wire reel, flat strip, and rod form. The wire classification (number) is located on a label on the side of the reel or container (*Figure 37*).

Stainless steel filler metal should be selected to closely match the alloy composition of the base metal. Stainless steel filler metals also require specific shielding gases or gas mixtures.

7.1.4 Aluminum and Aluminum Alloy Filler Metals

Aluminum filler metals are covered by *AWS Specification A5.10*. Aluminum filler metals usually contain magnesium, manganese, zinc, silicon, or copper for increased strength. Corrosion resistance and weldability are also considerations. Aluminum filler metals are designed to weld specific types of aluminum and should be selected for compatibility. The most widely used aluminum filler metals are ER4043 (contains silicon) and ER5356 (contains magnesium). The filler metal classification (number) is located on a label on the side of the reel or container (*Figure 38*).

7.1.5 Copper and Copper Alloy Filler Metals

Copper filler metals are covered by *AWS Specification A5.7*. Most copper filler metals contain other elements that increase strength, deoxidize the weld metal, and match the base metal composition. *Figure 39* shows the AWS classification for copper filler metal.

7.1.6 Nickel and Nickel Alloy Filler Metals

Nickel-based filler metals (*Figure 40*) are covered by *AWS Specification A5.14*. These filler metals contain other elements to match base metal appli-

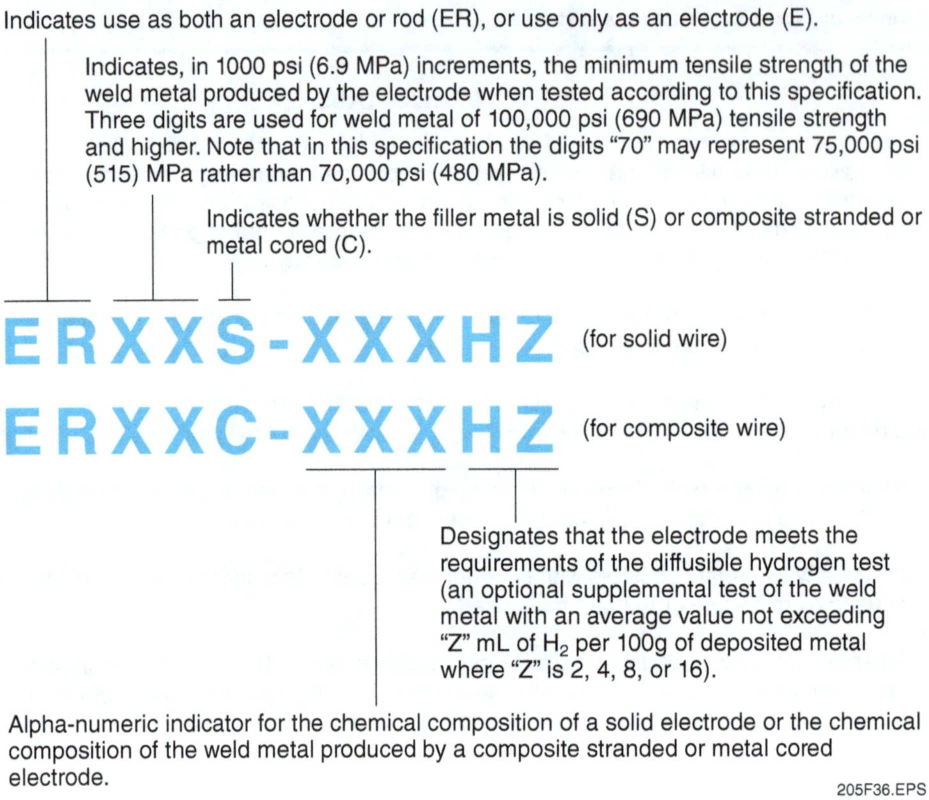

Figure 36 AWS classification for low-alloy steel filler metal.

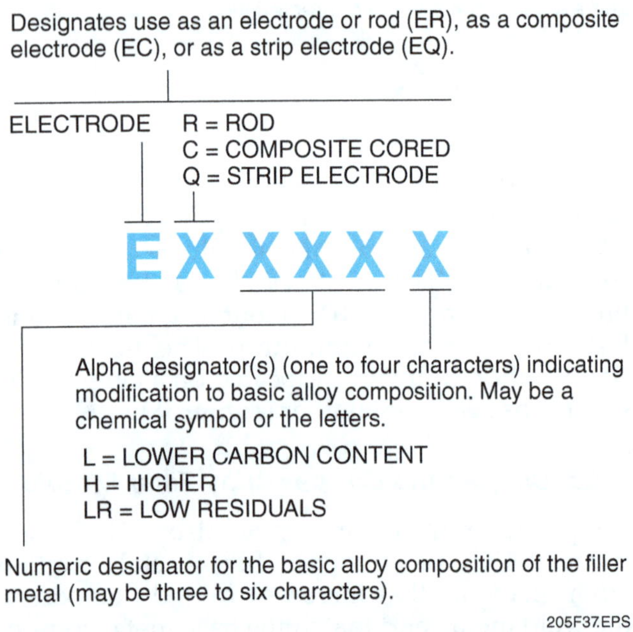

Figure 37 AWS classification for stainless steel filler metal.

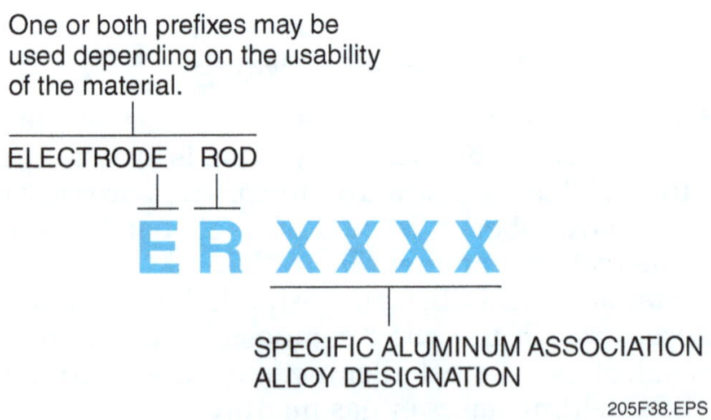

Figure 38 AWS classification for aluminum filler metal.

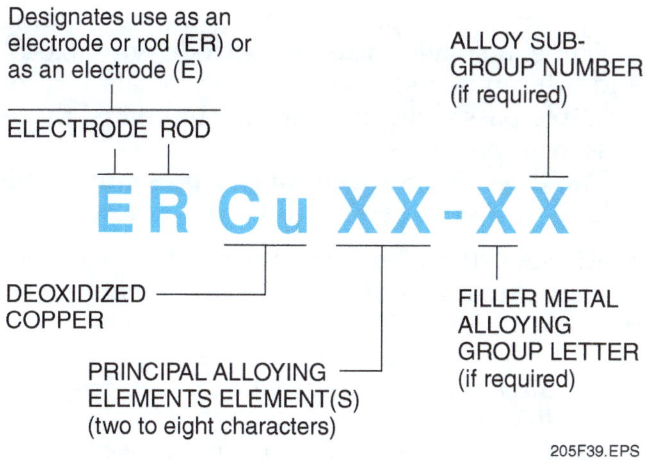

Figure 39 AWS classification for copper filler metal.

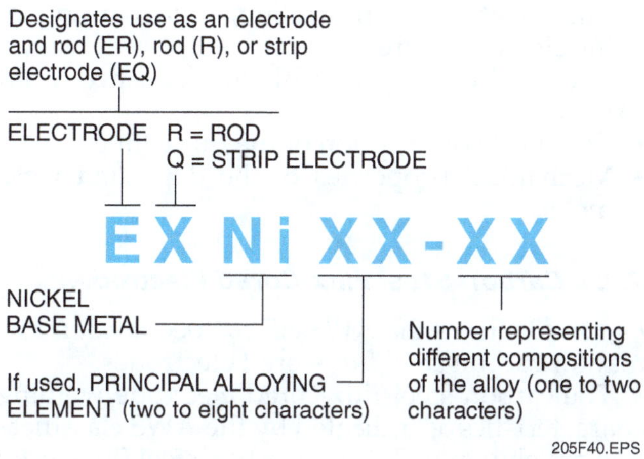

Figure 40 AWS classification for nickel filler metal.

cations and to increase the strength and quality of the weld metal. For GTAW, DCEN is used with high-purity argon, helium, or both as a shielding gas. For GMAW, DCEP is used. Argon shielding gas is often used, as are mixtures of argon and helium.

7.1.7 Magnesium Alloy Filler Metals

Magnesium alloy filler metals (*Figure 41*) are covered by *AWS Specification A5.19*. These filler metals are usually used with GTAW, GMAW, and PAW processes. Oxyfuel welding should be used only for temporary repair work. For GTAW welding, the techniques and equipment are similar to those for aluminum. Argon, helium, or both are used for shielding. Alternating current is preferred for arc cleaning and penetration. Direct current is also employed with DCEP, used for thin materials, and DCEN, used for mechanized welding with

Hot Tip

Magnesium Fires

While ignition of magnesium is a very remote possibility when welding, the ignition will cease when the heat source is removed. Ignition of the weld pool is also prevented by the gas shielding used in GTAW, GMAW, and PAW processes. Most magnesium fires occur with accumulations of fines from grinding or filing or chips from machining. Accumulations of fines on clothing should be avoided. Graphite-based or salt-based powders recommended for extinguishing magnesium fires should be stored in the work area. If large amounts of fines are produced, they should be collected in a waterwash-type of dust collector designed for use with magnesium. Special precautions for the handling of wet magnesium fines should be followed.

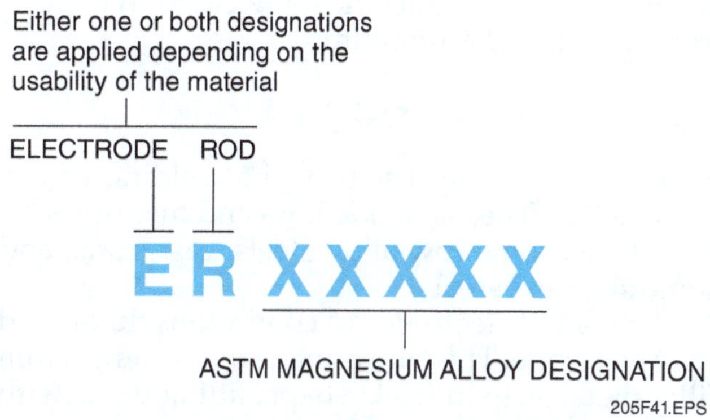

Figure 41 AWS classification for magnesium filler metal.

helium for deep penetration. GTAW is recommended for the defect repair of clean magnesium castings. GMAW of magnesium alloys is the same as for other metals. Argon is generally used for shielding; however, mixtures of argon and helium are occasionally used. Pulsed and short-circuit GMAW are used for magnesium alloys, and spray transfer without pulsing is also used; however, globular transfer is not suitable.

7.1.8 Titanium and Titanium Alloy Filler Metals

Titanium and titanium alloy filler metals (*Figure 42*) are covered by *AWS Specification A5.16*. These filler metals are generally used with GTAW, GMAW, SAW, and PAW processes. Titanium is sensitive to embrittlement by oxygen, nitrogen, and hydrogen at temperatures above 500°F. Like

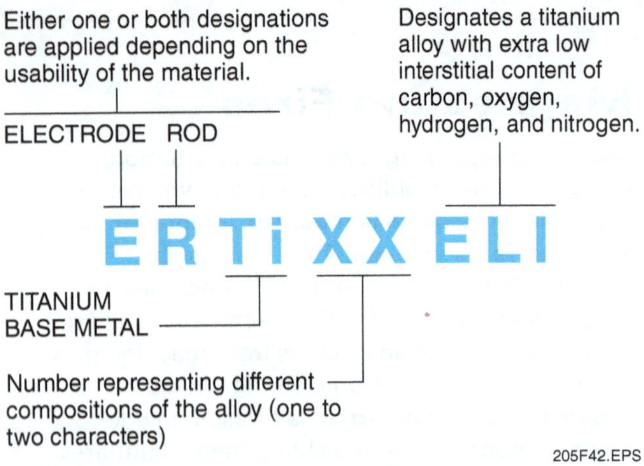

Figure 42 AWS classification for titanium filler metal.

aluminum, titanium requires weld cleaning and high-purity gas shielding, and an especially adequate postflow shielding to avoid embrittlement. Titanium can be successfully fusion welded to zirconium, tantalum, niobium, and vanadium. Titanium should not be fusion welded to copper, iron, nickel, and aluminum.

7.2.0 FCAW Filler Metals

FCAW is primarily used for the welding of ferrous metals. These include low- and medium-carbon steels, some low-alloy steels, cast irons, and some stainless steels.

The FCAW electrode is a continuous flux-cored wire electrode. The wire is made by forming a thin filler metal strip into a U-shape, filling the U with the flux material, squeezing the strip shut around the flux, and then drawing the tube through dies to size the wire and compact the flux. During manufacture, the wire is carefully inspected for defects. Flux-cored wires are manufactured in a range of sizes from 0.030" to ⅝₂". The finished flux-cored wire is coiled on all the standard spool sizes and forms.

> **NOTE**
> All flux cored wires are considered to be low hydrogen. For this reason, proper electrode control procedures must be followed to keep the wire dry.

By varying the power settings on the welding machine, a smaller size of electrode wire (0.045" and 1⁄16") can be used both for thin metals and for multiple-pass welds on metals of unlimited thickness in all positions.

The three AWS specifications presently available for flux-cored electrodes are as follows:

- *AWS A5.20/A5.20M:2005 Specification for Carbon Steel Electrodes for Flux Cored Arc Welding*
- *AWS A5.22:2008 Specification for Stainless Steel Electrodes for Flux Cored Arc Welding and Stainless Steel Flux Cored Rods for Gas Tungsten Arc Welding*
- *AWS A5.29/A5.29M:2005 Specification for Low-Alloy Steel Electrodes for Flux Cored Arc Welding*

Each specification defines standards for the following:

- Classification, identification, and marking of the electrode wire
- Chemical composition of the deposited weld metal
- Chemical composition of the flux filling
- Mechanical properties of the deposited weld metal

7.2.1 Carbon Steel Flux-Cored Electrodes

Figure 43 shows the *AWS A5.20* specification format for carbon steel flux-cored electrodes.

The carbon steel flux-cored electrode welding characteristics as indicated by the AWS classification are shown in *Table 6*. Carbon steel flux-cored electrode classes and uses are shown in *Table 7*.

7.2.2 Low-Alloy Steel Flux-Cored Electrodes

Figure 44 shows the *AWS A5.29* specification format for low-alloy steel flux-cored electrodes.

7.2.3 Stainless Steel Flux-Cored Electrodes and Rods

Figure 45 shows the *AWS A5.22* specification format for stainless steel flux-cored electrodes and rods.

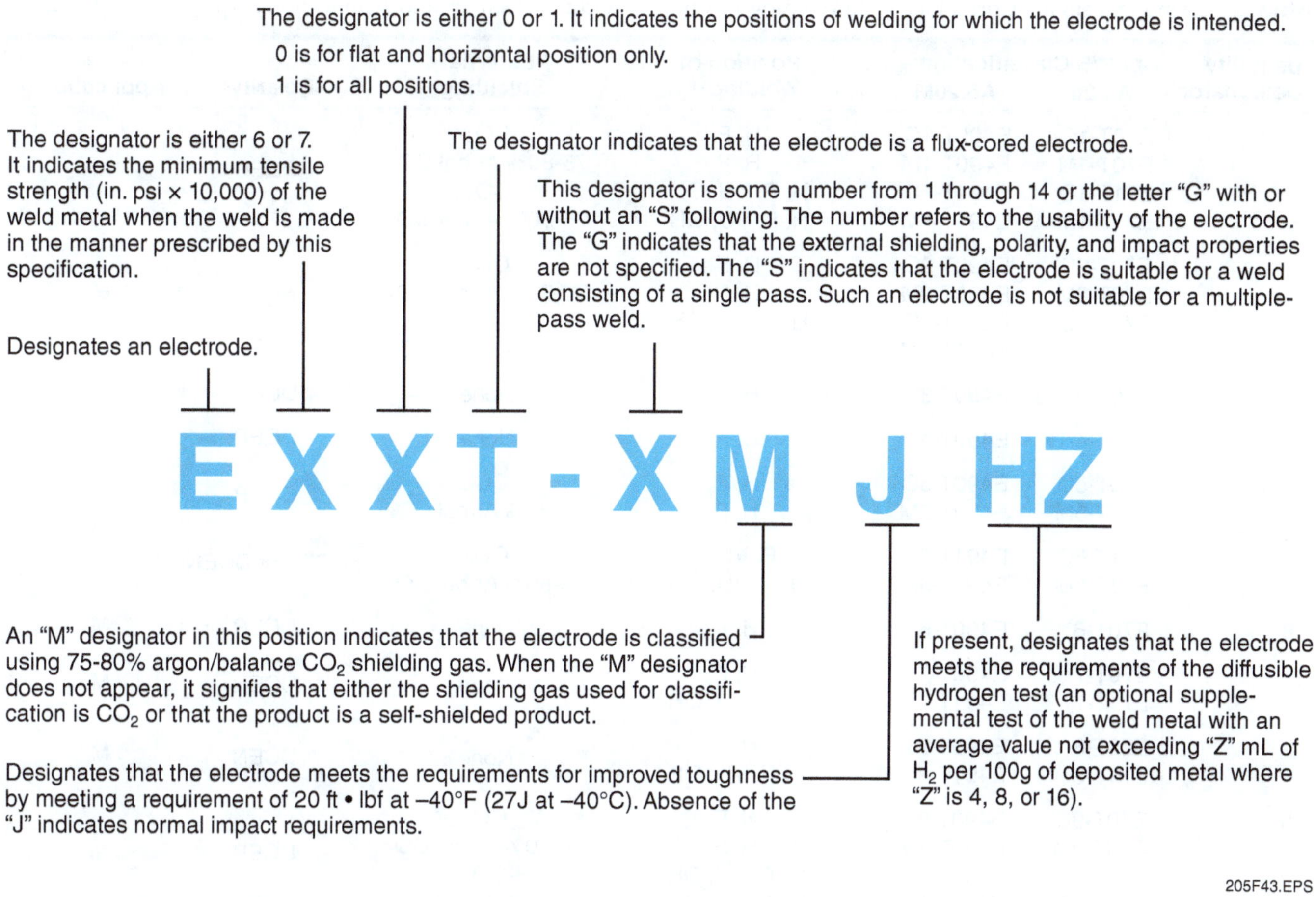

Figure 43 AWS classification for carbon steel flux-cored electrodes.

Hot Tip

FCAW of Structural Steel

For large structural welding projects, such as bridges and skyscrapers, many construction companies prefer carbon (or low-alloy) steel FCAW to meet the toughness requirements for new construction using the newer self-shielding E71T-8 or gas-shielded E71T-1 electrodes. Many ironworkers who use SMAW are reluctant to use FCAW, and those who have used FCAW predominantly relied upon E71T-11 for structural welds because it was easy to use in all positions. However, welds with –11 wire sometimes had cracking problems and had to be rewelded. When welders are properly retrained for FCAW using the newer electrodes, the cracking problems are eliminated and the deposition rates are higher because the newer electrodes run hotter and faster than the –11 electrodes or SMAW electrodes.

Table 6 Carbon Steel Flux-Cored Electrode Welding Position, Shielding, Polarity, and Application Requirements (1 of 2)

Usability Designator	AWS Classification A5.20	A5.20M	Position of Welding[a,b]	External Shielding[c]	Polarity[d]	Application[e]
1	E70T-1C	E490T-1C	H, F	CO_2	DCEP	M
	E70T-1M	E490T-1M	H, F	75-80% Ar/bal CO_2		
	E71T-1C	E491T-1C	H, F, VU, OH	CO_2		
	E71T-1M	E491T-1M	H, F, VU, OH	75-80% Ar/bal CO_2		
2	E70T-2C	E490T-2C	H, F	CO_2	DCEP	S
	E70T-2M	E490T-2M	H, F	75-80% Ar/bal CO_2		
	E71T-2C	E491T-2C	H, F, VU, OH	CO_2		
	E71T-2M	E491T-2M	H, F, VU, OH	75-80% Ar/bal CO_2		
3	E70T-3	E490T-3	H, F	None	DCEP	S
4	E70T-4	E490T-4	H, F	None	DCEP	M
5	E70T-5C	E490T-5C	H, F	CO_2	DCEP	M
	E70T-5M	E490T-5M	H, F	75-80% Ar/bal CO_2		
	E71T-5C	E491T-5C	H, F, VU, OH	CO_2	DCEP or DCEN[f]	
	E71T-5M	E491T-5M	H, F, VU, OH	75-80% Ar/bal CO_2		
6	E70T-6	E490T-6	H, F	None	DCEP	M
7	E70T-7	E490T-7	H, F	None	DCEN	M
	E71T-7	E491T-7	H, F, VU, OH			
8	E70T-8	E490T-8	H, F	None	DCEN	M
	E71T-8	E491T-8	H, F, VU, OH			
9	E70T-9C	E490T-9C	H, F	CO_2	DCEP	M
	E70T-9M	E490T-9M	H, F	75-80% Ar/bal CO_2		
	E71T-9C	E491T-9C	H, F, VU, OH	CO_2		
	E71T-9M	E491T-9M	H, F, VU, OH	75-80% Ar/bal CO_2		
10	E70T-10	E490T-10	H, F	None	DCEN	S

205T06A.EPS

Table 6 Carbon Steel Flux-Cored Electrode Welding Position, Shielding, Polarity, and Application Requirements (2 of 2)

Usability Designator	AWS Classification A5.20	A5.20M	Position of Welding[a,b]	External Shielding[c]	Polarity[d]	Application[e]
11	E70T-11 E71T-11	E490T-11 E491T-11	H, F H, F, VU, OH	None	DCEN	M
12	E70T-12C E70T-12M E71T-12C E71T-12M	E490T-12C E490T-12M E491T-12C E491T-12M	H, F H, F H, F, VU, OH H, F, VU, OH	CO_2 75-80% Ar/bal CO_2 CO_2 75-80% Ar/bal CO_2	DCEP	M
13	E61T-13 E71T-13	E431T-13 E491T-13	H, F, VU, OH	None	DCEN	S
14	E71T-14	E491T-14	H, F, VU, OH	None	DCEN	S
G	E60T-G E70T-G	E430T-G E490T-G	H, F	Not Specified	Not Specified	M
	E61T-G E71T-G	E431T-G E491T-G	H, F, VD or VU, OH	Not Specified	Not Specified	M
	E60T-GS E70T-GS	E430T-GS E490T-GS	H, F	Not Specified	Not Specified	S
	E61T-GS E71T-GS	E431T-GS E491T-GS	H, F, VD or VU, OH	Not Specified	Not Specified	S
	EX0T-G[g] EX1T-G[g] EX0T-GS[g] EX1T-GS[g]	EX0T-G[g] EX1T-G[g] EX0T-GS[g] EX1T-GS[g]	H, F H, F, VD or VU, OH H, F H, F, VD or VU, OH	Not Specified Not Specified Not Specified Not Specified	Not Specified Not Specified Not Specified Not Specified	S

Notes:

a. H = horizontal position; F = flat position; OH = overhead position; VD = vertical position with downward progression; VU = vertical position with upward progression.

b. Electrode sizes suitable for out-of-position welding, i.e., welding position other than flat or horizontal, usually are those sizes that are smaller than the 3⁄32" (2.4 mm) size. For that reason, electrodes meeting the requirements for the groove weld tests and the fillet weld tests may be classified as EX1T-X or EX1T-XM (where X represents the tensile strength and usability designators) regardless of their size.

c. Properties of weld metal from electrodes that are used with external gas shielding (EXXT-1, EXXT-1M, EXXT-2, EXXT-2M, EXXT-5, EXXT-5M, EXXT-9, EXXT-9M, EXXT-12, and EXXT-12M) vary according to the shielding gas employed. Electrodes classified with the specified shielding gas should not be used with other shielding gases without first consulting the manufacturer of the electrode.

d. The term "DCEP" refers to direct current electrode positive (dc, reverse polarity). The term "DCEN" refers to direct current electrode negative (dc, straight polarity).

e. M = single or multiple pass; S = single pass only. (Single-pass only classes are highly oxidizing and are used on metals that are rusty or have mill scale.)

f. Some E71T-5 and E71T-5M electrodes may be recommended for use on DCEN for improved out-of-position welding. Consult the manufacturer for the recommended polarity.

g. The letter "X" can be replaced with either a "6" or "7" to designate the weld metal's tensile strength.

205T06B.EPS

Table 7 Carbon Steel Flux-Cored Electrode Classes and Uses (1 of 2)

AWS Classification	Intended Uses
EXXT-1 EXXT-1M	Electrodes designated as –1 use CO_2 shielding. However, other gas mixtures can be used to improve arc stability if recommended by the manufacturer. Electrodes designated as –1M use an argon-CO_2 mixture to increase manganese and silicon in the weld metal to increase yield and tensile strengths. Reducing argon in a gas mixture or using CO_2 shielding only with –1M electrodes may cause deterioration of arc and out-of-position welding characteristics and loss of strength. Both electrodes have spray transfer characteristics and low spatter. Most electrodes of this class have rutile-base slag and produce high disposition rates.
EXXT-2 EXXT-2	These electrodes are similar to –1 and –1M types with higher manganese or silicon or both. The higher levels of deoxidizers allow single-pass welding of heavily oxidized or rimmed steel. The –2M electrodes have an even higher manganese content and can be used on metals with heavier mill scale, rust, or foreign matter that cannot be tolerated by some –1 or –1M electrodes.
EXXT-3	These self-shielding electrodes are designed for single-pass, very high-speed welding and use spray transfer. Besides the listed positions, they can be used on 20° inclined downhill welds of sheet metal. They are not recommended for T or lap joints in materials thicker than 3⁄16" or for butt, edge, or corner joints in materials thicker than 1⁄4".
EXXT-4	These self-shielding electrodes use globular-type transfer and are designed for very high deposition rates. They produce low-sulfur welds that resist hot cracking. They are also designed for low penetration beyond the root of the weld because of poor joint fitup.
EXXT-5 EXXT-5M	These electrodes have a globular transfer and a lime-fluoride base slag that produces welds that have better impact strength and hot and/or cold cracking resistance than other similar classes with rutile-based slags. The –5 electrodes are used with a CO_2 shielding gas, however, argon-CO_2 mixes can be used to reduce spatter if recommended by the manufacturer. Reducing argon in a gas mixture or using CO_2 shielding only with –5M electrodes may cause deterioration of arc an out-of-position welding characteristics and loss of strength. Operator appeal for both types of electrodes is not as good as other similar classes with rutile slag.
EXXT-6	These self-shielding electrodes use spray-type transfer. The slag system is designed for good penetration into the root of the weld and excellent slag removal, even in deep grooves.
EXXT-7	These self-shielding electrodes use a droplet/spray-type transfer and produce a low-sulfur weld that is very resistant to cracking. The slag system is designed to produce high deposition rates for larger size electrodes in the flat or horizontal positions.
EXXT-8	These self-shielded electrodes have small droplet/spray-type transfer. They are suitable for all positions. The weld metal has good toughness, low temperature performance, and crack resistance.
EXXT-9 EXXT-9M	These electrodes are essentially the same as the –1 and –M electrodes except the deposit weld metal with improved impact properties. Some electrodes in this class require joints that are relatively clean of oil, excess oxide, and scale.
EXXT-10	These self-shielding electrodes have small droplet transfer. They are used for high speed, single-pass welds on any thickness material in the flat, horizontal, and 20° inclined vertical positions.
EXXT-11	These self-shielding electrodes have a smooth spray-type transfer and are general-purpose electrodes for all positions. Their use on material thicknesses greater than 3⁄4" is not generally recommended unless preheat and interpass temperatures are maintained. Consult the manufacturer for specific recommendations.

205T07A.EPS

Table 7 Carbon Steel Flux-Cored Electrode Classes and Uses (2 of 2)

AWS Classification	Intended Uses
EXXT-12 EXXT-12M	These electrodes are essentially the same as –1 and –1M electrodes except they are modified to improve impact toughness and to lower manganese requirements of the A-1 Analysis Group in the *ASME Boiler and Pressure Code* Section IX. As a result, tensile strength and hardness will be decreased from –1 and –1 electrodes. Users should check hardness when a required hardness level is specified or required.
EXXT-13	These self-shielding electrodes are usually welded with a short-circuiting transfer. The slag system allows an all-position root pass on pipe welds. They are recommended only for the root pass and are not generally recommended for multiple passes.
EXXT-14	These self-shielding electrodes have a smooth spray-type transfer. The slag system allows high speed welds in all positions. They are used to make welds on sheet metal up to 3⁄16" thick and are often designed especially for galvanized, aluminized or other coated steels. They are generally not recommended for T or lap joints in materials thicker than 3⁄16" or butt, edge, or corner joints in materials thicker than 1⁄4". The manufacturer should be consulted for specific recommendations.
EXXT-G	These are multipass electrodes not covered by other classifications. Except for requirements to assure carbon steel deposits and tensile strengths, other requirements are agreed upon by the purchaser and manufacturer.
EXXT-GS	These are single-pass electrodes not covered by other classifications. Except for tensile strength, other requirements are agreed upon by the purchaser and the manufacturer.

205T07B.EPS

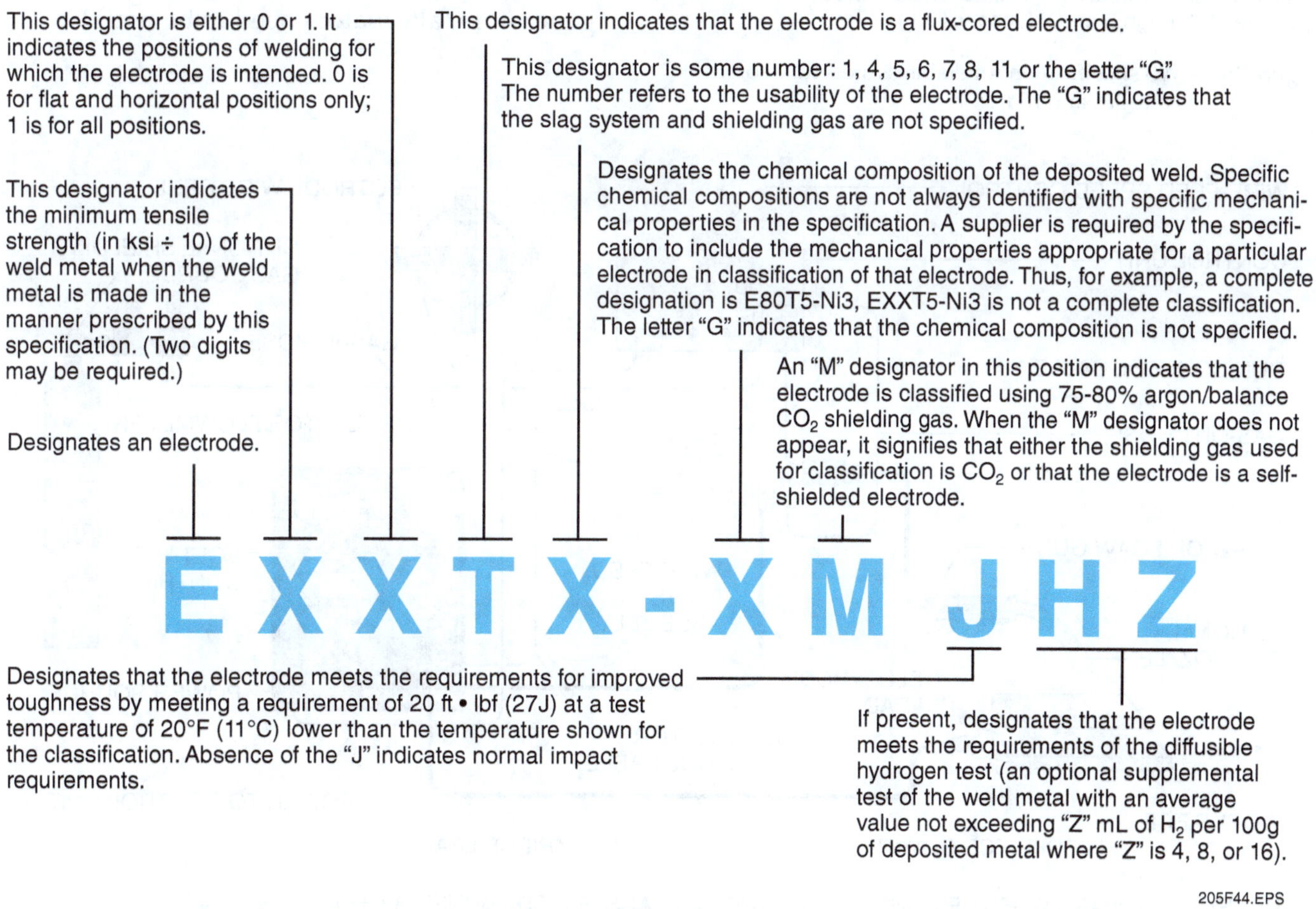

Figure 44 AWS specification for low-alloy steel flux-cored electrodes.

8.0.0 Welding Equipment Setup

To weld safely and efficiently, the welding equipment must be properly set up (*Figure 46*). The following sections explain the steps for setting up welding equipment.

8.1.0 Selecting a GMAW Power Source

To select a GMAW power source (welding machine), the following factors must be considered:

- GMAW requires a constant-voltage or constant voltage/constant current power source.
- A DCEP-type welding current is required.
- The maximum amperage required must be available.
- The primary power requirements must be met; for example, are there electrical receptacles to plug a welding machine into, or do engine-driven generators/alternators need to be used?

8.2.0 Selecting an FCAW Power Source

To select an FCAW welding machine, the following factors must be considered:

- FCAW requires a constant-voltage power source.
- The welding current required for FCAW is DC.
- The required operating amperage ranges from 150 amps with a 0.045" electrode to 650 amps or more with a ⅛" electrode; larger electrodes require even higher amperage.
- Is the primary power source adequate? Are there enough AC electrical receptacles to plug a welding machine into, or do engine-driven generators/alternators need to be used?

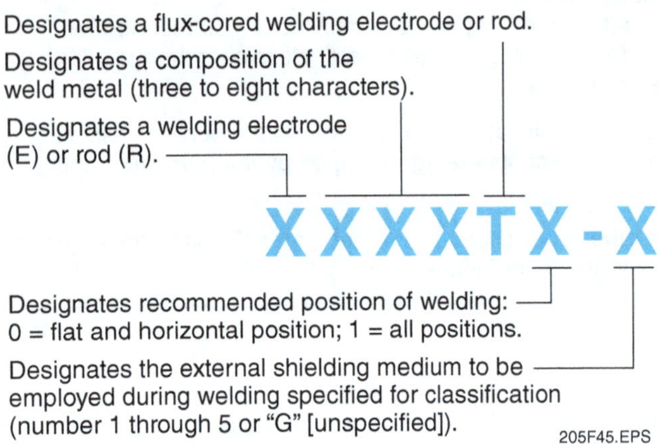

Figure 45 AWS specification for stainless steel flux-cored electrodes and rods.

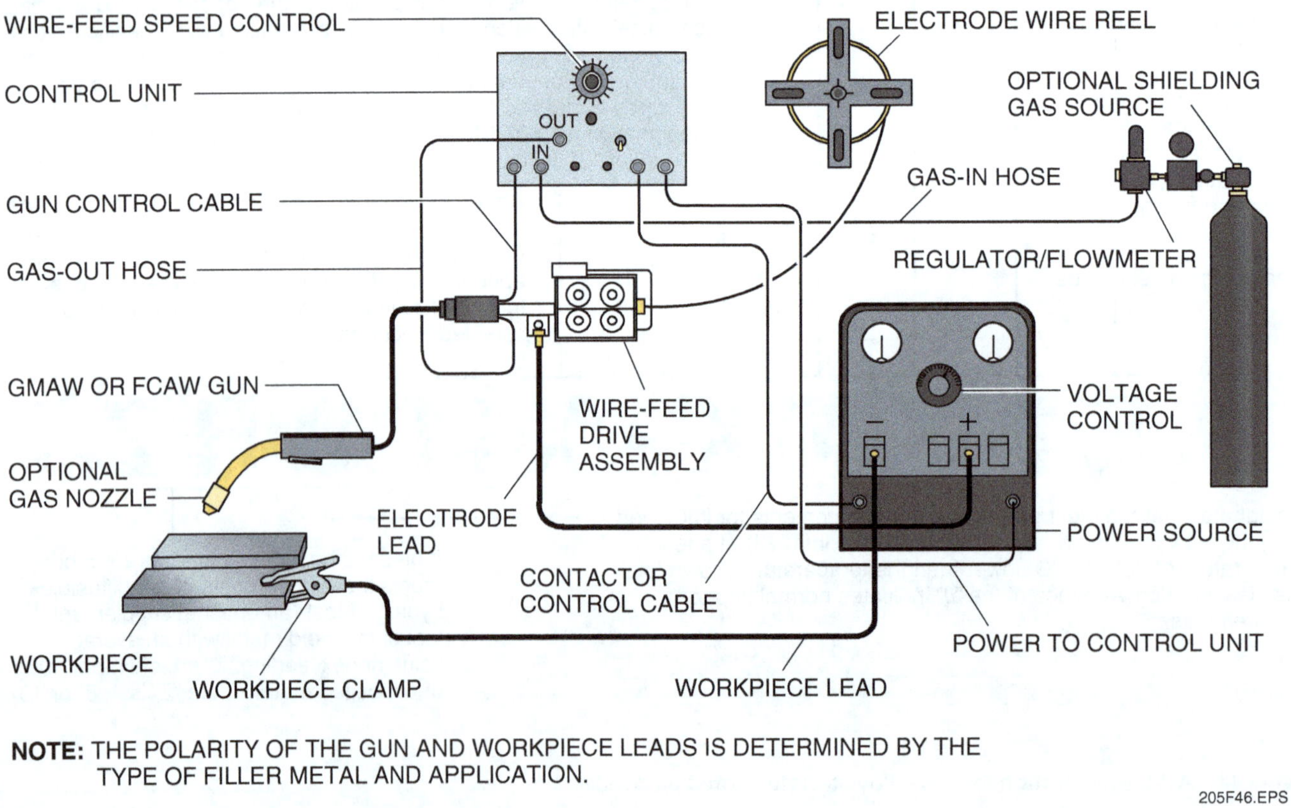

Figure 46 Diagram of a GMAW/FCAW system.

8.3.0 Positioning the Equipment

Because the gun cables are short, the wire feeder must be positioned close to the work to be performed. The power source must also be located near the wire feeder because of the limited length of the contactor/control cable that runs from the power source to the wire feeder. Normally, the wire feeder is located on top of the power source, on the floor, or a cart near the power source.

Select a site where the equipment will not be in the way but will be protected from welding or cutting sparks. There should be good air circulation to keep the machine cool. The environment should be free from explosive or corrosive fumes and as free as possible from dust and dirt. Welding machines have internal cooling fans that will pull these materials into the welding machine if they are present. Also, dust and dirt can collect on the wire, causing weld contamination and clogging of the liner in the gun cable. The site should be free of standing water or water leaks. If an engine-driven generator or alternator is used, position it so that it can be easily refueled and serviced.

There should be easy access to the site so that the equipment can be started, stopped, or adjusted as needed. If the machine will be plugged into an outlet, be sure that the outlet has been properly installed by a licensed electrician to ensure that it is grounded. Also, be sure to identify the location of the electrical disconnect before plugging the welding machine into the outlet.

8.4.0 Moving Welding Power Sources

Large engine-driven generators are mounted on a trailer frame, and they can easily be moved by a pickup truck or tractor with a trailer hitch. Other types of welding machines may have a skid base or be mounted on steel or rubber wheels. Be careful when moving wheel-mounted welding machines by hand. Some machines are top-heavy and could fall over in a tight turn or if the floor or ground is uneven or soft.

CAUTION

Secure or remove the wire feeder and gas cylinder before attempting to move the power source.

WARNING!

If a welding machine starts to fall over, do not attempt to hold it. Welding machines are very heavy, and severe crushing injuries can occur if a welding machine falls on you.

Most welding machines have a lifting eye that is used to move machines mounted on skids or to lift any machine. Before lifting a welding machine, check the equipment specifications for the weight. Be sure the lifting device and tackle can handle the weight of the machine. Always use a sling and/or a shackle. Never attempt to lift a machine by placing the lifting hook directly in the machine's lifting eye. This is because the safety latch on the hook cannot be closed. Also, before lifting or moving a welding machine, be sure the welding cables are secure. *Figure 47* shows the proper way to lift a welding machine.

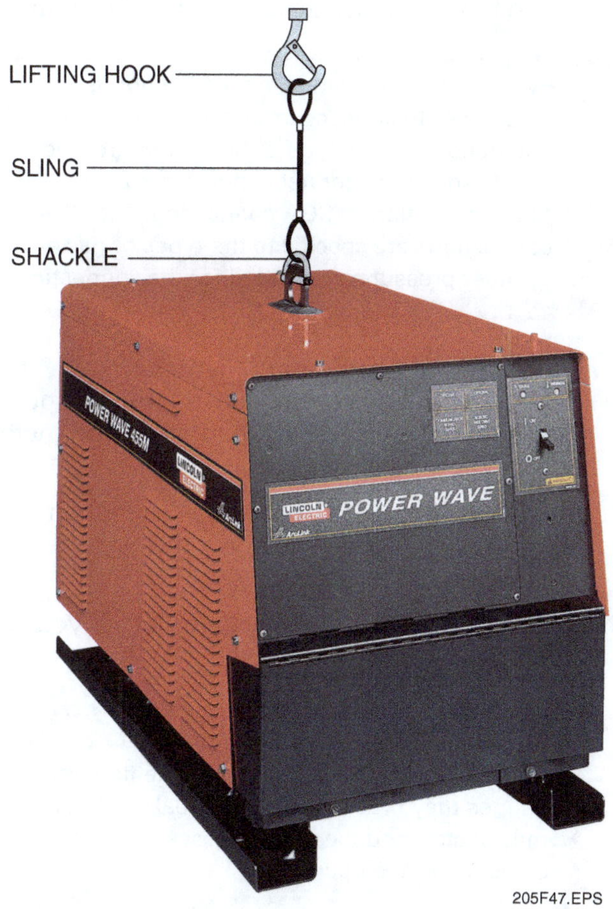

Figure 47 Lifting a welding machine.

8.5.0 Connecting the Shielding Gas

The shielding gas connects to the wire feeder unit or to the power source if the wire feeder is integrated with the power source.

Follow these steps to connect the shielding gas:

Step 1 Identify the required shielding gas by referring to the WPS or site quality standards.

Step 2 Locate a cylinder of the correct gas or mixture, and secure it near the wire feeder.

> **CAUTION**
> Be sure to secure the cylinder so that it cannot fall over.

Step 3 Remove the cylinder's protective cap, and momentarily crack open the cylinder valve to blow out any dirt, then close it again.

> **WARNING!**
> Even though they are available, do not use an adapter to connect a regulator/flowmeter equipped with one type of CGA (Compressed Gas Association) connection to a gas cylinder that has a different CGA connection. The CGA connections are specific to the types of gas and cylinder pressures permitted for the connection.

Step 4 Using a regulator/flowmeter with the correct CGA connection for the cylinder, install the pressure regulator/flowmeter on the cylinder.

Step 5 Connect the gas hose to the flowmeter and to the gas connection on the wire feeder (or power source).

Step 6 Slowly crack open the cylinder valve, and then open it completely.

> **WARNING!**
> Do not open the valve quickly or the flow meter gauges may break. If a gauge breaks, the glass may shatter and the gas may escape, which could injure the operator.

Hot Tip

Adjusting Flowmeters

In most cases, the top of the floating flowmeter ball is used as the reference point for setting the gas flow. However, always check the manufacturer's instructions to determine the correct adjustment method. Use the lowest flow rate as often as possible to conserve gas and protect the environment.

205SA05.EPS

Step 7 Adjust the gas flow to the specified flow rate using the flowmeter valve.

> **NOTE**
> Some flowmeters are equipped with several scales of different calibrations around the same sight tube, for monitoring the flows of different types (densities) of gases. Be sure to rotate the scales or read the correct side for the type of gas being used.

Hot Tip

Gas Flow Rates

More gas flow is not always better. Too much flow can deflect off the workpiece. The resultant swirling gas can create turbulence that pulls in air, which contaminates the weld and creates pinholes. If black flecks occur in the weld puddle or pinholes occur in the weld, try reducing the gas flow.

8.6.0 Selecting and Installing Filler Wire

Follow these steps to select and install the filler wire:

Step 1 Identify the filler wire required by referring to the WPS or site quality standard.

Step 2 Locate a spool of the wire, and mount it on the wire feeder spool holder. Lock the holder, and set the drag brake for a slight drag. The brake control is usually located on the end of the spool axle.

Step 3 Check (and change if necessary) the wire feed drive wheels, liner, and gun contact tube to make sure that they are the correct size for the wire selected. Adjust the wheel tension if necessary. For detailed instructions on changing or adjusting the wire feeder drive wheels or changing the gun's contact tube or liner, refer to the specific manufacturer's instructions.

Step 4 Feed the wire into the feeder wheels, and then use the JOG control to feed it through the gun cable and the gun contact tube.

8.7.0 Placing the Workpiece Clamp

The workpiece clamp must be properly placed to prevent damage to surrounding equipment. If the electrical welding current travels through a bearing, seal, valve, or contacting surface, it could cause severe damage from heat and arcing, which would require that these items be replaced. Carefully check the area to be welded, and position the workpiece clamp so the welding current will not pass through any contacting surface. When in doubt, ask your supervisor for assistance before proceeding.

CAUTION

Welding current passing through electrical or electronic equipment will cause severe damage to the equipment. Before welding on any type of mobile equipment, the ground lead at the battery must be disconnected to protect the electrical system. If welding done is near the battery, the battery must be removed. Batteries produce hydrogen gas, which is extremely explosive. A welding spark could cause the battery to explode.

WARNING!

Do not weld near batteries. A welding spark could cause a battery to explode, showering the area with battery acid.

Workpiece clamps must never be connected to pipes carrying flammable or corrosive materials. The welding current could cause overheating or sparks, resulting in an explosion or fire.

The workpiece clamp must make a good electrical contact when it is connected. Dirt and paint will inhibit the connection and cause arcing, resulting in overheating of the workpiece clamp. Dirt and paint also affect the welding current and can cause defects in the weld. Clean the surface before connecting the workpiece clamp. If the workpiece clamp is damaged or does not close securely onto the surface, replace it.

8.8.0 Energizing the Power Source

Electrically powered welding machines are energized by plugging them into an electrical outlet. The electrical requirements (primary current) will be located on the equipment specification tag displayed prominently on the machine. Most machines require single-phase 240V power, 208V three-phase power, or 460V three-phase power. Most machines requiring single-phase 240V power have a three-pronged plug. Machines requiring three-phase 208V or 460V power have a four-pronged plug. Both types of plugs are shown in *Figure 48*.

If a welding machine does not have a power plug, an electrician must connect it. The electrician will add a plug or hard-wire the machine directly into an electrical box.

WARNING!

Never use a welding machine until you identify the electrical disconnect for that machine. In the event of an emergency, you must be able to quickly turn off the power to the welding machine at the disconnect.

205F48.EPS

Figure 48 Typical three- and four-pronged plugs.

8.9.0 Starting Engine-Driven Generators

Before welding can take place with an engine-driven generator, the engine must be checked and then started. As with a car engine, the engine powering the generator must also have routine maintenance performed.

8.9.1 Performing Pre-Start Checks

Many sites have pre-start checklists that must be completed and signed prior to starting or operating an engine-driven generator. Check with your supervisor. If your site has such a checklist, complete and sign it. If your site does not have a pre-start checklist, perform the following checks before starting the engine:

- Check the oil, using the engine oil dipstick. If the oil is low, add the appropriate grade of oil for the time of year.
- Check the coolant level in the radiator if the engine is liquid-cooled. If the coolant level is low, add coolant.

CAUTION

Do not add plain water to radiators that contain antifreeze. Antifreeze not only protects radiators from freezing in cold weather, it also has rust inhibitors and additives to aid cooling. If the antifreeze is diluted, it will not function properly. If the weather turns cold, the system may freeze, causing damage to the radiator, engine block, and water pump.

- Check the fuel. The unit may have a fuel gauge or a dipstick. If the fuel is low, add the correct fuel (diesel or gasoline) to the fuel tank. The type of fuel required should be marked on the fuel tank. If it is not marked, contact your supervisor to verify the fuel required, and have the tank marked.

CAUTION

Adding gasoline to a diesel engine or diesel to a gasoline engine will cause severe engine problems and can also cause a fire hazard. Always be sure to add the correct fuel to the fuel tank.

- Check the battery water level unless the battery is sealed. Add demineralized water if the battery water level is low.
- Check the electrode holder to be sure it is not grounded. If the electrode holder is grounded, it will arc and overheat the welding system when the welding machine is started. This is a fire hazard and will cause damage to the equipment.
- Open the fuel shut-off valve if the equipment has one. If there is a fuel shut-off valve, it should be located in the fuel line between the fuel tank and the carburetor.
- Record the hours from the hour meter if the equipment has one. An hour meter records the total number of hours that the engine runs. This information is used to determine when the engine needs to be serviced. The hours will be displayed on a gauge similar to an odometer.
- Clean the unit. Use a compressed air hose to blow off the engine and generator or alternator. Use a rag to remove heavier deposits that cannot be removed with the compressed air.

WARNING!

Always wear eye protection when using compressed air to blow dirt and debris from surfaces. Never point the nozzle at yourself or anyone else.

NOTE

Cleaning may not be required on a daily basis. Clean the unit as required.

8.9.2 Starting the Engine

Most engines have an On/Off ignition switch and a starter. They may be combined into a key switch similar to the ignition on a car. To start the engine, turn on the ignition switch and press the starter. Release the starter when the engine starts. The engine speed is controlled by the governor. If the governor switch is set for idle, the engine will slow to an idle after a few seconds. If the governor is set to welding speed, the engine will continue to run at the welding speed.

Small engine-driven generators may have an On/Off switch and a pull cord. These are started by turning on the ignition switch and pulling the cord, in much the same way as a lawn mower is started.

Engine-driven generators should be started about 5 to 10 minutes before they are needed for welding. This will allow the engine to warm up before a welding load is applied.

8.9.3 Stopping the Engine

If no welding is required for thirty minutes or more, stop the engine by turning off the ignition switch. If you are finished with the welding machine for the day, close the fuel valve if there is one.

8.9.4 Performing Preventive Maintenance

Engine-driven generators require regular preventive maintenance to keep the equipment operating properly. Most sites have a preventive maintenance schedule based on the hours that the engine operates. In severe conditions, such as in very dusty environments or cold weather, maintenance may have to be performed more frequently.

CAUTION

To prevent damage to the equipment, perform preventive maintenance as recommended by the site procedures or by the manufacturer's maintenance schedule in the equipment manual.

The responsibility for performing preventive maintenance varies by site. Check with your supervisor to determine who is responsible for performing preventive maintenance.

When performing preventive maintenance, follow the manufacturer's guidelines in the equipment manual. Typical procedures to be performed as a part of preventive maintenance include the following:

Hot Tip

Dust Removal

Blow out the welding machine and drive unit with compressed air occasionally to remove dust. A significant buildup of dust inside the machine can provide an alternate path for the electricity and could result in some of the components shorting out.

- Changing the oil
- Changing the gas filter
- Changing the air filter
- Checking/changing the antifreeze
- Greasing the undercarriage
- Repacking the wheel bearings

9.0.0 Hand Tools for Weld Cleaning

The tools used for weld cleaning include handheld tools, such as chipping hammers, wire brushes, and pliers, and power tools, such as pneumatic weld-flux chippers and needle scalers.

Wire brushes are used to clean welds and to remove paint or surface corrosion. Wire brushing removes light to medium corrosion, but does not remove tight corrosion. Tight corrosion must be removed by filing, grinding, or sandblasting.

Chipping hammers are used to remove welding slag. The head of a chipping hammer has a point at one end and a chisel at the other. When chipping hammers become dull, they can be sharpened on a grinder. When sharpening chipping hammers, use care not to overheat the head. The head is hardened by tempering, and overheating removes the temper, causing the head to become soft. If the temper is removed from a chipping hammer head, it will mushroom and wear out much faster than a tempered head. Prevent overheating by plunging the head of the chipping hammer into a pail of water every few seconds while grinding.

Welders should also have clamping pliers for handling hot metal or for holding and aligning metal to be welded. Hot metal should not be handled with leather welding gloves. The heat will cause the leather to shrivel and become hard. *Figure 49* shows a wire brush, chipping hammers, various clamping pliers, and a common file. Before they are used, files should have handles installed on the tang to prevent injury.

Figure 49 Wire brush, chipping hammers, clamping pliers, and file.

SUMMARY

Both gas metal arc welding and flux-cored arc welding can be dangerous if the proper safety precautions are not followed. Before proceeding, make sure that you understand the equipment and filler metal requirements and the safety precautions presented in this module. Before GMAW or FCAW can begin, the appropriate equipment and filler metals must be selected and the equipment set up. By following the recommendations in this module, you will be able to select and safely set up the appropriate welding equipment. In addition, this module will provide you with the information you need to understand the types of filler metals used for GMAW and FCAW.

Review Questions

1. In GMAW, the arc is stabilized and the molten filler and base metals protected from oxidation by _____.

a. the application of pressure
b. use of self-fluxing alloy
c. use of intense heat
d. shielding gas

2. Of the two basic FCAW processes, one is self-shielding and the other one uses _____ shielding gas.

a. intermittent
b. internal
c. external
d. constant

3. When working near oil, grease, and other petroleum products, keep large amounts of _____ away from them.

a. argon
b. helium
c. oxygen
d. hydrogen

4. Depending on the range selected, for each 100 amperes of current, the arc voltage of a welding machine is typically lower than the open-circuit voltage by _____ volt(s).

a. 0 to 1
b. 1 to 2
c. 2 to 3
d. 4 to 5

5. The welding arc generates intense heat that ranges from _____.

a. 1,000°F to 5,000°F
b. 6,000°F to 10,000°F
c. 5,000°F to 12,500°F
d. 8,500°F to 12,500°F

6. The advantage of engine-driven generators is that they are _____.

a. portable
b. easy to operate
c. less expensive
d. easy to maintain

7. In FCAW, the flux inside the filler core contains ingredients that _____.

a. cause arc flutter
b. oxidize the weld
c. generate shielding gas
d. prevent slag from forming

8. In FCAW, edge preparation is seldom needed for double-welded butt joints on plate up to 3/8" thick, because welds made with CO_2 shielding possess _____.

a. deionization
b. a smooth profile
c. a lower temperature
d. deeper penetration

9. When using short-circuiting welding in GMAW, the current passing through the wire electrode heats and then melts the wire in a process called _____.

a. pinch effect
b. inductance
c. arc blow
d. slope

10. The duty cycle of a GMAW gun is determined and rated with _____.

a. argon – CO_2
b. helium
c. argon
d. CO_2

11. Light-duty GMAW guns are cooled by _____.
   a. domestic water
   b. shielding gas flow
   c. special cooling fluid
   d. demineralized water

12. To meter the shielding gas to the welding gun at the proper flow rate, a(n) _____ is required.
   a. auxiliary gun control unit
   b. gas regulator/flowmeter
   c. cylinder valve
   d. gate valve

13. Because of its current transfer and ionization properties, the commonly used shielding gas that can be used alone or in combination with other gases is _____.
   a. argon
   b. helium
   c. nitrogen
   d. carbon dioxide

14. FCAW is primarily used for the welding of _____.
   a. copper
   b. aluminum
   c. ferrous metals
   d. nonferrous metals

15. After first checking the equipment specification for the weight of the machine, a welding machine can be lifted by _____ the lifting eye.
   a. attaching a sling and/or a shackle to
   b. attaching a heavy-duty chain to
   c. attaching the hook directly to
   d. running a stout rope through

Trade Terms Introduced in This Module

Axially: In a straight line along an axis.

Globular transfer: A GMAW process in which the electrode wire does not touch the base metal, and the filler metal is transferred randomly in the arc as large, irregular globules.

Pulsed transfer: A GMAW welding process in which the welding power is cycled from a low level (background level) to a higher level (welding level), resulting in lower average voltage and current; this is a variation of the spray transfer process.

Short-circuiting transfer: A GMAW welding process in which the electrode wire shorts against the base metal. The welding current repeatedly and rapidly melts the wire and establishes an arc, which is extinguished when the wire again touches the base metal.

Spray transfer: A GMAW welding process in which the electrode wire does not touch the base metal, and the filler metal is transferred axially (in a straight line) in the arc as fine droplets.

Stickout: The length of electrode that projects from the contact tube, measured from the arc tip to the contact tube.

Additional Resources

This module is intended to present thorough resources for task training. The following reference works are suggested for further study. These are optional materials for continued education rather than for task training.

AWS B1.10: 1999 Guide for the Nondestructive Examination of Welds. Miami, FL: American Welding Society.

AWS B1.11: 2000 Guide for the Visual Examination of Welds. Miami, FL: American Welding Society.

AWS D3.5-93R: Guide for Steel Hull Welding. Miami, FL: American Welding Society, 1993.

Lincoln Electric website
offers sources for products and training:
http://www.lincolnelectric.com

Modern Welding Technology. Howard B. Cary. Englewood Cliffs, NJ: Prentice Hall, Inc.

OSHA 1910.269, Appendix C, Protection from Step and Touch Potentials. Current edition. Washington, DC: Occupational Safety and Health Administration (OSHA).

OSHA 1926.351, Arc Welding and Cutting. Current edition. Washington, DC: Occupational Safety and Health Administration (OSHA).

Welding Handbook. Volume 1. *Welding Science & Technology*. Miami, FL: American Welding Society, 2001.

Welding Handbook. Volume 2. Part 1: *Welding Processes*. Miami, FL: American Welding Society, 2004.

NCCER CURRICULA — USER UPDATE

NCCER makes every effort to keep its textbooks up-to-date and free of technical errors. We appreciate your help in this process. If you find an error, a typographical mistake, or an inaccuracy in NCCER's curricula, please fill out this form (or a photocopy), or complete the online form at **www.nccer.org/olf**. Be sure to include the exact module ID number, page number, a detailed description, and your recommended correction. Your input will be brought to the attention of the Authoring Team. Thank you for your assistance.

Instructors – If you have an idea for improving this textbook, or have found that additional materials were necessary to teach this module effectively, please let us know so that we may present your suggestions to the Authoring Team.

NCCER Product Development and Revision

13614 Progress Blvd., Alachua, FL 32615

Email: curriculum@nccer.org

Online: www.nccer.org/olf

❑ Trainee Guide ❑ Lesson Plans ❑ Exam ❑ PowerPoints Other ____________

Craft / Level: Copyright Date:

Module ID Number / Title:

Section Number(s):

Description:

Recommended Correction:

Your Name:

Address:

Email: Phone:

GMAW and FCAW: Plate

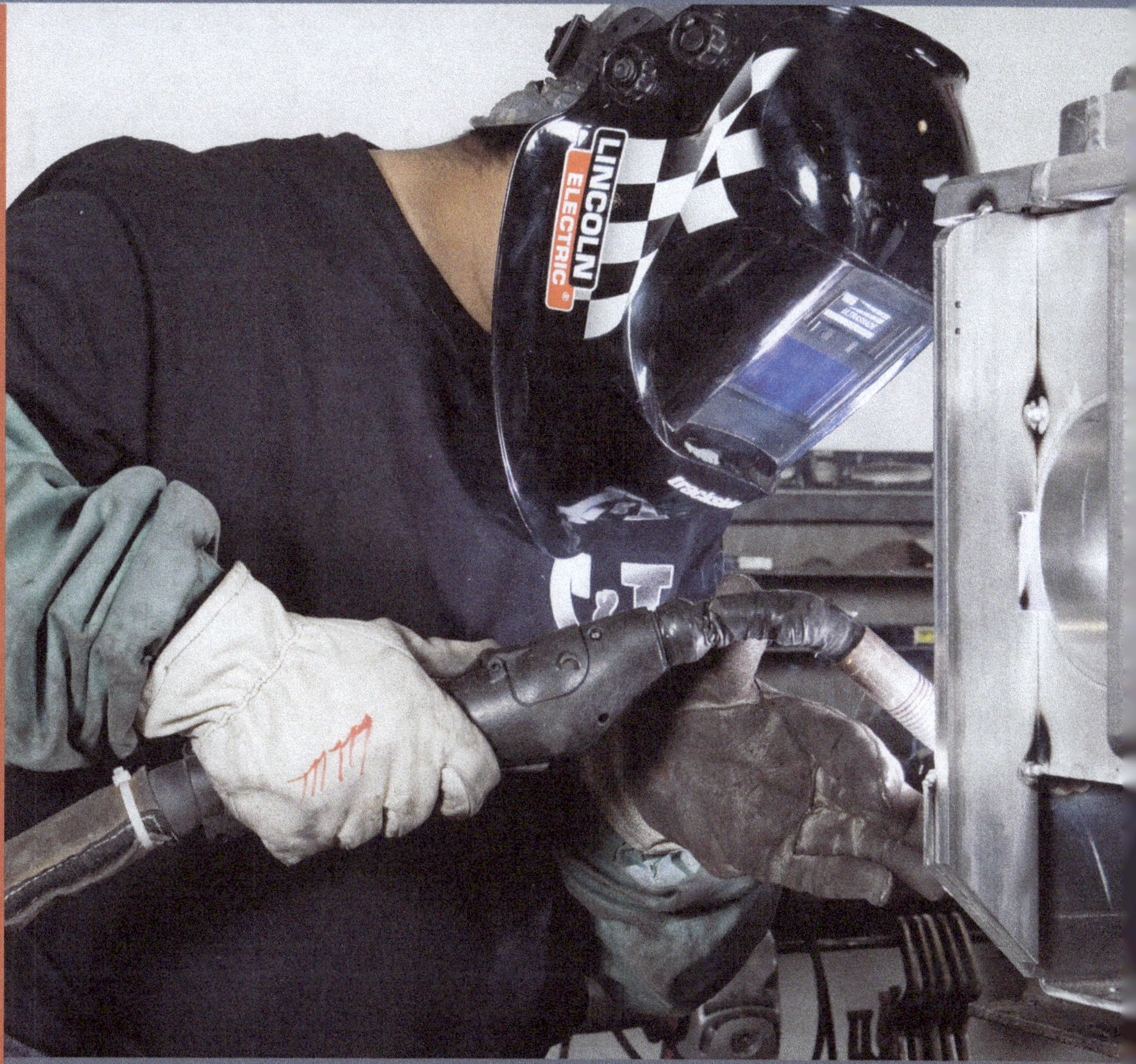

29206-09

29206-09
GMAW and FCAW: Plate

Objectives

When you have completed this module, you will be able to do the following:

1. Perform GMAW-S (short-circuit) multiple-pass fillet welds on carbon steel plate coupons in multiple positions, using solid or composite wire and shielding gas.
2. Perform GMAW-S (short-circuit) multiple-pass V-groove welds on carbon steel plate coupons in multiple positions (with or without backing), using solid or composite wire and shielding gas.
3. Perform GMAW spray fillet and V-groove welds on carbon steel plate coupons in multiple positions (with or without backing), using solid or composite wire and shielding gas.
4. Perform FCAW multiple-pass fillet welds on carbon steel plate coupons in multiple positions, using flux-cored wire and, if required, shielding gas.
5. Perform FCAW multiple-pass V-groove welds on carbon steel plate coupons in multiple positions (with or without backing), using flux-cored wire and, if required, shielding gas.

Trade Terms

Axis
Degassing

Required Trainee Materials

1. Pencil and paper
2. Appropriate personal protective equipment, including:
   - Safety glasses with a full face shield or a helmet
   - Flame-retardant clothing
   - Welding gloves
   - 8-inch or taller high-top safety shoes or boots
   - Earmuffs or earplugs

Prerequisites

Before you begin this module, it is recommended that you successfully complete *Core Curriculum; Maritime Structural Fitter Level One;* and *Maritime Structural Fitter Level Two.*

Industry Recognized Credentials

If you are training through an NCCER-accredited sponsor, you may be eligible for credentials from NCCER's Registry. The ID number for this module is 29206-09. Note that this module may have been used in other NCCER curricula and may apply to other level completions. Contact NCCER's Registry at 888.622.3720 or go to **www.nccer.org** for more information.

Contents

Topics to be presented in this module include:

Figures and Tables

1.0.0 Introduction

This module will teach trainees how to perform fillet and open V-groove welds on carbon steel plate using the gas metal arc welding process and the flux-cored arc welding process in all positions as directed by the instructor.

1.1.0 The GMAW Process

In the GMAW arc welding process, a continuous, consumable, solid wire or composite cored wire electrode is used for the filler metal and shielding gas to protect the weld zone. GMAW is a fast and effective method of making high-quality welds (*Figure 1*). The GMAW process is commonly used to make welds on carbon steel plate as well as on aluminum and other metals.

> **NOTE**
> Composite cored wire electrodes are a subset of GMAW solid wire carbon steel electrodes as defined in *AWS Specification A5.18.*

1.2.0 The FCAW Process

In the FCAW arc welding process, a continuous, consumable, flux-cored wire is used as a filler metal to form a weld (*Figure 2*). Shielding gas protects the molten weld metal from contamination by the atmosphere. The shielding gas can be generated entirely from the flux in the wire core or from both the flux in the wire core and an external shielding gas. In all types of flux-cored wires, the flux contains agents for degassing the weld, stabilizing the arc, and forming a protective slag over the weld bead. Other elements are often added to the flux-cored wire to alloy with the base metal and improve its chemical and mechanical characteristics. The FCAW process is basically limited to welding low-carbon steels, some low-alloy steels, and some stainless steels. FCAW is an excellent process for making fillet and groove welds on carbon steel plate.

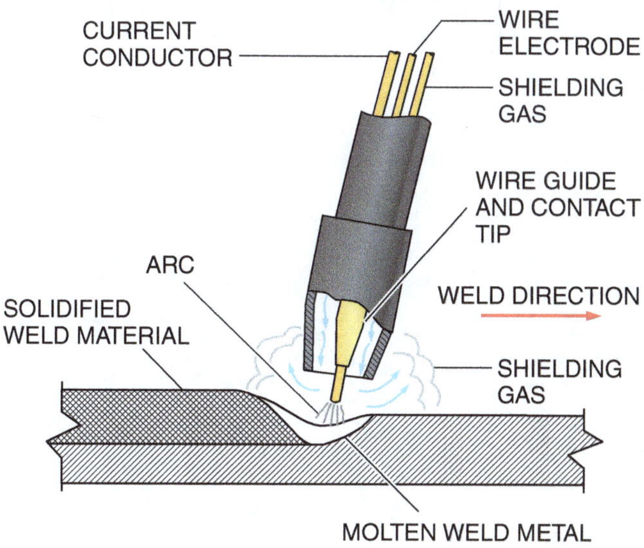

206F01.EPS

Figure 1 GMAW process.

1.3.0 GMAW and FCAW Equipment

The equipment used for gas-shielded GMAW and FCAW or self-shielded FCAW is basically the same except for the components needed for the gas used by the gas-shielding process. This additional equipment includes guns equipped with a gas nozzle, gas hoses, a gas control solenoid valve, a gas regulator/flowmeter, and a shielding gas cylinder (gas source). *Figures 3* and *4* are examples of typical GMAW and FCAW equipment.

Hot Tip

Wind Resistance

Self-shielded FCAW is particularly useful in situations with high winds, because the shielding gas is created inside the welding arc as the flux vaporizes. The molten puddle has a constant supply of shield gas displacing the air from the arc outward, and the puddle is immediately covered by the residual slag. An external shield gas, in contrast, attempts to force gas down from the nozzle to displace the air, and the gas can be blown away from the weld puddle by strong winds.

206SA01.EPS

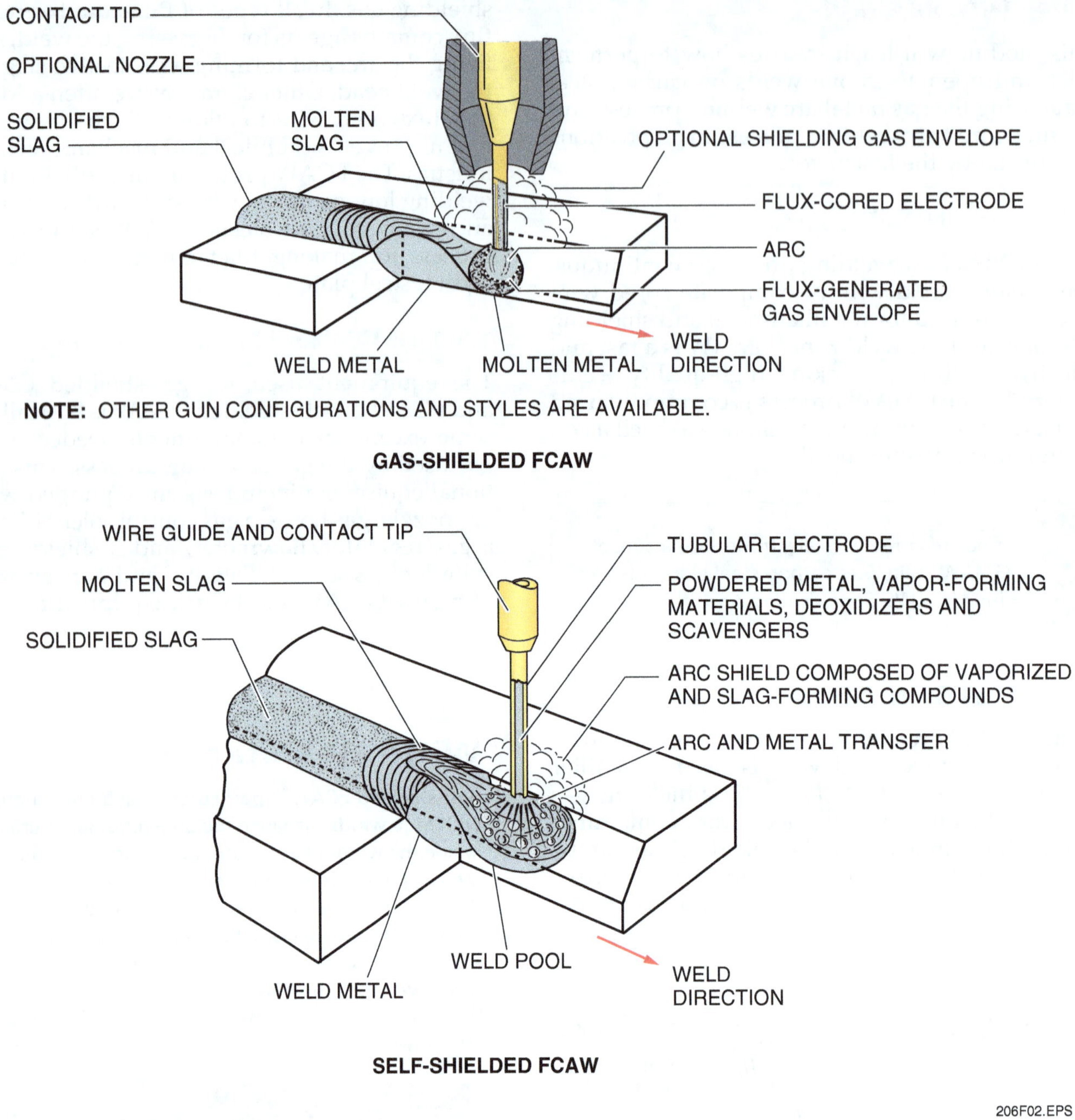

Figure 2 FCAW process.

206F03.EPS

Figure 3 Gas-shielded GMAW equipment.

206F04.EPS

Figure 4 Self-shielded FCAW equipment.

2.0.0 Safety Summary

The following is a summary of safety procedures and practices that must be observed while cutting or welding. Keep in mind that this is only a summary. Complete safety coverage is provided in the *Welding Level One* module, *Welding Safety*. If you have not completed that module, do so before continuing. Above all, be sure to wear appropriate protective clothing and equipment when welding or cutting.

2.1.0 Protective Clothing and Equipment

- Always use safety glasses with a full face shield or a helmet. The glasses, face shield, or helmet lens must have the proper light-reducing shade for the type of welding or cutting to be performed. Never directly or indirectly view an electric arc without using a properly tinted lens.
- Wear proper protective leather and flame-retardant clothing along with welding gloves that will protect you from flying sparks, molten metal, and heat.
- Wear 8-inch or taller high-top safety shoes or boots. Make sure that the tongue and lace area of the footwear will be covered by a pant leg. If the tongue and lace area is exposed or if the footwear must be protected from burn marks, wear leather spats under the pants or chaps over the front and top of the footwear.
- Wear a solid material (non-mesh) hat with a bill pointing to the rear or, if much overhead cutting or welding is required, a full leather hood with a welding face plate and the correct tinted lens. If a hard hat is required, use one that allows the attachment of rear deflector material and a face shield.
- Wear earmuffs or at least earplugs to protect your ear canals from sparks.

2.2.0 Fire/Explosion Prevention

- Never carry matches or gas-filled lighters in your pockets. Sparks can cause the matches to ignite or the lighter to explode, causing serious injury.

- Never perform any type of heating, cutting, or welding until a hot-work permit has been obtained and an approved fire watch has been established. Most work-site fires caused by these types of operations are started by cutting torches.
- Never use oxygen to blow dust or dirt from clothing. The oxygen can remain trapped in the fabric for a time. If a spark hits the clothing during this time, the clothing can burn rapidly and violently out of control.
- Make sure that any flammable material in the work area is moved or shielded by a fire-resistant covering. Approved fire extinguishers must be available before attempting any heating, welding, or cutting operations.
- Never release a large amount of oxygen or use oxygen as compressed air. Its presence around flammable materials or sparks can cause rapid and uncontrolled combustion. Keep oxygen away from oil, grease, and other petroleum products.
- Never release a large amount of fuel gas, especially acetylene. Methane and propane tend to concentrate in and along low areas and can ignite at a considerable distance from the release point. Acetylene is lighter than air, but is even more dangerous than methane. When mixed with air or oxygen, acetylene will explode at much lower concentrations than any other fuel.
- To prevent fires, maintain a neat and clean work area, and make sure that any metal scrap or slag is cold before disposal.
- Before cutting or welding containers, such as tanks or barrels, check to see if they have contained any explosive, hazardous, or flammable materials, including petroleum products, citrus products, or chemicals that decompose into toxic fumes when heated. As a standard practice, always clean and then fill any tanks or barrels with water or purge them with an appropriate purging gas to displace any oxygen.

2.3.0 Work Area Ventilation

- Make sure to follow confined space procedures before conducting any welding or cutting in the confined space.
- Never use oxygen to provide ventilation in confined spaces.
- Always perform cutting or welding operations in a well-ventilated area. Cutting or welding operations involving zinc or cadmium materials or coatings result in toxic fumes. For long-term cutting or welding of such materials, always wear an approved full-face, supplied-air respirator that uses breathing air supplied from outside the work area. For occasional, very short-term exposure, a HEPA-rated or metal-fume filter may be used on a standard respirator.
- Make sure confined spaces are properly ventilated for cutting or welding purposes.

Figure 5 shows mobile and portable extraction equipment used to remove cutting or welding fumes and vapors from work areas.

Figure 5 Mobile and portable extraction equipment.

3.0.0 Welding Equipment Setup

Before welding can take place, the area must be made ready, the welding equipment must be set up, and the metal to be welded must be prepared. The following sections will explain how to set up equipment for welding.

3.1.0 Preparing the Welding Area

To practice welding, a welding table, bench, or stand is needed. The welding surface must be steel, and provisions must be made for placing weld coupons out of position (*Figure 6*).

Follow these steps to set up the area for welding:

Step 1 Make sure that the area is properly ventilated. Make use of doors, windows, and fans.

Step 2 Check the area for fire hazards. Remove any flammable materials before proceeding.

Step 3 Know the location of the nearest fire extinguisher. Do not proceed unless the extinguisher is charged and you know how to use it.

Step 4 Set up flash shields around the welding area.

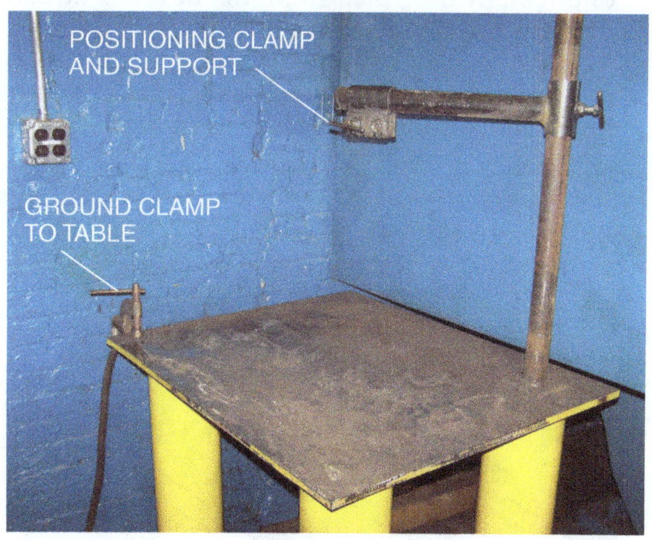

Figure 6 Welding station.

3.2.0 Preparing the Practice Weld Coupons

If possible, the practice weld coupons should be ⅜" thick carbon steel. If this size is not readily available, ¼" to ¾" thick steel can be used to practice welding. Use a wire brush or grinder to remove

Hot Tip

Machine Bevels and Root Faces for GMAW or FCAW

For V-groove joints made in thick materials, machined bevels and root faces provide the cleanest and most uniform way to prepare materials for satisfactory GMAW or FCAW welds. While SMAW, FCAW, and GTAW can tolerate some poor fitup for root passes, controlling penetration of root passes with poor fitup using GMAW is difficult.

heavy mill scale or corrosion. Prepare weld coupons for practicing the following:

- *Running beads* – The coupons can be any size or shape that can be easily handled.
- *Overlapping beads* – The coupons can be any size or shape that can be easily handled.
- *Fillet welds* – Cut the metal into 4" × 5" rectangles for the base and 3" × 5" rectangles for the web. *Figure 7* shows the weld coupons for fillet welding.
- *Open V-groove welds* – As shown in *Figure 8*, cut the metal into 3" × 5" rectangles with one (or both) of the 5" lengths beveled at 22½ degrees for stainless steel or 30 to 37½ degrees for carbon steel or low-alloy plate. Grind a 0" to ⅛" root face on the bevel as directed by your instructor.

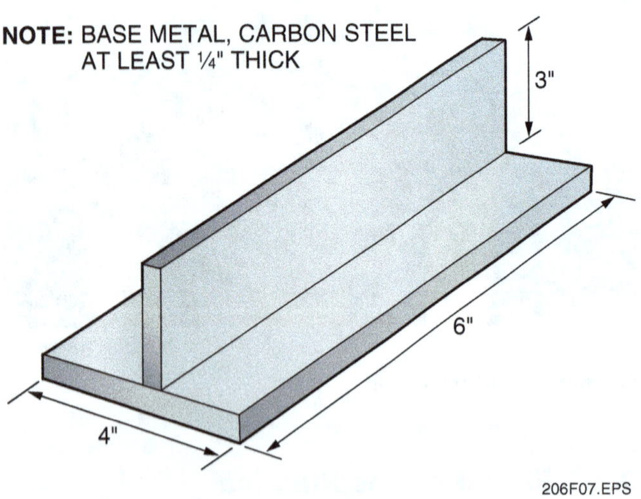

Figure 7 Fillet weld coupons.

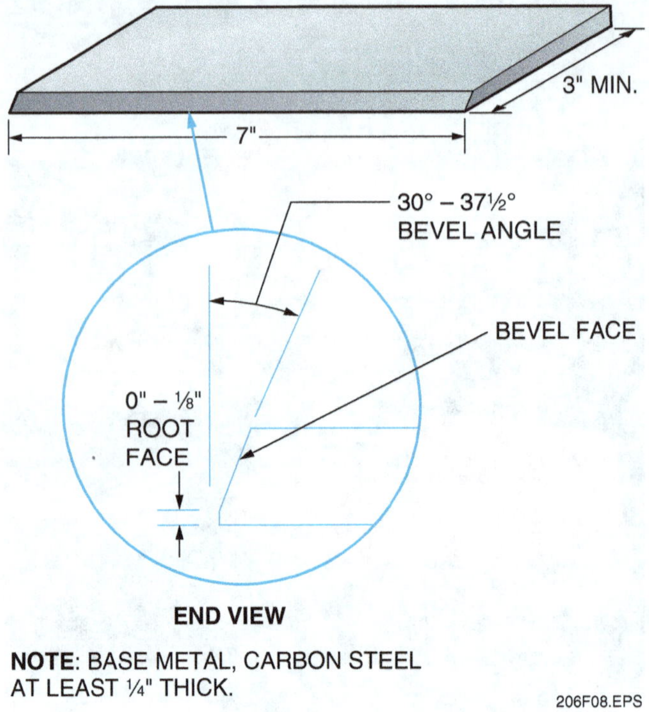

Figure 8 Metal cut for open V-groove weld coupons.

> **NOTE**
> Check with your instructor to determine which bevel angles should be used.

> **NOTE**
> The welding codes allow the root face and opening on open root welds to be 0" to ⅛". Adjust the root face and root opening as needed when you start your welding practices.

Follow these steps to prepare each V-groove weld coupon:

Step 1 Check the bevel face of each coupon. There should be no dross and a 0" to ⅛" root face. The bevel angle should be 30 to 37½ degrees. An example of this is shown in *Figure 9*.

Step 2 Align the beveled coupons on a flat surface with a root opening determined by your instructor. Place tack welds on the back side of the joint. Use from three to four ¼" tack welds.

> **NOTE**
> When tack welding open V-groove practice coupons together, use a piece of wire or rod of the correct diameter to space the coupons for the desired root opening.

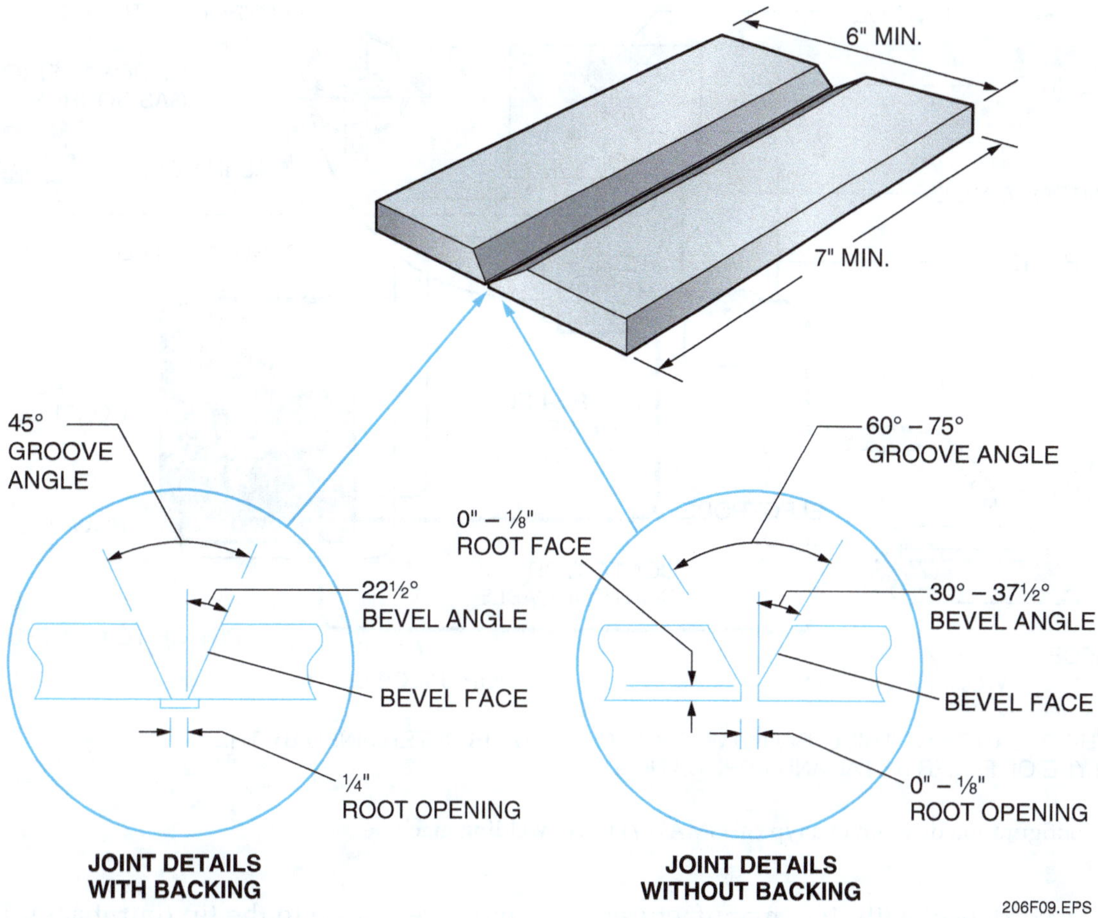

Figure 9 Open V-groove weld coupon.

3.3.0 The Welding Machine

Identify the proper welding machine for GMAW/FCAW use, and follow these steps to set it up for welding:

Step 1 Verify that the welding machine can be used for GMAW/FCAW, with or without gas shielding for FCAW as required.

Step 2 Verify the location of the primary disconnect.

Step 3 Configure the welding machine for gas-shielded GMAW/FCAW or self-shielding FCAW as directed by your instructor (*Figure 10*). Configure the gun polarity, and equip the gun with the correct liner material and contact tube for the diameter of the wire being used and the correct nozzle for the application.

Hot Tip

Conserving Material

Steel for practice welding is expensive and difficult to obtain. Every effort should be made to conserve the material that is available. Reuse weld coupons until all surfaces and joints have been used for welding, then cut the weld coupon apart and reuse the pieces. Use material that cannot be cut into weld coupons to practice running beads.

GOING GREEN

Recycling Metal

When metals are no longer usable in the welding shop or on the job, they should be collected and sold to a recycling company rather than sent to a landfill. Selling the scrap metal can make money for the company or welding school, and it protects the environment.

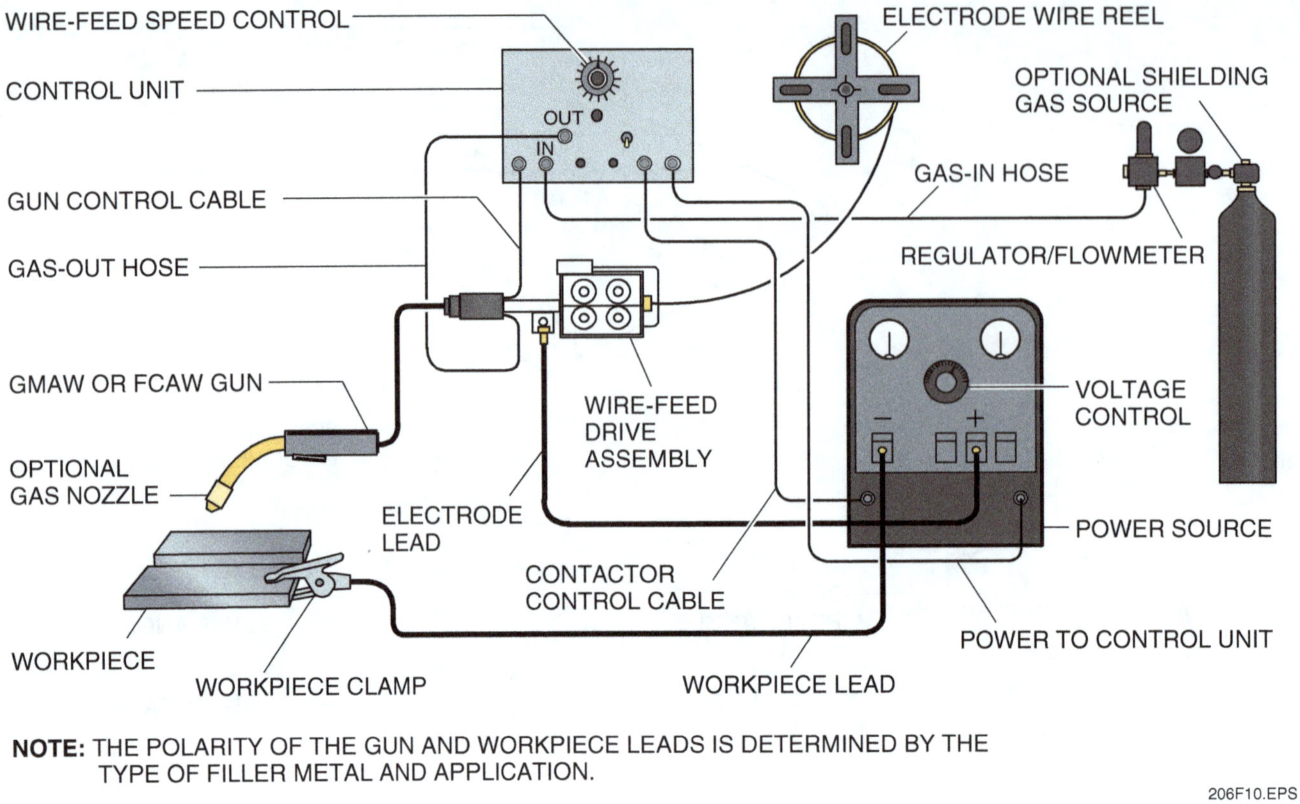

Figure 10 Configuration diagram of a typical GMAW/FCAW welding machine.

Step 4 In accordance with the manufacturer's instructions, load the wire feeder and gun with a solid, composite, or flux-cored electrode wire of the proper diameter as directed by your instructor.

Step 5 If required, connect the proper shielding gas for the application as specified by the wire electrode manufacturer, the WPS, the site quality standards, or your instructor.

Step 6 Connect the clamp of the workpiece lead to the workpiece.

Step 7 Turn on the welding machine and, if necessary, purge the gun as directed by the gun manufacturer's instructions.

Step 8 Set the initial welding voltage and wire feed speed for the type and size of electrode wire as recommended by the manufacturer, directed by the instructor, or provided in the WPS.

3.4.0 Welding Voltage

Arc length is determined by voltage, which is set at the power source. Arc length is the distance from the wire electrode tip to the base metal or molten pool at the base metal. If voltage is set too high, the arc will be too long, which could cause the wire to fuse to the tip (burnback). Too high a voltage also causes porosity and excessive spatter. Voltage must be increased or decreased as wire feed speed is increased or decreased. Set the voltage so that the arc length is just above the surface of the base metal (*Figure 11*).

3.5.0 Welding Amperage

With a standard constant-voltage power source, the electrode feed speed control on the power source controls the welding amperage after the welder has set the initial recommended setting.

Hot Tip

Wire Electrode Manufacturer's Recommendations

Always obtain and follow the manufacturer's recommendations or WPS for shielding gas (if required) and for initially setting the welding voltage, current (if variable), and wire feed speed parameters. These balanced parameters are critical and are based on the welding position, size, and composition of the solid, composite, or flux-cored wired being used.

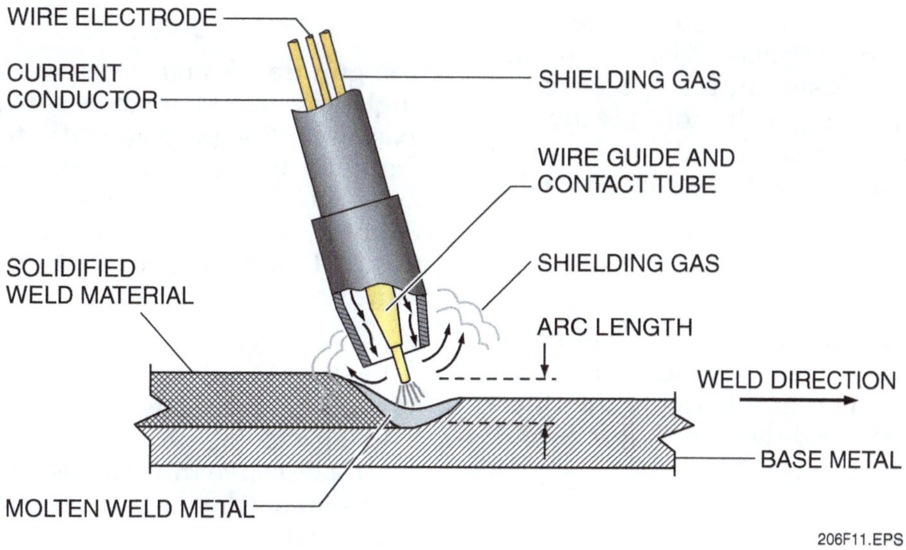

Figure 11 Arc length.

Hot Tip

Arc Voltage

A minimum arc voltage is needed to maintain spray transfer. However, penetration is not directly related to voltage. Penetration will increase with voltage for a time, but beyond an optimum voltage, it will actually decrease with further voltage increases.

The welding power source provides the amperage necessary to melt the wire electrode while maintaining the selected welding voltage. Within limits, when the wire electrode feed speed is increased, the welding amperage and deposition rate also increase. This results in higher welding heat, deeper penetration, and higher beads. When the wire electrode feed speed is decreased, the welding amperage automatically decreases. With lower welding amperage and less heat, the deposition rate drops and the weld beads are smaller and have less penetration.

Note that some constant-voltage power sources used for GMAW/FCAW provide varying degrees of current modification, such as slope or induction adjustments. Power sources with a slope adjustment allow the amount of amperage change to vary in relation to voltage range of the unit. This allows the correct short-circuiting current to be established for certain welding applications and wire electrode requirements. Standard constant-voltage GMAW/FCAW units have current slope that is fixed by the manufacturer for general welding applications and conditions. Other units allow adjustment of the response time for short-circuit current by using an induction control. In an instantaneous short-circuit current draw mode of operation, the machine draws a high level of current as soon as the electrode touches the workpiece. In the adjustable induction mode of operation, the machine gradually increases the welding current after the initial contact of the electrode to the workpiece. This smooths the arc characteristics during short-circuiting transfer. Pulse transfer power sources allow the adjustment of the peak current for the pulse and the background current between pulses to match specific welding applications and wire electrode requirements. In other cases, where a voltage-sensing wire feeder is used with a constant-current power source, the current is adjustable and the wire feeder varies the wire feed speed to maintain the proper arc voltage at the wire electrode tip.

3.6.0 Welding Travel Speed

Weld travel speed is the speed of the electrode tip across the base metal in the direction of the weld, measured in inches per minute (ipm). Travel

Hot Tip

Travel Speed and Wire Feed Speed

Beginners are tempted to turn down the wire feed speed if they experience difficulty controlling the weld puddle. Wire electrodes must be run at certain balanced parameters that cannot be changed individually. Voltage, wire feed speed, and travel speed are adjusted and balanced together to control the weld puddle.

speed has a great effect on penetration and bead size. Slower travel speeds build higher beads with deeper penetration. Faster travel speeds build smaller beads with less penetration. Ideally, the welding parameters should be adjusted so that the electrode tip is positioned at the leading edge of the weld puddle during travel.

3.7.0 Gun Position

Gun position for GMAW/FCAW has an effect on weld penetration along with voltage and travel speed. The gun position in relation to the direction of the weld is defined the same as for SMAW welding.

3.7.1 Work Angle

The gun work angle (*Figure 12*) is an angle less than 90 degrees between a line perpendicular to the major workpiece surface at the point of electrode contact, and a plane determined by the electrode axis and the weld axis. For a T-joint or corner joint, the line is perpendicular to the nonbutting member. For pipe, the plane is determined by the electrode axis and a line tangent to the pipe surface at the same point.

3.7.2 Travel Angle

The gun travel angle (*Figure 12*) is defined as an angle less than 90 degrees between the electrode axis and a line perpendicular to the weld axis at the point of electrode contact in a plane determined by the electrode axis and the weld axis. For pipe, the plane is determined by the electrode axis and a line tangent to the pipe surface at the same point.

- *Neutral angle* – The term *neutral angle* describes the travel angle when the center line of the gun electrode is exactly perpendicular (90 degrees) to the welding surface. The term is nonstandard for a 0-degree travel angle. Weld penetration is moderate when the gun electrode is used in this position.
- *Push angle* – A push angle is generally used for welding thin materials or when shallow penetration is required. A push angle is the travel angle created when the welding gun is tilted back so that the electrode tip precedes the gun in the direction of the weld. In this position, the electrode tip and shielding gas are being directed ahead of the weld bead. Welding with this angle produces a smaller weld bead with less penetra-

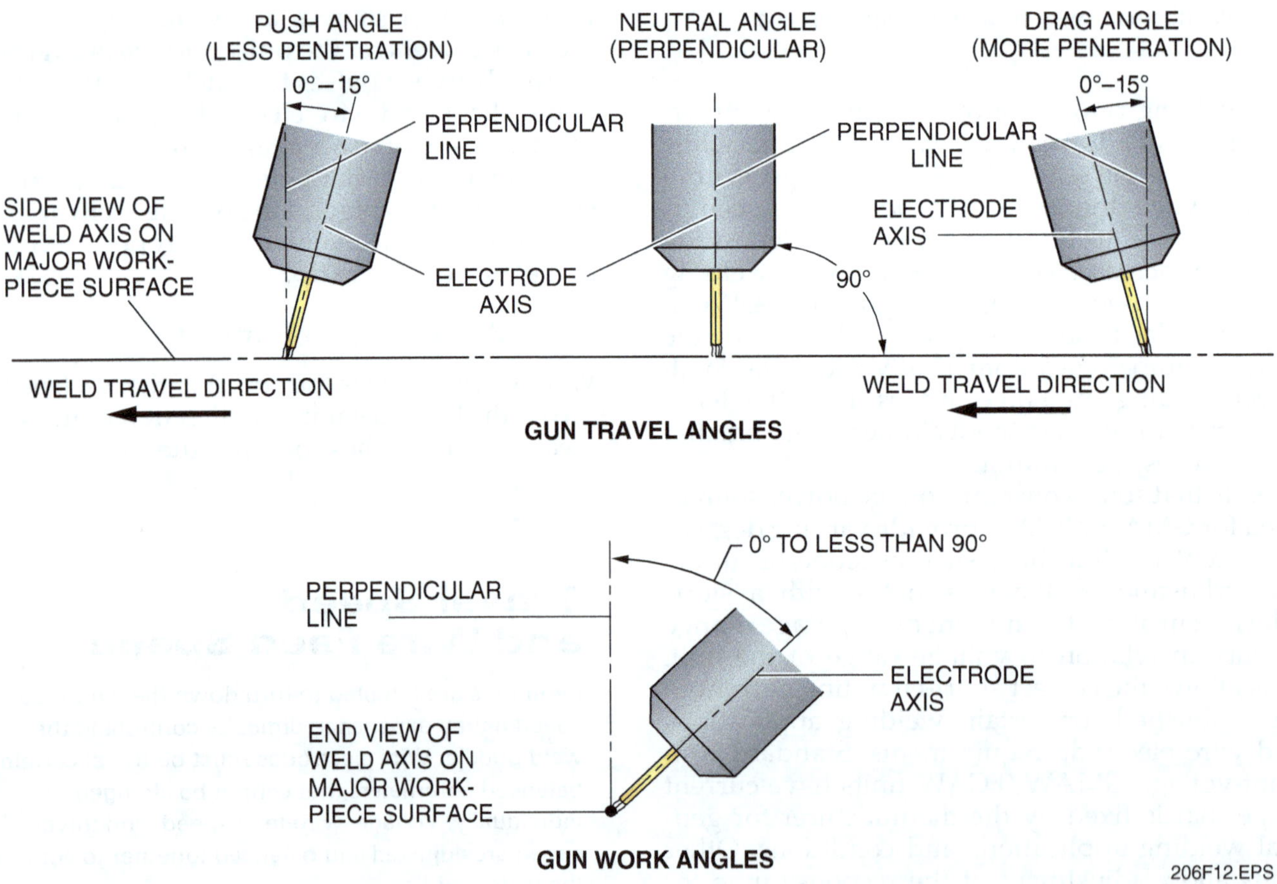

Figure 12 Gun travel and work angles.

tion. Push angles greater than 15 degrees are not recommended for manual welding.

- *Drag angle*– A drag angle is the travel angle created when the gun is tilted forward so that the electrode tip trails the gun in the direction of the weld. In this position, the electrode tip and shielding gas are being directed back toward the weld bead. Welding with this angle produces a higher and wider weld bead with more penetration. Maximum penetration is achieved with a drag angle of approximately 25 degrees.

3.8.0 Electrode Extension, Stickout, and Standoff Distance

Electrode extension is the length of the wire that extends beyond the contact tip of the welding gun. Changing the extension influences the welding current for low-conductivity metal wires because it changes the preheating of the wire caused by the resistance of the wire to current flow. As the extension increases, preheating of the wire increases. Since the welding power source is self-regulating, it does not have to furnish as much welding current to melt the wire, so the current output automatically decreases. This results in less penetration and increased deposition rates. Increasing extension is useful for bridging gaps and compensating for mismatch, but it can cause overlap or lack of fusion and a ropy bead appearance. When the extension is decreased, the power source is forced to increase its current output to burn off the wire. Too little extension can cause the wire to weld to the contact tip and develop porosity in the weld. For high-conductivity metal wires, the preheating effect of wire resistance is minimal, and wire speed and voltage settings have a more direct effect.

GMAW and FCAW-G extensions vary between ¼" to 1", depending on the transfer mode. The extension for self-shielding FCAW and an insulating nozzle is ¾" to 1", depending on the welding position and wire size. *Figure 13* shows typical electrode extensions for various GMAW and FCAW gun configurations. In addition, other gun nomenclature is defined. Stickout is the distance from the gas nozzle or insulating nozzle to the end of the electrode.

Standoff distance is the distance from the gas nozzle or insulating nozzle to the workpiece. Contact tip extension or setback is usually dependent on the transfer mode for the GMAW application.

Hot Tip

GMAW of Stainless Steel

For stainless steel, a drag angle is normally used for fillet welds and a push angle for square or grooved butt joints. Stainless steel is usually welded in the flat and horizontal positions with spray transfer. Vertical and overhead welds usually employ short-circuiting transfer for at least the root passes, if not the fill and cover passes. Pulsed-spray transfer can also be used for fill and cover passes. Remember that all open joints in stainless steel materials must be backed with copper, stainless steel, or inert gas to prevent air from reaching the weld puddle until the puddle solidifies. When practicing welding using stainless steel electrodes and carbon steel coupons, use the travel angles and transfer methods outlined, along with carbon steel backing strips or a backing fixture for open or butt joints, when required. When welding plate that is thicker than ¼", the gun should be moved back and forth in the travel direction, with a slight side-to-side motion. On thinner materials, only the back and forth motion is necessary.

Hot Tip

Arc Blow

When current flows, magnetic fields are created. The magnetic fields tend to concentrate in corners, deep grooves, or the ends of the base metal. When the arc approaches these concentrated magnetic fields, it is deflected from its intended path. Arc blow can cause defects such as excessive weld spatter and porosity. If arc blow occurs, try one or more of these methods to control it:

- Change the position of the workpiece clamp. This will change the flow of welding current, affecting the way the magnetic fields are created.
- Change the angle of the gun. Reducing or even reversing the angle of the gun can compensate for the arc blow.

3.9.0 Gas Nozzle Cleaning

As the welding machine is being used, weld spatter accumulates on the gas nozzle and the contact tip. If the gas nozzle is not properly cleaned, it will restrict the shielding gas flow and cause porosity in the weld.

Clean the gas nozzle with a reamer or round file or with the tang of a standard file. After cleaning, the nozzle can be sprayed or dipped in a special antispatter compound. The antispatter compound helps prevent the spatter from sticking to the nozzle.

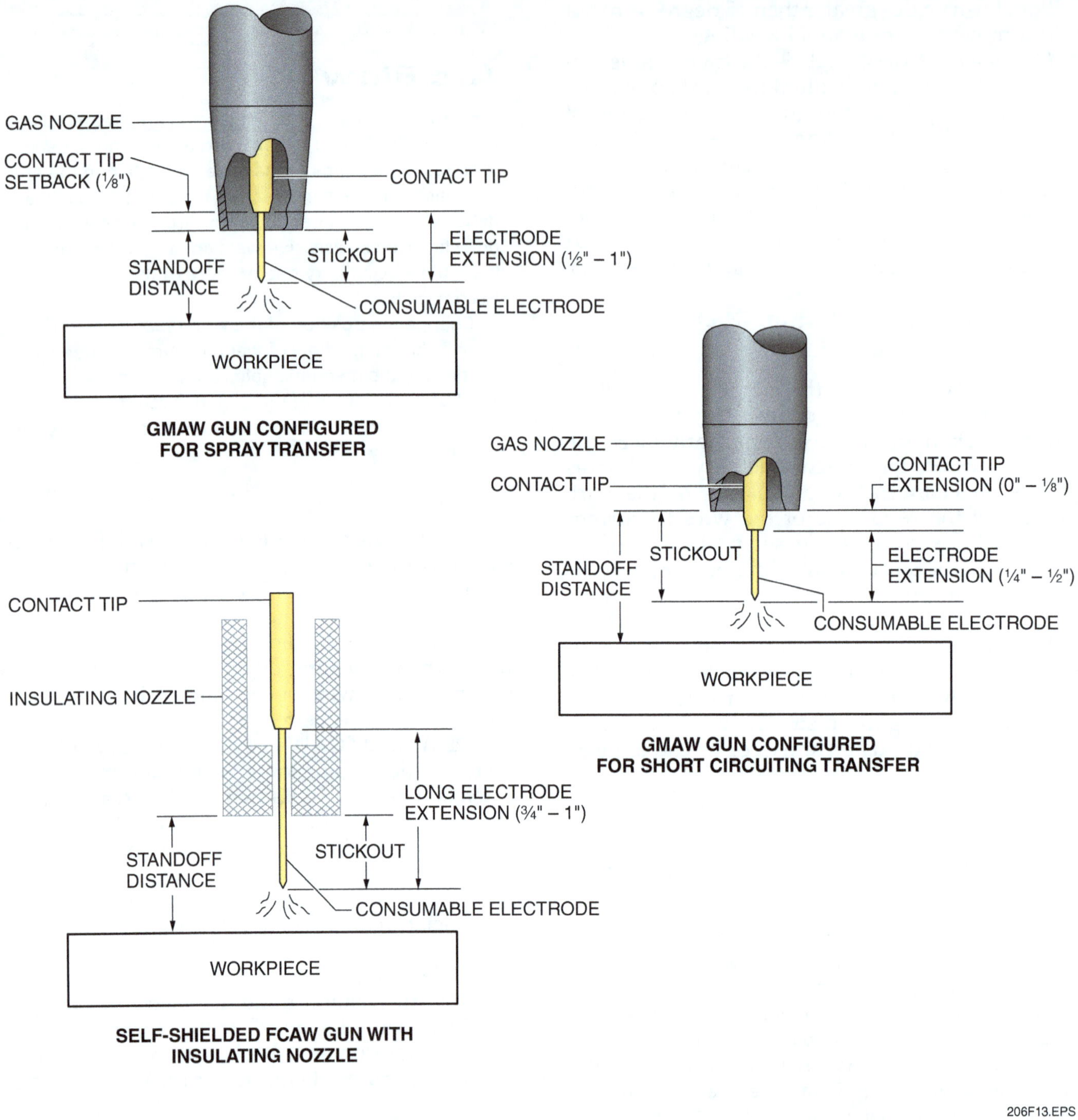

Figure 13 GMAW and FCAW gun configurations.

4.0.0 Bead Types

The following sections will explain how to create beads and welds using GMAW and FCAW equipment. The two basic bead types are stringer beads and weave beads.

4.1.0 Stringer Beads

Stringer beads are made with little or no side-to-side motion of the electrode. The width of a stringer bead is generally no more than six times the wire electrode diameter for smaller diameter (0.045" to $\frac{1}{16}$") wire or four times the wire electrode diameter for larger diameter ($\frac{3}{32}$" to $\frac{1}{8}$") wire. *Figure 14* shows a stringer bead made with 0.023" diameter wire.

Hot Tip

Antispatter Materials

Use only antispatter material specifically designed for GMAW and FCAW-G gas nozzles. Water-based products may cause hydrogen embrittlement in hydrogen-sensitive steel. Always refer to material safety data sheets for specific safety considerations.

206SA02.EPS

206F14.EPS

Figure 14 Close-up view of a GMAW stringer bead.

CAUTION

The width of stringer beads is specified in the welding code or WPS being used at your site. Do not exceed the widths specified for your site.

4.1.1 Practicing Stringer Beads

Practice running stringer beads in the flat position. Experiment with different voltages, travel angles, and stickouts. Change the wire feed speed and observe the bead size. Try different travel angles and observe the bead shape and size.

NOTE

Practice stringer bead welds using GMAW and FCAW as directed by your instructor. Use solid, composite, or flux-cored electrode wire sizes and, if required, appropriate shielding gas as directed by your instructor.

Follow these steps to run stringer beads:

Step 1 Hold the gun at the desired travel angle (drag or push), with the electrode tip directly over the point where the weld is to begin, and pull the gun trigger.

Step 2 Hold the arc in place until the weld puddle begins to form.

Step 3 Slowly advance the arc while maintaining the gun angle.

NOTE

Stay in the leading edge of the puddle to prevent incomplete fusion.

Step 4 Continue to weld until a bead about 2" to 3" long is formed.

Step 5 Stop, raise the gun to a neutral (0-degree) travel angle, pause until the crater is filled, and then release the trigger.

Step 6 Inspect the bead for the following:
- Straightness of the bead
- Uniform appearance of the bead face
- Smooth, flat transition with complete fusion at the toes of the weld
- No porosity
- No excessive undercut
- No overlap
- No lack of fusion
- Complete penetration
- No pinholes (fisheyes)
- No inclusions
- Crater filled
- No cracks

Step 7 Continue practicing stringer beads until you can make acceptable welds every time.

4.2.0 Weave Beads

Weave beads are made with wide, side-to-side motions of the electrode. The width of a weave bead is determined by the amount of side-to-side motion. *Figure 15* shows weave beads made with 0.023" diameter wire.

When making a weave bead, use care at the toes to be sure there is proper tie-in to the base metal. To ensure proper tie-in at the toes, slow down or pause slightly at the edges. The pause at the edges will also flatten out the weld, giving it the proper profile.

206F15.EPS

Figure 15 Close-up view of GMAW weave beads.

4.2.1 Practicing Weave Beads

Practice running weave beads about ⅝" wide in the flat position. Experiment with different weave motions, travel angles, and stickouts. Change the wire feed speed and observe the bead size. Try different gun angles and observe the bead shape and size.

> **NOTE**
> Practice weave bead welds using GMAW and/or FCAW as directed by your instructor. Use solid, composite, or flux-cored electrode wire sizes and, if required, appropriate shielding gas as directed by your instructor.

Follow these steps to run weave beads:

Step 1 Hold the gun at the desired travel angle with the electrode tip directly over the point where the weld is to begin.

Step 2 Pull the gun trigger.

Step 3 Hold the arc in place until the weld puddle begins to form.

Step 4 Slowly advance the arc in a weaving motion while maintaining the gun angle (*Figure 16*).

Step 5 Continue to weld until a bead about 2" to 3" long is formed.

Hot Tip

Weave Bead Width

The width of weave beads is often specified in the welding code or WPS being used at your site. Don't try to exceed the widths specified for your site. *AWS D1.1:2008* limits the width of a weave bead on prequalified joints to ⅝" in flat, horizontal, and overhead positions and 1" in the vertical positions.

Hot Tip

Weave Beads and Stainless Steel

Weave beads are not normally used with stainless steel because of gas shielding problems and excessive heat input to the base metal. However, some fully automatic processes can make reasonable weave beads using pulse transfer techniques.

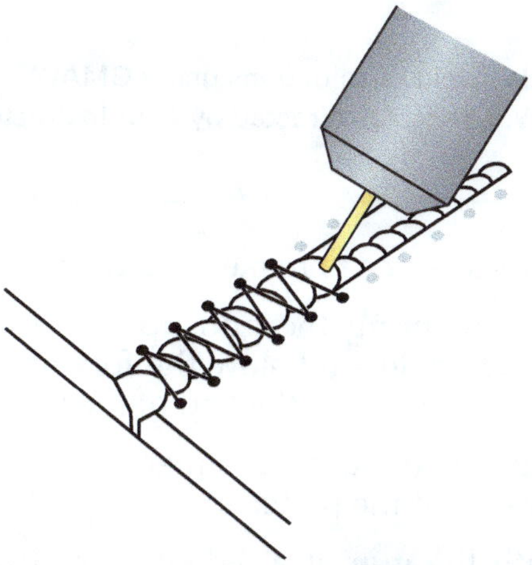

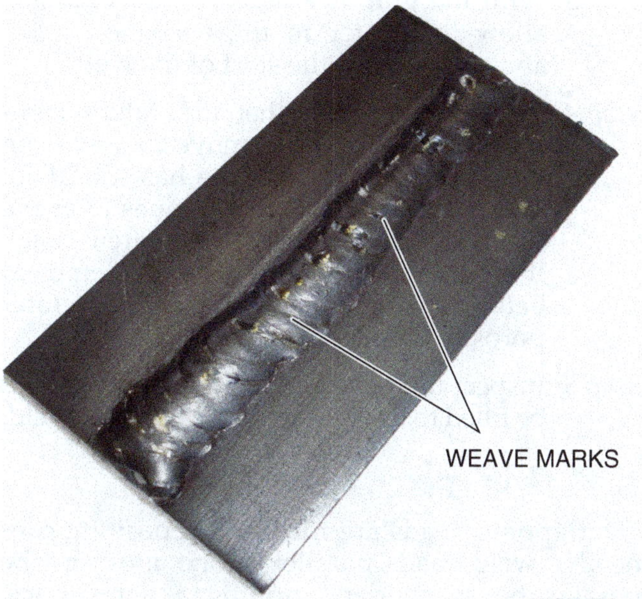

206F16.EPS

Figure 16 Weave motion.

Step 6 Stop, raise the gun to a neutral (0-degree) travel angle, pause until the crater is filled, and then release the trigger.

Step 7 Inspect the bead for the following:
- Straightness of the bead
- Uniform appearance of the bead face
- Smooth, flat transition with complete fusion at the toes of the weld
- No porosity
- No excessive undercut
- No overlap
- No lack of fusion
- Complete penetration
- No pinholes (fisheyes)
- No inclusions
- Crater filled
- No cracks

Step 8 Continue practicing weave beads until you can make acceptable welds every time.

Hot Tip

Bead Cleaning

There is no such thing as a self-cleaning electrode or shield gas. Welders must make sure that surfaces to be welded are clean and free of contaminants. It is important that multiple-pass welds are cleaned before each successive bead is laid down. Attempting to lay multiple-pass beads without properly cleaning the previous bead will result in inclusions in the welded joint. Shown here is a radiograph of a butt weld that has two slag lines easily visible in the weld root.

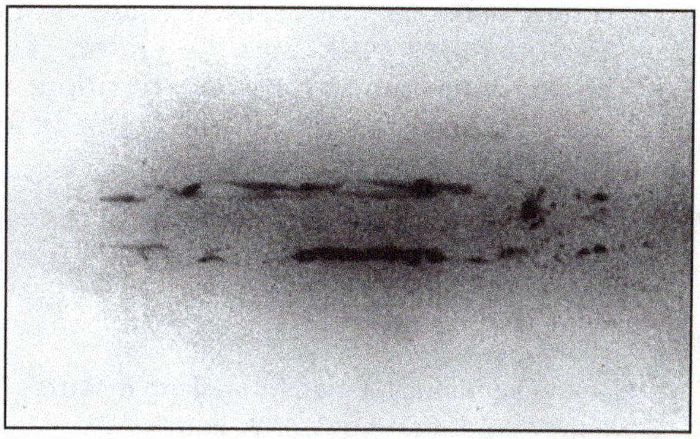

206SA03.EPS

4.3.0 Weld Restarts

A restart is the junction where a new weld connects to and continues the bead of a previous weld. Restarts are important because an improperly made restart creates a weld discontinuity. One advantage of the semiautomatic GMAW and FCAW process is that restarts can be minimized. Plan the weld to eliminate restarts, if possible. If a restart has to be made, it must be made so that it blends smoothly with the previous weld and does not stand out. The technique for making a restart is the same for both stringer and weave beads. Practice weld restarts using GMAW and FCAW as directed by your instructor.

Follow these steps to make a restart:

Step 1 With the weld clean, hold the torch at the proper travel angle while restarting the arc directly over the center of the crater.

NOTE

The welding codes do not allow arc strikes outside the area of permanent welds.

Step 2 Pull the gun trigger.

Step 3 Move the electrode tip over the crater in a small circular motion to fill the crater with a molten puddle.

Step 4 Move to the leading edge of the puddle; as soon as the puddle fills the crater, continue the stringer or weave bead pattern being used.

Step 5 Inspect the restart. A properly made restart will blend into the bead, making it hard to detect. If the restart has undercut, not enough time was spent in the crater to fill it. If the undercut is on one side or the other, use more of a side-to-side motion as you move back into the crater. If the restart has a lump, it was overfilled; too much time was spent in the crater before resuming the forward motion.

Step 6 Continue to practice restarts until they are correct.

> **NOTE**
> Use the same techniques for making restarts whenever performing GMAW and FCAW.

4.4.0 Weld Terminations

A weld termination is made at the end of a weld. A termination normally leaves a crater. When making a termination, the welding codes require that the crater must be filled to the full cross section of the weld. This can be difficult, since most terminations are at the edge of a plate, where welding heat tends to build up, making filling the crater more difficult. *Figure 17* shows the proper technique for terminating a weld.

> **NOTE**
> Practice weld terminations using GMAW, FCAW, or both as directed by your instructor.

Follow these steps to make a weld termination:

Step 1 Start to bring the gun up to a 0-degree travel angle, and slow the forward travel as you approach the end of the weld.

Step 2 Stop the forward movement about ⅛" from the end of the plate.

Step 3 With the gun at a 0-degree travel angle, allow the crater to fill, and then move about ⅛" toward the start of the weld.

Step 4 Release the trigger when the crater is completely filled. Do not remove the gun from the weld until the puddle has solidified. The gas postflow that continues after the welding stops protects the molten metal. If the gun and shielding gas are removed before the crater has solidified, crater porosity or cracks can occur.

Step 5 Inspect the termination. The crater should be filled to the full cross section of the weld.

4.5.0 Overlapping Beads

Overlapping beads are made by depositing connective weld beads parallel to one another. The parallel beads overlap, forming a flat surface. This is also called padding. Overlapping beads are used to build up a surface and to make multiple-pass welds. Both stringer and weave beads can be overlapped. When viewed from the end, properly overlapped beads will have a relatively flat surface. *Figure 18* shows proper and improper overlapping string and weave beads.

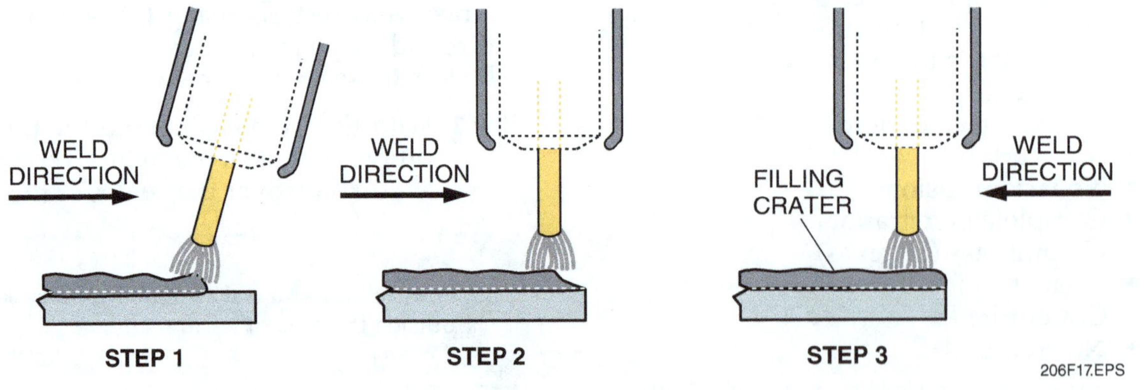

Figure 17 Terminating a weld.

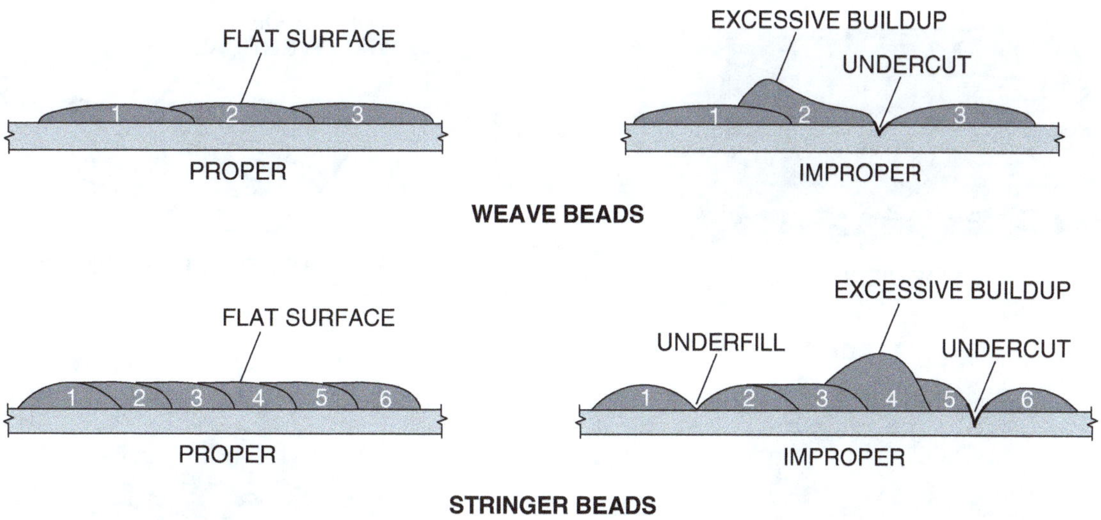

Figure 18 Proper and improper overlapping beads.

4.5.1 Practicing Overlapping Stringer or Weave Beads

Follow these steps to weld overlapping stringer or weave beads, using a specific diameter solid, composite, or flux-cored electrode wire and the proper shielding gas as directed by your instructor.

> **NOTE**
>
> Practice overlapping beads using GMAW and/or FCAW as directed by your instructor. Avoid making weave beads on stainless steel.

Step 1 Mark out a square on a piece of steel.

Step 2 Weld a stringer or weave bead along one edge at a 0-degree work angle.

Step 3 Clean the weld.

Step 4 Position the gun at a work angle of 10 to 15 degrees toward the side of the previous bead to get proper tie-in (*Figure 19*), and pull the trigger.

Step 5 Continue running overlapping stringer or weave beads until the square is covered.

> **NOTE**
>
> Chip or brush each bead before running the new bead.

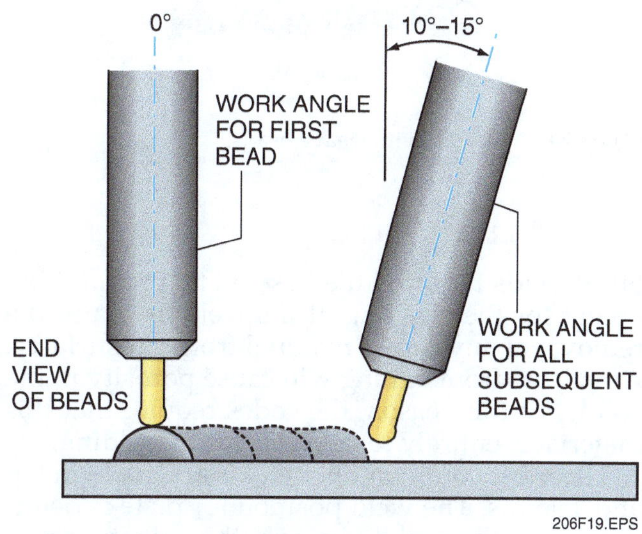

Figure 19 Work angles for overlapping beads.

Step 6 Continue building layers of stringer or weave beads, one on top of the other, until your technique is perfected.

> **WARNING!**
>
> The base metal will get very hot as it is being built up.

4.5.2 Building a Pad with Stringer or Weave Beads

Build a pad using stringer beads. Repeat building the pad using weave beads, as shown in *Figure 20*.

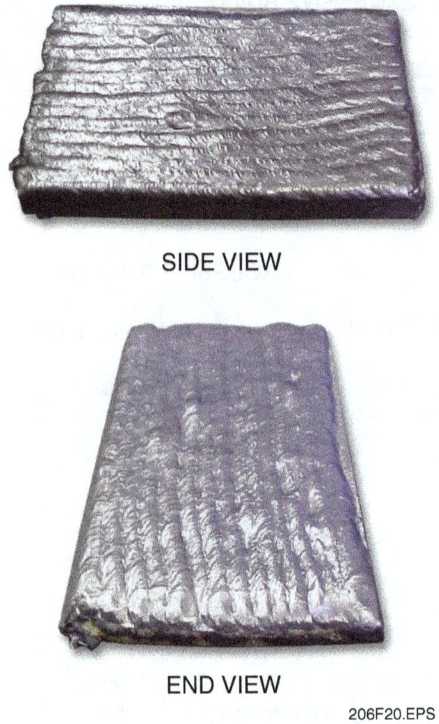

Figure 20 Pad built with weave beads.

5.0.0 Fillet Welds

Fillet welds require little base metal preparation, except for the cleaning of the weld area and the removal of any excess material from cut surfaces. Any dross from cutting will cause porosity in the weld. For this reason, the codes require that this material is entirely removed prior to welding.

The most common fillet welds are made in lap and T-joints. The weld position for plate is determined by the weld axis and the orientation of the workpiece. The positions for fillet welding on plate are flat, or 1F (the letter F stands for fillet); horizontal, or 2F; vertical, or 3F; and overhead, or 4F. In the 1F and 2F positions, the weld axis can be inclined up to 15 degrees. Any weld axis inclination for the other positions varies with the rotational position of the weld face as specified in AWS standards. *Figure 21* shows the four fillet welding positions for plate.

Fillet welds can be concave or convex, depending on the WPS or site quality standards. The welding codes require a fillet weld to have a uniform concave or convex face, although a slightly non-uniform face is acceptable. Convexity is the distance the weld extends above a line drawn between the toes of the weld. The convexity of a fillet weld or individual surface bead must not exceed that permitted by the applicable code or standard. A fillet weld must be repaired if the profile has defects.

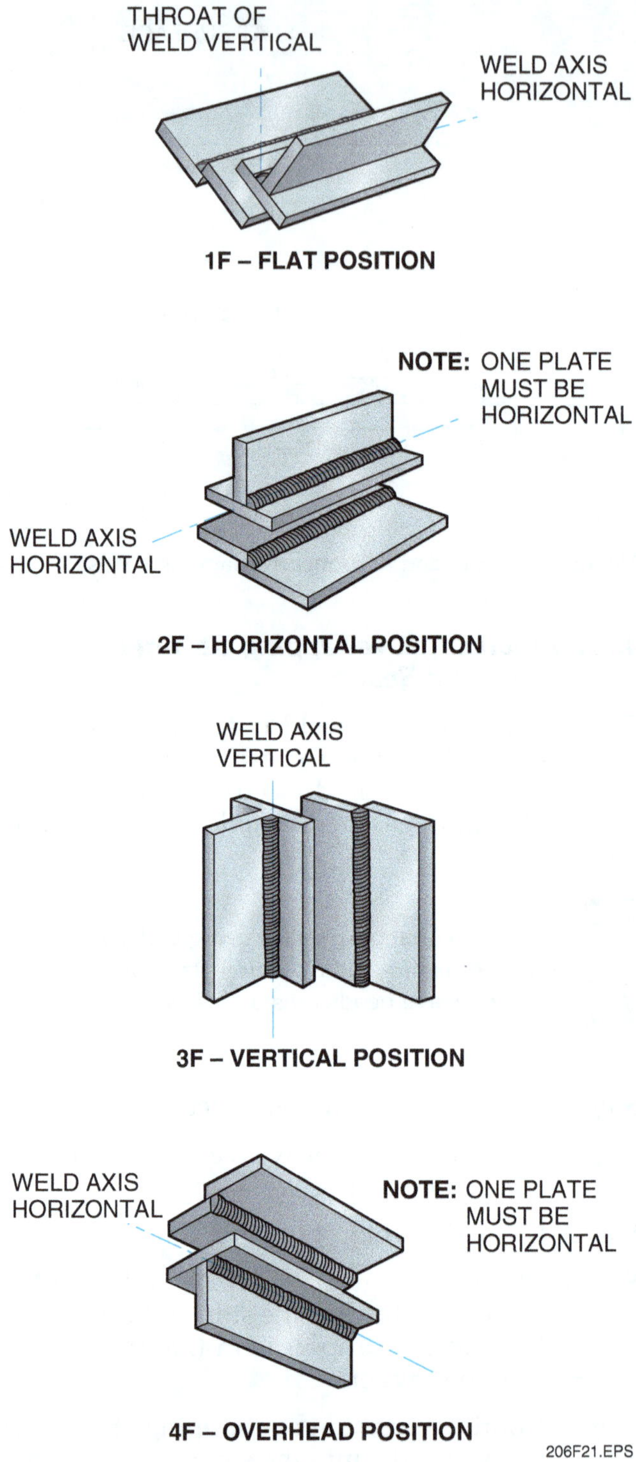

Figure 21 Fillet weld positions.

A fillet weld is unacceptable and must be repaired if the profile has insufficient throat, excessive convexity, excessive undercut, overlap, insufficient leg, or incomplete fusion.

Figure 22 shows ideal, acceptable, and unacceptable fillet weld profiles.

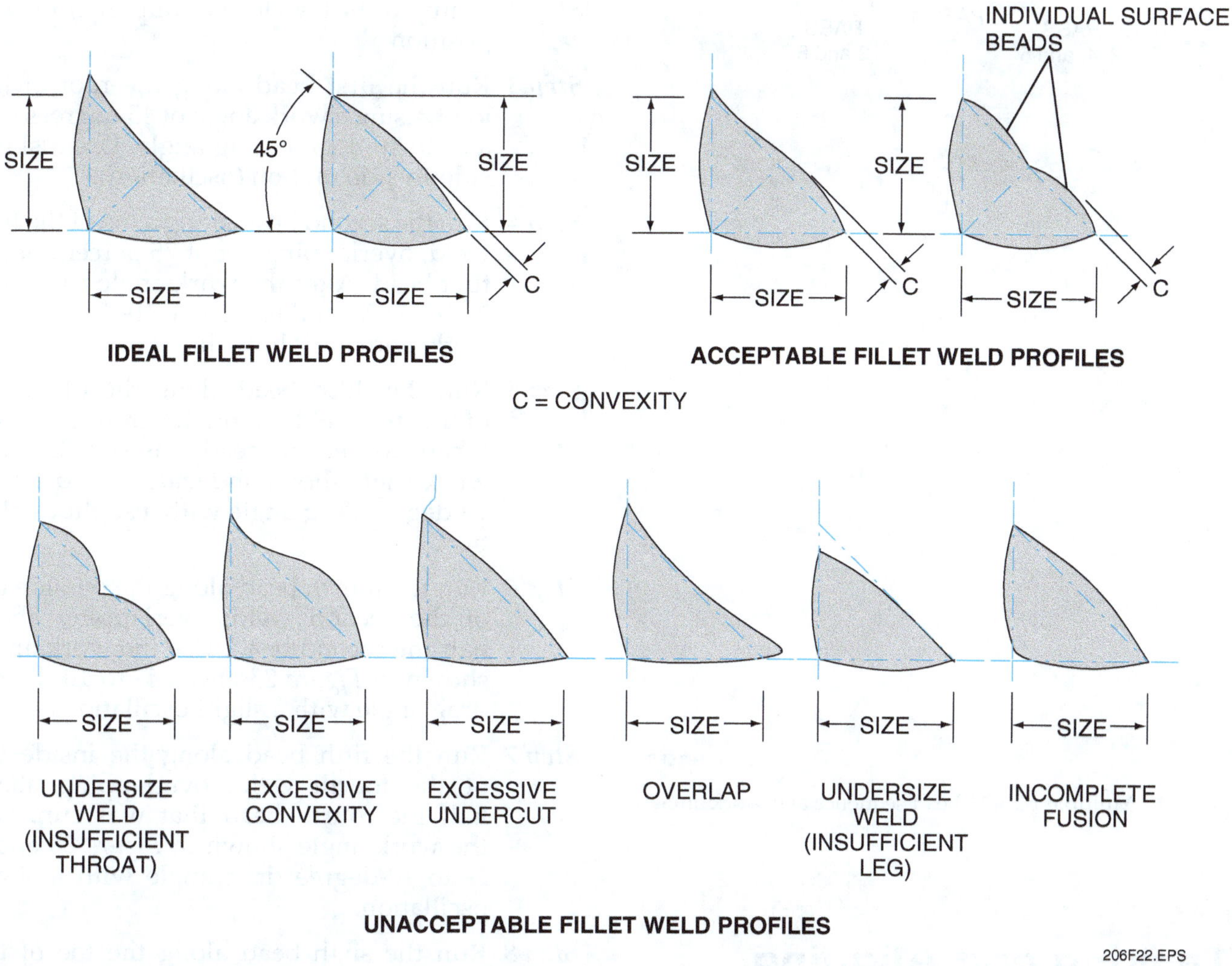

Figure 22 Ideal, acceptable, and unacceptable fillet weld profiles.

5.1.0 Practicing Flat (1F) Position Fillet Welds

Practice flat (1F) position fillet welds by welding multiple-pass (six-pass) convex fillet welds in a T-joint. Use an instructor-specified size of solid, composite, or flux-cored electrode wire with appropriate shielding gas as required. When making flat fillet welds, pay close attention to the work angle and travel speed. For the first bead, the work angle is vertical (45 degrees to both plate surfaces). The angle is adjusted for all subsequent beads. Increase or decrease the travel speed to control the amount of weld metal buildup. *Figure 23* shows the work angles for 1F position fillet welds.

Hot Tip

Preferred Fillet Weld Contours

In single-pass, weave-bead fillet welds where two workpieces are being joined at an angle (not lap joints), flat or slightly convex faces are usually preferred because weld stresses are more uniformly distributed through the fillet weld and workpieces.

NOTE

Practice 1F fillet welds using GMAW (spray and short-circuiting transfer) and FCAW as directed by your instructor.

Follow these steps to make a flat fillet weld. Please note in the following steps to clean all weld beads before beginning the next bead.

Step 1 Tack two plates together to form a T-joint for the fillet weld coupon (*Figure 24*). Clean the tack welds.

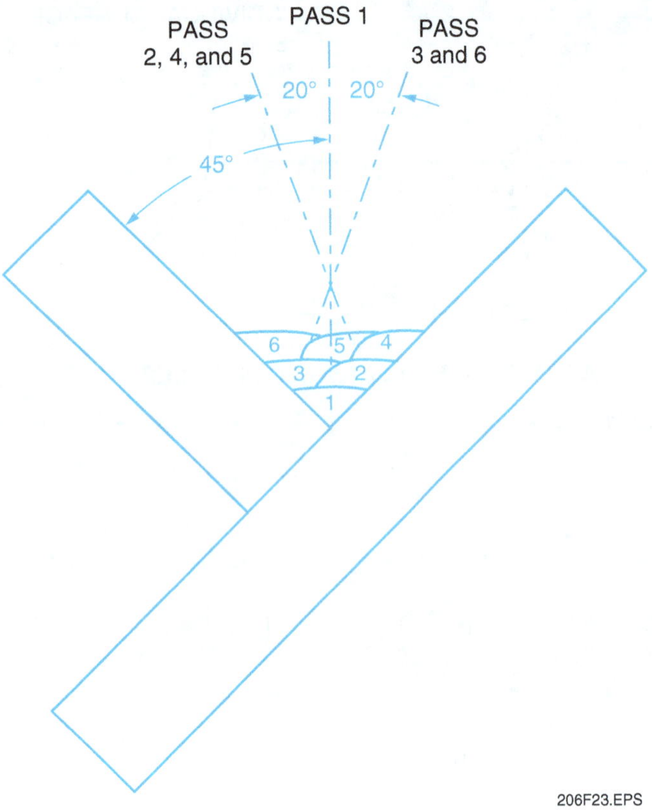

Figure 23 Multiple-pass 1F bead sequence and work angles.

Hot Tip

Tacking and Aligning Workpieces

When tacking workpieces together, both sides of the workpieces are usually tacked with welds that are about ½" long. This helps to position the workpieces and to minimize distortion when the final welds are made. After the first tack weld, use a hammer or other tool to align the workpieces side to side and end to end; then tack the opposite side. Tack the far ends of the workpieces in the same manner. Intermediate tack welds can be made every 5" to 6" as necessary to minimize lengthwise distortions.

NOTE: BASE METAL, CARBON STEEL AT LEAST ¼" THICK

Figure 24 Fillet weld coupon.

Step 2 Clamp or tack weld the coupon in the flat position.

Step 3 Run the first bead along the root of the joint, using a work angle of 45 degrees and a 5- to 10-degree drag angle. Use a slight side-to-side motion (oscillation).

Step 4 Run the second bead along a toe of the first weld, overlapping about 75 percent of the first bead. Alter the work angle as shown in *Figure 23*, and use a 5- to 10-degree drag angle with a slight oscillation.

Step 5 Run the third bead along the other toe of the first weld, filling the groove created when the second bead was run. Use the work angle shown in *Figure 23* and a 5- to 10-degree drag angle with a slight oscillation.

Step 6 Run the fourth bead along the outside toe of the second weld, overlapping about half the second bead. Use the work angle shown in *Figure 23* and a 5- to 10-degree drag angle with a slight oscillation.

Step 7 Run the fifth bead along the inside toe of the fourth weld, overlapping about half the fourth bead that was run. Use the work angle shown in *Figure 23* and a 5- to 10-degree drag angle with a slight oscillation.

Step 8 Run the sixth bead along the toe of the fifth weld, filling the groove created when the fifth bead was run. Use the work angle shown in *Figure 23* and a 5- to 10-degree drag angle with a slight oscillation.

Step 9 Inspect the weld. The weld is acceptable if it has the following features:

- Uniform appearance on the bead face
- Craters and restarts filled to the full cross section of the weld
- Uniform weld size ±1⁄16"
- Acceptable weld profile in accordance with the applicable code or standard
- Smooth transition with complete fusion at the toes of the weld
- No porosity
- No excessive undercut
- No overlap
- No inclusions
- No cracks

5.2.0 Practicing Horizontal (2F) Position Fillet Welds

Practice horizontal (2F) fillet welding by placing multiple-pass fillet welds in a T-joint using solid,

Hot Tip

T-Joint Heat Dissipation

In T-joints, the welding heat dissipates more rapidly in the thicker or nonbutting member. On various bead passes, the arc may have to be concentrated slightly more on the thicker or the nonbutting member to compensate for the heat loss.

composite, or flux-cored electrode wire sizes and, if required, appropriate shielding gas as directed by your instructor. When making horizontal fillet welds, pay close attention to the gun angles and travel speed. For the first bead, the electrode work angle is 45 degrees. The work angle is adjusted for all other welds. Increase or decrease the travel speed to control the amount of weld metal buildup.

NOTE

Practice 2F fillet welds using GMAW (spray and short-circuiting transfer) and FCAW as directed by your instructor.

Follow these steps to make a horizontal fillet weld:

Step 1 Tack two plates together to form a T-joint for the fillet weld coupon. Clean the tack welds.

Step 2 Clamp or tack weld the coupon in the horizontal position.

Step 3 Run the first bead along the root of the joint using a work angle of approximately 45 degrees with a 5- to 10-degree drag angle and a slight side-to-side oscillation (*Figure 25*).

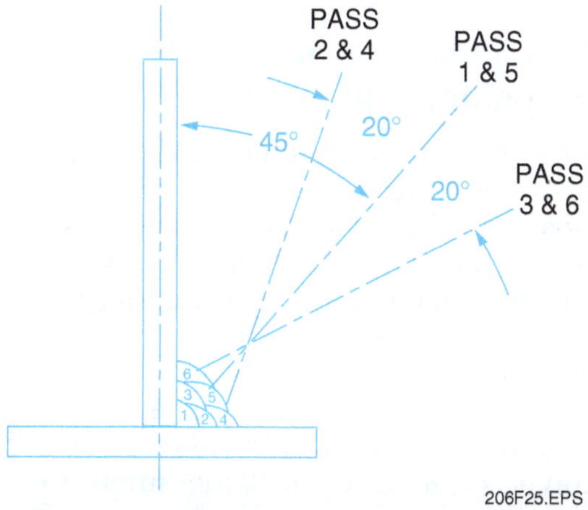

Figure 25 Multiple-pass 2F bead sequence and work angles.

Step 4 Clean the weld.

Step 5 Run the remaining passes at the appropriate work angles using a 5- to 10-degree drag angle and a slight oscillation at the appropriate work angles (*Figure 25*). Overlap each previous pass and clean the weld after each pass.

Step 6 Have your instructor inspect the weld. The weld is acceptable if it has the following features:

- Uniform appearance on the bead face
- Craters and restarts filled to the full cross section of the weld
- Uniform weld size ±1⁄16"
- Acceptable weld profile in accordance with the applicable code or standard
- Smooth transition with complete fusion at the toes of the weld
- No porosity
- No excessive undercut
- No overlap
- No inclusions
- No cracks

5.3.0 Practicing Vertical (3F) Position Fillet Welds

Practice vertical (3F) fillet welding by placing multiple-pass fillet welds in a T-joint using solid, composite, or flux-cored electrode wire sizes and, if required, appropriate shielding gas as directed by your instructor. Normally, vertical welds are made welding uphill (also called vertical-up welding) from the bottom to the top, using a gun push angle (up-angle). However, downhill welding (also called vertical-down) can also be accomplished using a slight drag angle. When vertical welding, either stringer or weave beads can usually be used. However, weave beads are avoided on stainless steel or in downhill welding. On the job, the site WPS or quality standard will specify which technique to use.

Hot Tip

T-Joint Vertical Plate Undercut

The most common defect for T-joints is an undercut on the vertical plate of the joint. If this problem occurs, angling the arc slightly more toward the vertical plate and the bead at the top of the weave will force more metal into the bead at the top edge of the weld.

Hot Tip

Vertical Fillet Welds

When making vertical fillet welds, pay close attention to the gun angles and travel speed. For the first bead, the work angle is approximately 45 degrees. The work angle is adjusted for all other welds. Increase or decrease the travel speed to control the amount of weld metal buildup.

NOTE

Practice 3F fillet welds using GMAW and/or FCAW as directed by your instructor. Check with your instructor to see if you should run stringer or weave beads, or practice both techniques.

5.3.1 Vertical Weave Bead Fillet Welds

Follow these steps to make an uphill fillet weld:

Step 1 Tack two plates together to form a T-joint for the fillet weld coupon.

Step 2 Clamp or tack weld the coupon in the vertical position.

Step 3 Starting at the bottom, run the first bead along the root of the joint using a work angle of approximately 45 degrees and a 10- to 15-degree push angle. Pause in the weld puddle to fill the crater.

Step 4 Clean the weld.

Step 5 Run the remaining passes using a 5- to 10-degree push angle and a side-to-side weave technique with a 45-degree work angle (*Figure 26*). Use a slow motion across the face of the weld, pausing at each toe for penetration and to fill the crater. Adjust the travel speed across the face of the weld to control the buildup. Overlap each previous pass as required, and clean the weld after each pass.

Step 6 Have the instructor inspect the weld. The weld is acceptable if it has the following features:

- Uniform appearance on the bead face
- Craters and restarts filled to the full cross section of the weld
- Uniform weld size ±1⁄16"
- Acceptable weld profile in accordance with the applicable code or standard
- Smooth transition with complete fusion at the toes of the weld
- No porosity

Hot Tip

Vertical Weave Bead

To help control undercut, an alternate pattern, such as a small triangular weave, can be used in vertical welds. By pausing the rod at the edges, the previous undercut can be filled. This action also creates undercut at the existing weld pool, but that will be filled in by the next weave.

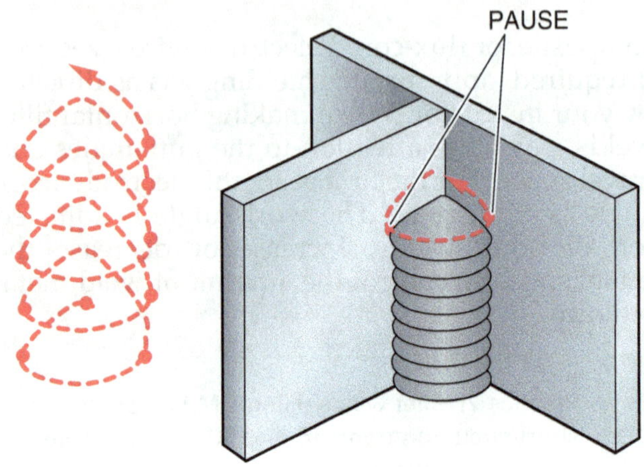

TRIANGULAR WEAVE PATTERN

206SA04.EPS

- No excessive undercut
- No overlap
- No inclusions
- No cracks

5.3.2 Vertical Stringer Bead Fillet Welds

Repeat vertical fillet (3F) welding using stringer beads. Use a slight oscillation and side-to-side motion, pausing slightly at each toe to prevent undercut. For stringer beads, use a 0- to 10-degree push angle and the work angles shown in *Figure 26* as required.

5.4.0 Practicing Overhead (4F) Position Fillet Welds

Practice overhead (4F) fillet welding by welding multiple-pass fillet welds in a T-joint using solid, composite, or flux-cored electrode wire sizes and, if required, appropriate shielding gas as directed by your instructor. When making overhead fillet welds, pay close attention to the gun angles and travel speed. For the first bead, the work angle is approximately 45 degrees. The work angle is adjusted for all other welds. Increase or decrease the travel speed to control the amount of weld metal buildup.

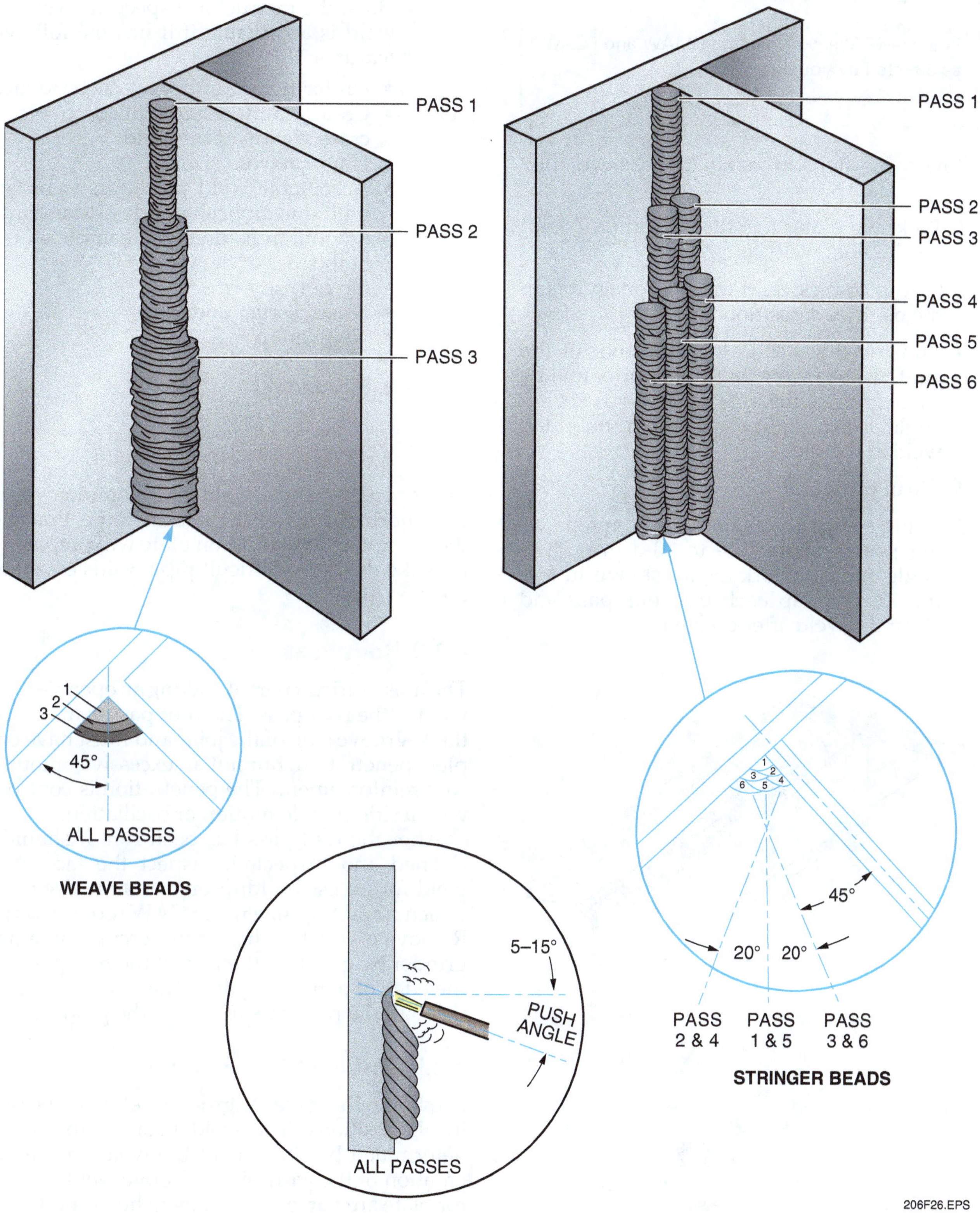

Figure 26 Multiple-pass 3F bead sequences and work angles for weave and stringer beads.

NOTE

Practice 4F fillet welds using GMAW and FCAW as directed by your instructor.

Follow these steps to make an overhead fillet weld:

Step 1 Tack two plates together to form a T-joint for the fillet weld coupon.

Step 2 Clamp or tack weld the coupon so it is in the overhead position.

Step 3 Run the first bead along the root of the joint, using a work angle of approximately 45 degrees with a 5- to 10-degree drag angle. Use a slight oscillation to tie-in the weld at the toes.

Step 4 Clean the weld.

Step 5 Using a slight oscillation, run the remaining passes using a 5- to 10-degree drag angle and the work angles shown in *Figure 27*. Overlap each previous pass and clean the weld after each pass.

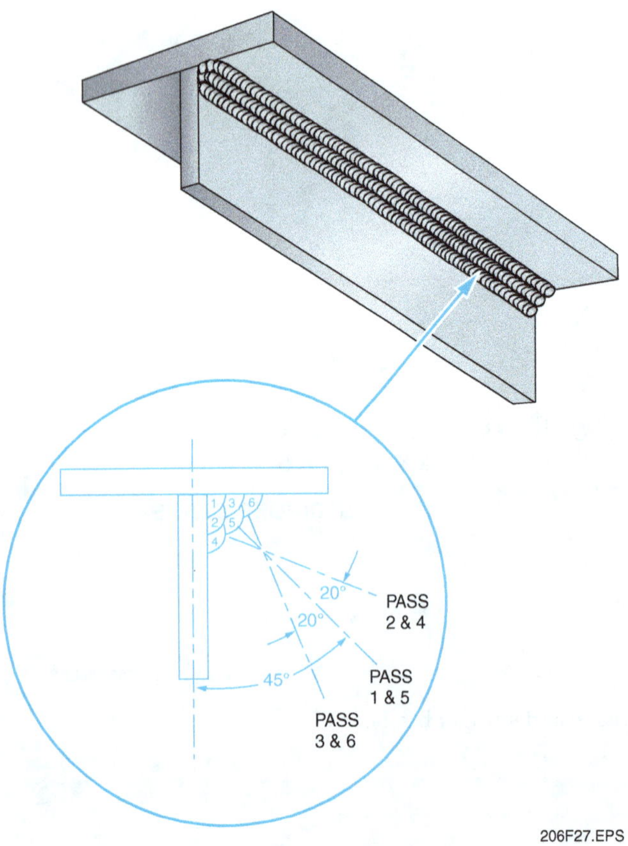

Figure 27 Multiple-pass 4F bead sequence and work angles.

Step 6 Have the instructor inspect the weld. The weld is acceptable if it has the following features:

- Uniform appearance on the bead face
- Craters and restarts filled to the full cross section of the weld
- Uniform weld size ±1⁄16"
- Acceptable weld profile in accordance with the applicable code or standard
- Smooth transition with complete fusion at the toes of the weld
- No porosity
- No excessive undercut
- No overlap
- No inclusions
- No cracks

6.0.0 Open V-Groove Welds

The open V-groove weld is a common groove weld normally made on plate and pipe. Practicing the open V-groove weld on plate will prepare you to make the more difficult pipe welds covered in the next level.

6.1.0 Root Pass

The most difficult part of making an open V-groove weld is the root pass. The root pass is made from the V-groove side of the joint and must have complete penetration, but not an excessive amount of root reinforcement. The penetration is controlled with a side-to-side motion or oscillation.

After the root pass has been run, it should be cleaned and inspected. Inspect the face of the weld for excess buildup or undercut (*Figure 28*), which could trap slag if an FCAW root pass is run. Remove excess buildup or undercut with a hand grinder by grinding the face of the root pass with the edge of a grinding disk. Use care not to grind through the root pass or widen the groove.

6.2.0 Groove Weld Positions

As shown in *Figure 29*, groove welds can be made in all positions. The weld position for plate is determined by the axis of the weld and the orientation of the workpiece. Groove weld positions for plate are flat, or 1G (groove); horizontal, or 2G; vertical, or 3G; and overhead, or 4G.

In the 1G and 2G positions, the weld axis can be inclined up to 15 degrees. Any weld axis inclination for the other positions varies with the rotational position of the weld face as specified in the AWS.

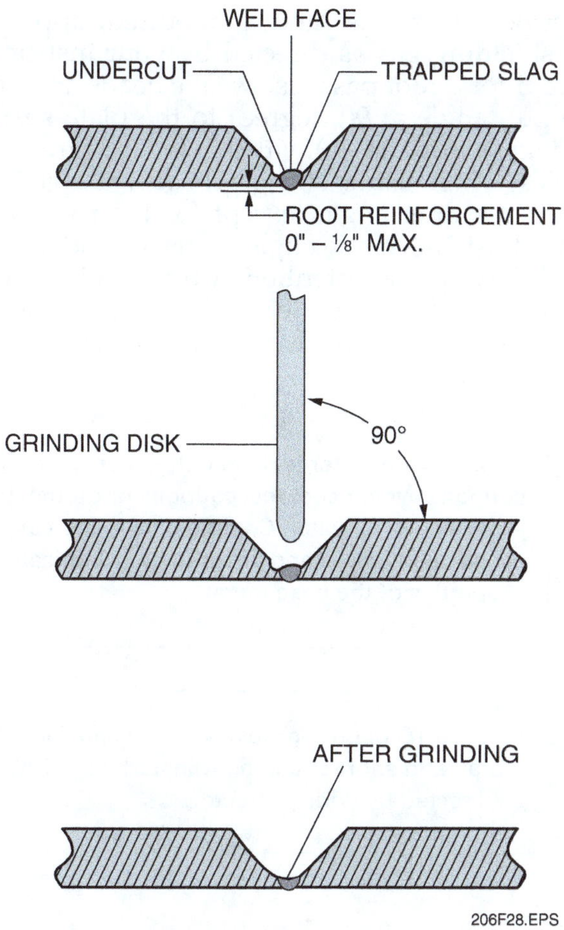

Figure 28 Grinding a root pass.

Hot Tip

V-Groove Welds for Stainless Steel

When practicing V-groove welds on stainless steel, remember to practice using a push angle instead of a drag angle. Also, practice using a backing for the weld, either with base metal strips or with a purging plate.

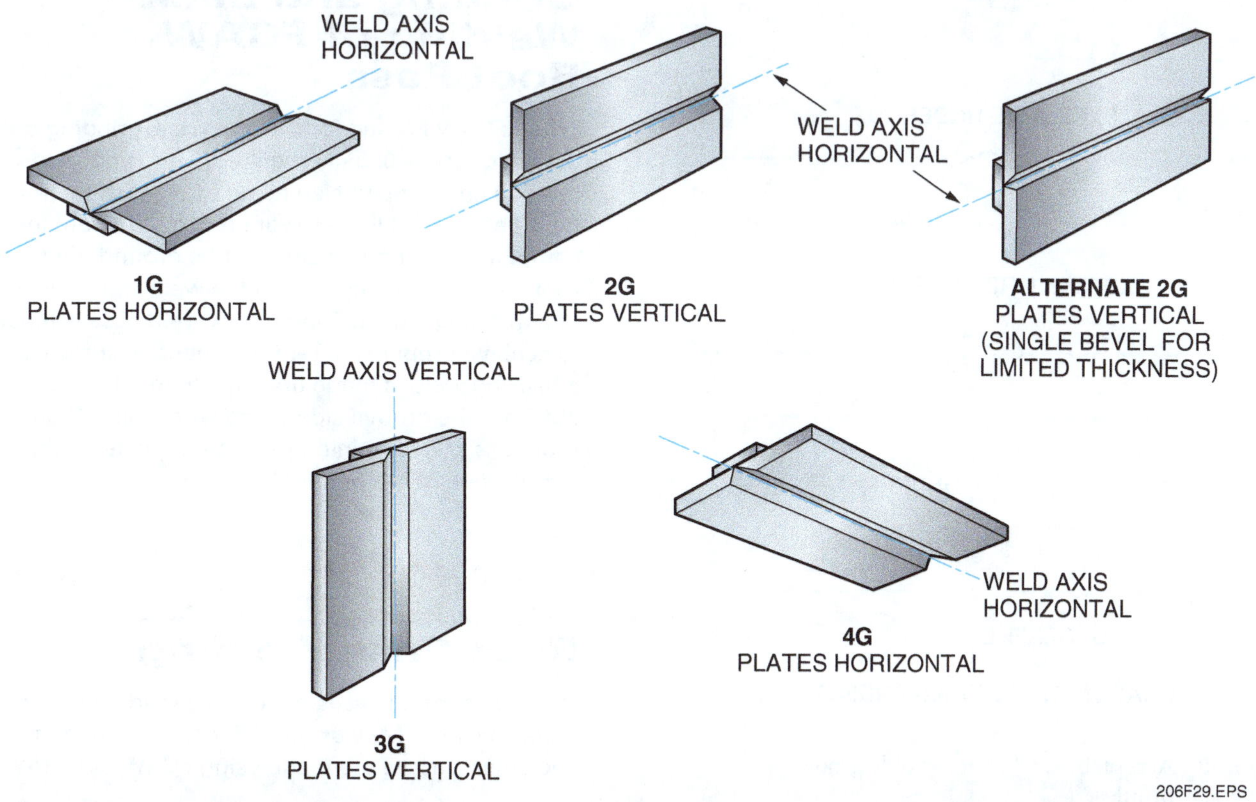

Figure 29 Grove weld positions for plate.

6.3.0 Acceptable and Unacceptable Groove Weld Profiles

Groove welds should be made with slight reinforcement and a gradual transition to the base metal at each toe. As shown in *Figure 30*, groove welds must not have excessive reinforcement or undercut, underfill, or overlap. If a groove weld has any of these defects, it must be repaired.

6.4.0 Practicing Flat (1G) Position Open V-Groove Welds

Practice flat (1G) open V-groove welds, as shown in *Figure 31*. Use solid, composite, or flux-cored electrode wire sizes and, if required, appropriate shielding gas as directed by your instructor. Except for a root pass, use weave beads and keep the gun angle at 90 degrees to the plate surface (0-degree work angle) with a 5- to 10-degree drag angle. When using stringer beads, use a 5- to 10-degree drag angle (except for the root pass), but adjust the work angle to tie in and fill the bead. Pay special attention at the termination of the weld to fill the crater. If the practice coupon gets too hot between passes, you may have to cool it in water.

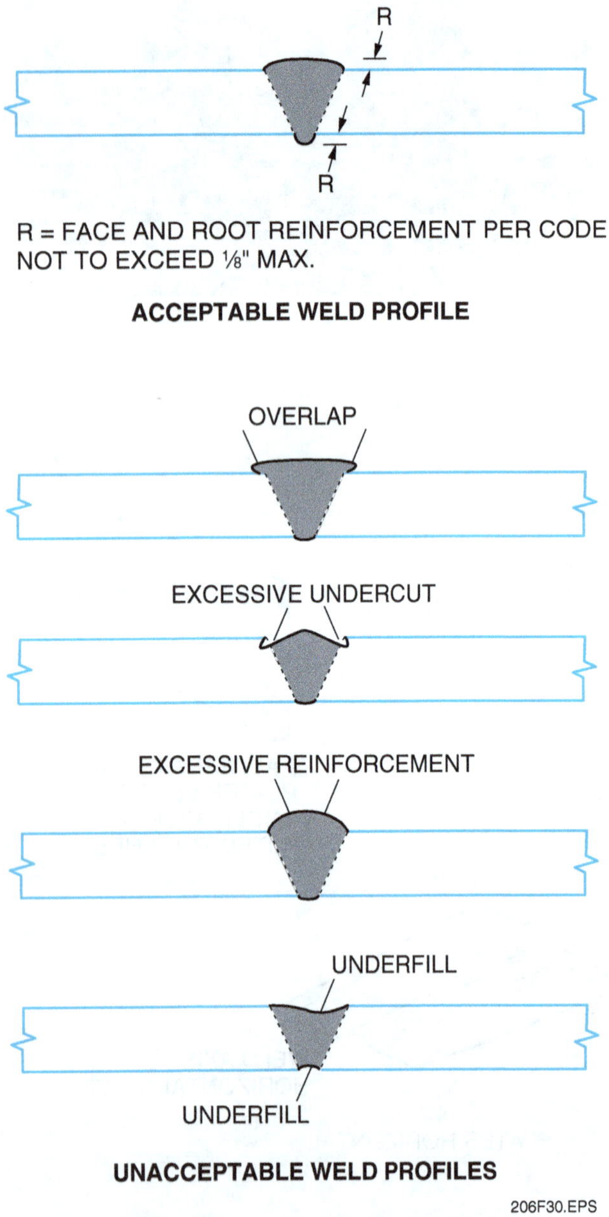

Figure 30 Acceptable and unacceptable groove weld profiles.

WARNING!

Cooling with water is done only on practice coupons. Never cool test coupons or on-the-job welds with water. Cooling with water can cause weld cracks and affect the mechanical properties of the base metal.

NOTE

Practice 1G open V-groove welds using GMAW (spray and short-circuiting transfer) and FCAW as directed by your instructor.

Hot Tip

Grinding and Back Welding of FCAW Root Pass

When FCAW is used for the root pass, grinding of the weld face is usually required. This is done to remove any trapped slag along the toes of the root pass before the fill passes are run. In addition, the completed V-groove weld must be ground out to the root pass on the back side of the weld with the edge of a grinding wheel. Then a back weld pass is made to achieve proper root reinforcement. In some cases, a backing weld is made first and its root is ground out from the groove side of the workpiece. Then the root pass and remaining passes are made on the groove side of the material.

Hot Tip

Root Pass Welding

In practice, root passes are usually made on various metals using GTAW or SMAW methods; then the fill and cover passes are made using FCAW or GMAW.

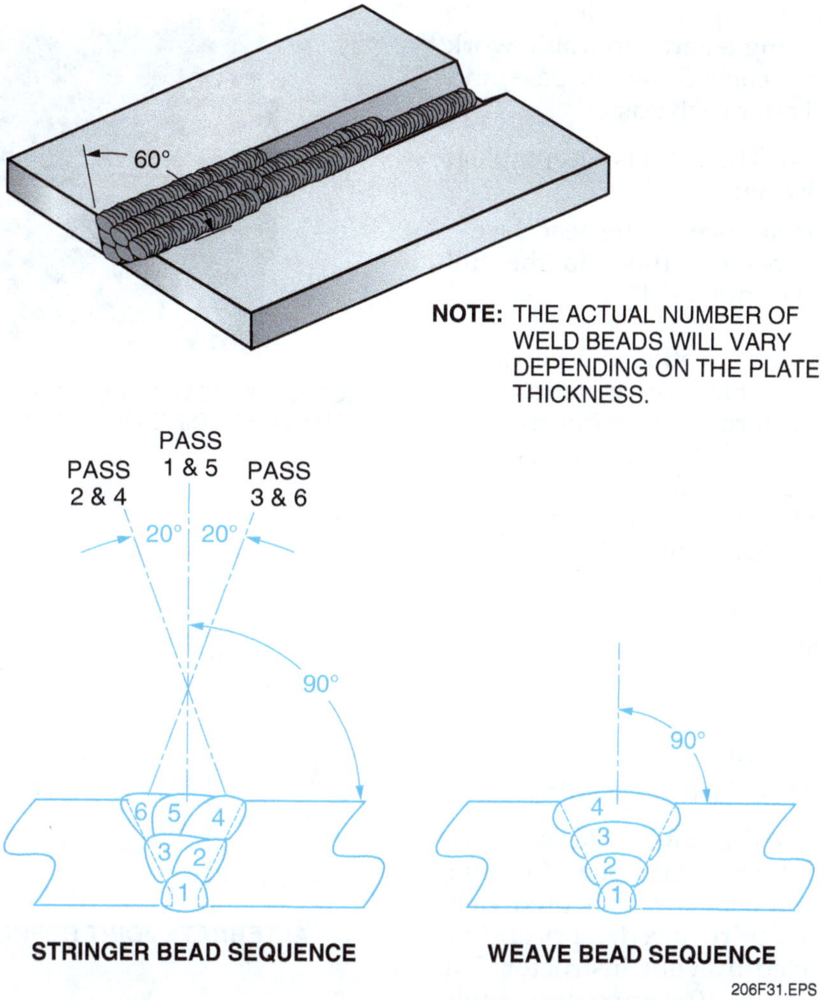

Figure 31 Multiple-pass 1G bead sequences and work angles.

WARNING!

Use pliers to handle the hot practice coupons. Wear gloves when placing the practice coupon in water. Steam will rise off the coupon and can burn or scald unprotected hands.

Hot Tip

Whiskers

When making a root pass, make sure that the wire electrode does not get ahead of the leading edge of the weld puddle. If it does, a portion of the wire may penetrate through the joint and be melted off and fused into the back side of the weld. This will cause pieces of wire, sometimes called whiskers, to protrude from the back of the weld.

Follow these steps to practice open V-groove welds in the flat position:

NOTE

Avoid using weave beads for stainless steel.

Step 1 Tack weld the practice coupon together as explained earlier.

Step 2 Clamp or tack weld the weld coupon in the flat position above the welding table surface.

Step 3 Run the root pass using a 5- to 10-degree drag angle and a slight side-to-side motion to control penetration.

Step 4 Clean the root pass and grind if required.

Step 5 Run the remaining passes using a 5- to 10-degree drag angle and applicable work angles. Overlap each previous pass and clean the weld after each pass.

Step 6 Inspect the weld. The weld is acceptable if it has the following:

- Uniform appearance on the bead face
- Craters and restarts filled to the full cross section of the weld
- Uniform weld width ±1⁄16"
- Acceptable weld profile in accordance with the applicable code or standard
- Complete uniform root reinforcement at least flush with the base metal to a maximum buildup of 1⁄8"
- Smooth, flat transition with complete fusion at the toes of the weld
- No porosity
- No excessive undercut
- No inclusions
- No cracks

6.5.0 Practicing Horizontal (2G) Position Open V-Groove Welds

A horizontal (2G) open V-groove weld is shown in *Figure 32*. Practice horizontal open V-groove welds using solid, composite, or flux-cored electrode wire sizes and, if required, appropriate shielding gas as directed by your instructor. For the gun angles, use a 5- to 10-degree drag angle (except for the root pass) and adjust the work angle as required. Pay particular attention to fill the crater at the termination of the weld.

> **NOTE**
> Practice 2G open V-groove welds using GMAW and/or FCAW as directed by your instructor.

Follow these steps to practice welding open V-groove welds in the horizontal position:

Step 1 Tack weld the practice coupon together as explained earlier. Use the standard or alternate horizontal weld coupon as directed by your instructor.

Step 2 Tack weld the flat plate weld coupon into the horizontal position.

Step 3 Run the root pass using a 5- to 15-degree drag angle and the appropriate work angle. Use a slight side-to-side motion to control penetration.

NOTE: THE ACTUAL NUMBER OF WELD BEADS WILL VARY DEPENDING ON THE PLATE THICKNESS.

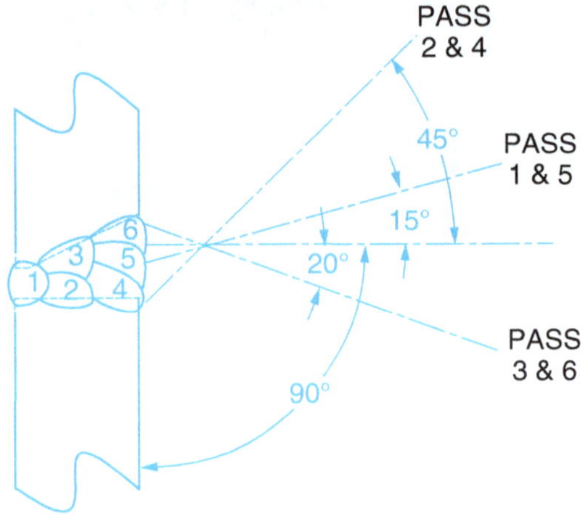

ALTERNATE JOINT REPRESENTATION

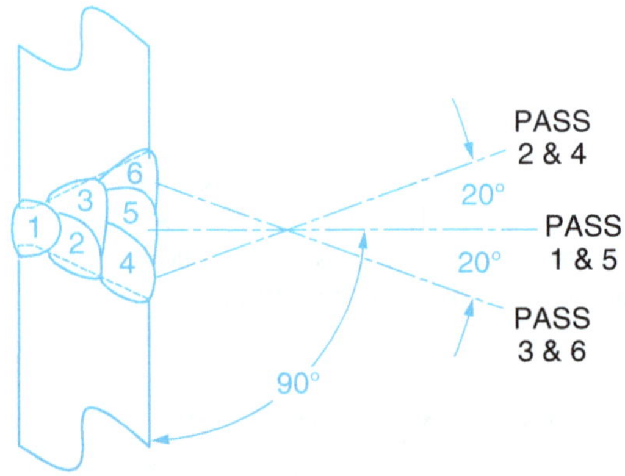

STANDARD JOINT REPRESENTATION

206F32.EPS

Figure 32 Multiple-pass 2G bead sequences and work angles.

Step 4 Clean the weld.

Step 5 Run the remaining passes using a 5- to 10-degree drag angle and appropriate work angles. Overlap each previous pass and clean the weld after each pass.

Step 6 Inspect the weld. The weld is acceptable if it has the following:

- Uniform appearance on the bead face
- Craters and restarts filled to the full cross section of the weld
- Uniform weld width ±1⁄16"
- Acceptable weld profile in accordance with the applicable code or standard
- Complete uniform root reinforcement at least flush with the base metal to a maximum buildup of 1⁄8"
- Smooth, flat transition with complete fusion at the toes of the weld
- No porosity
- No excessive undercut
- No inclusions
- No cracks

6.6.0 Practicing Vertical (3G) Position Open V-Groove Welds

Practice vertical (3G) multiple-pass groove welds using stringer beads or weave beads. Set the welding parameters at the lower ends of the suggested ranges. Out-of-position welds are made at lower welding voltages and currents than flat or horizontal welds to prevent the molten weld metal from sagging or running.

> **NOTE**
>
> When welding vertically, either stringer or weave beads can be used. However, weave beads are not normally used on stainless steel. On the job, the WPS or site quality standards will specify which technique to use. Check with your instructor to see if you should run stringer or weave beads or practice both techniques. The root bead can be run either uphill or downhill. The fill and cover beads can be run uphill or downhill as well. However, downhill weave beads are not normally used. Use the direction preferred at your site or as specified in the WPS or site quality standards.

Run the root bead, using a work angle of 90 degrees (nozzle center line bisects the V-groove) and a 5- to 15-degree push angle. The work angle is adjusted for all other welds.

6.6.1 Vertical Open V-Groove Weld with Stringer Beads

Practice 3G open V-groove welds using GMAW and FCAW as directed by your instructor. Use solid, composite, or flux-cored electrode wire sizes and appropriate shielding gas as directed by your instructor.

Follow these steps to make a vertical groove weld with stringer beads:

Step 1 Tack weld the practice coupon together as explained earlier.

Step 2 Clamp or tack weld the coupon in the vertical position.

Step 3 Run the root pass uphill with a work angle of 0 degrees (nozzle center line bisects the V-groove) and a 5- to 15-degree push angle.

Step 4 Clean the weld.

Step 5 Run the remaining passes at the work angles shown in *Figure 33*, using a 5- to 15-degree push angle. Overlap each pre-

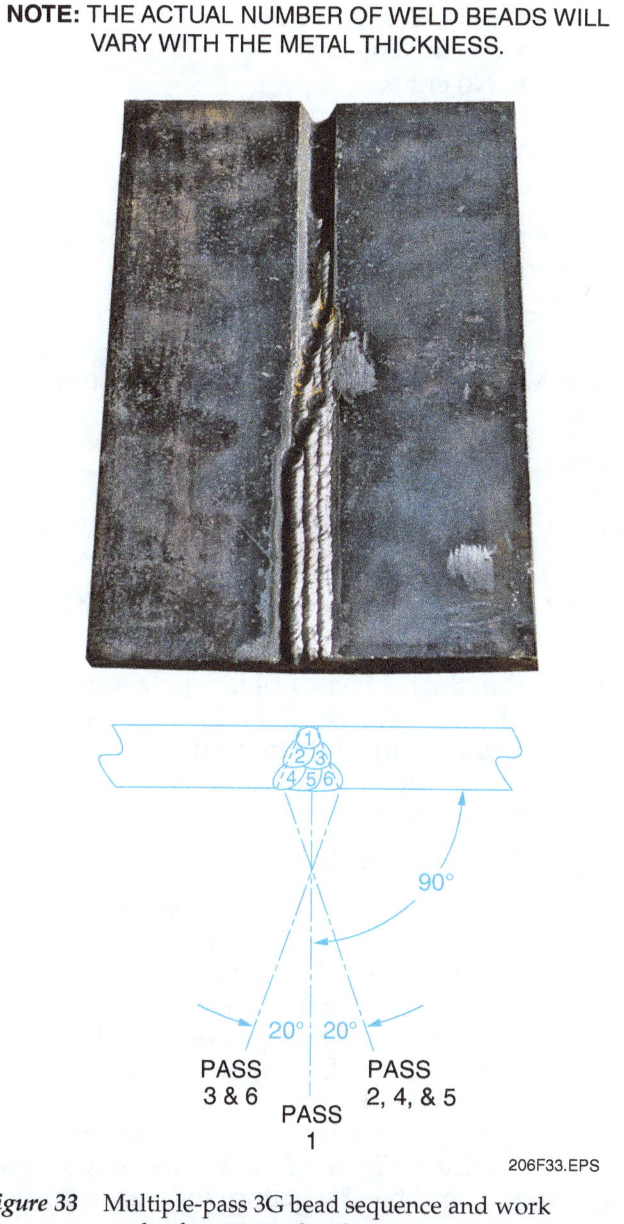

Figure 33 Multiple-pass 3G bead sequence and work angles for stringer beads.

vious pass and use a slight side-to-side oscillation to tie-in the welds at the toes. Clean the weld after each pass.

Step 6 Inspect the weld. The weld is acceptable if it has the following features:

- Uniform appearance on the bead face
- Craters and restarts filled to the full cross section of the weld
- Uniform weld width ±1/16"
- Acceptable weld profile in accordance with the applicable code or standard
- Complete uniform root reinforcement at least flush with the base metal to a maximum buildup of 1/8"
- Smooth, flat transition with complete fusion at the toes of the weld
- No porosity
- No excessive undercut
- No inclusions
- No cracks

6.6.2 Vertical Open V-Groove Weld with Weave Beads

Follow these steps to make a vertical (3G) groove weld with weave beads:

> **NOTE**
> Avoid using weave beads for stainless steel.

Step 1 Tack weld the practice coupon together as explained earlier.

Step 2 Clamp or tack weld the coupon in the vertical position.

Step 3 Run the root pass uphill with a work angle of 0 degrees (90 degrees from the plate) and a 5- to 15-degree push angle. Use a slight side-to-side oscillation to control the penetration.

Step 4 Clean the weld.

Step 5 Run the remaining passes as uphill weave beads across the root bead at a 0-degree work angle. Clean the weld after each pass. Use a zigzag weave pattern across the root bead and pause at each toe for good fusion. Use a push angle of 5 to 15 degrees. Increase or decrease the travel speed to control the amount of metal buildup. *Figure 34* shows the zigzag pattern and bead sequence for weave beads.

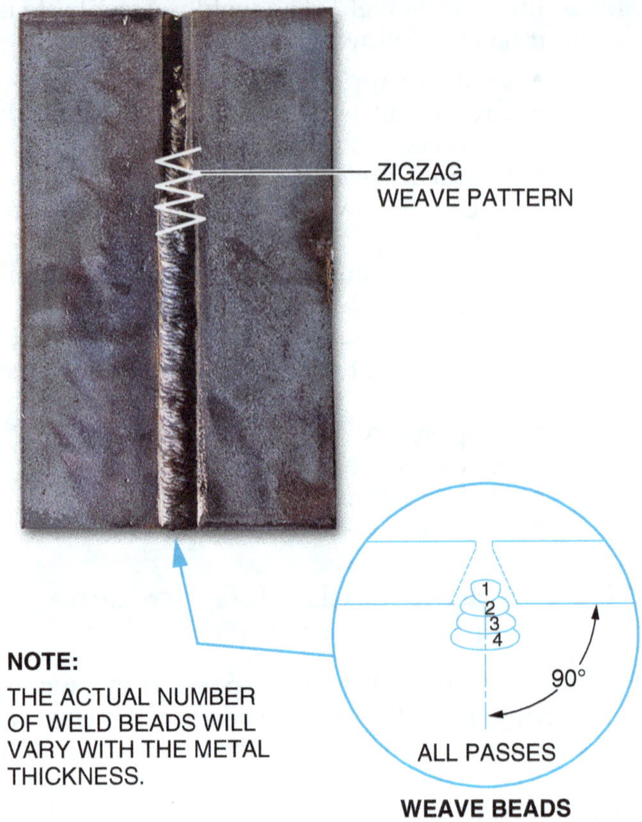

Figure 34 Multiple-pass 3G bead sequence and work angle for weave beads.

Step 6 Inspect the weld. The weld is acceptable if it has the following features:

- Uniform appearance on the bead face
- Craters and restarts filled to the full cross section of the weld
- Uniform weld width ±1/8"
- Acceptable weld profile in accordance with the applicable code or standard
- Complete uniform root reinforcement at least flush with the base metal to a maximum buildup of 1/8"
- Smooth, flat transition with complete fusion at the toes of the weld
- No porosity
- No excessive undercut
- No inclusions
- No cracks

6.7.0 Practicing Overhead (4G) Position Open V-Groove Welds

Practice overhead (4G) open V-groove welds using GMAW and FCAW as directed by your instructor. Use solid, composite, or flux-cored electrode wire sizes and appropriate shielding gas as directed by your instructor.

> **NOTE**
> To reduce heat and prevent the molten weld metal from sagging or running, GMAW and FCAW overhead welds are made with lower welding voltages and currents than flat or horizontal welds.

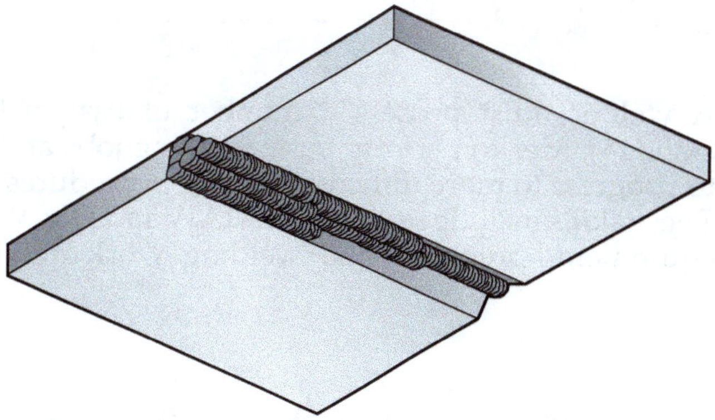

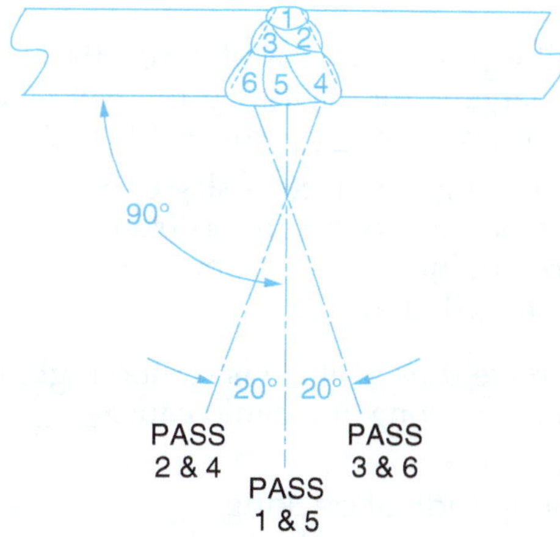

STRINGER BEAD SEQUENCE

206F35.EPS

Figure 35 Multiple-pass 4G bead sequence and work angles.

Use a work angle of 0 degrees (90 degrees to plate) and a 5- to 15-degree drag angle for the root bead.

Follow these steps to make an overhead groove weld with stringer beads:

Step 1 Tack weld the practice coupon together as explained earlier.

Step 2 Clamp or tack weld the coupon in the overhead position.

Step 3 Run the root pass, using a work angle of 0 degrees and a 5- to 15-degree drag angle. Use a slight side-to-side oscillation to control the penetration.

Step 4 Clean the weld.

Step 5 Run the remaining passes at the work angles shown in *Figure 35* using a 5- to 10-degree drag angle. Overlap each previous pass, and use a slight side-to-side motion to tie in the welds at the toes. Each bead should be cleaned before running the next bead.

Step 6 Inspect the weld. The weld is acceptable if it has the following features:

- Uniform appearance on the bead face
- Craters and restarts filled to the full cross section of the weld
- Uniform weld width ±1⁄16"
- Acceptable weld profile in accordance with the applicable code or standard
- Complete uniform root reinforcement at least flush with the base metal to a maximum buildup of 1⁄8"
- Smooth, flat transition with complete fusion at the toes of the weld
- No porosity
- No excessive undercut
- No inclusions
- No cracks

SUMMARY

A welder must possess a number of essential skills in order to perform basic welding jobs and to progress to more difficult welding procedures. These skills include setting up GMAW and FCAW equipment, preparing the welding work area, running stringer and weave beads, and making acceptable fillet welds and open V-groove welds on plate in all positions. Practice these welds until you can consistently produce acceptable welds as defined in the criteria for acceptance.

Review Questions

1. In all types of flux-cored wires, the flux contains agents for degassing the weld, stabilizing the arc, and _____ the weld bead.

a. forming a protective slag over
b. removing contaminants from
c. removing the slag from
d. strengthening

2. If the welding voltage is set too high, the arc will be too long and could cause _____.

a. overlap
b. longitudinal cracking
c. the wire to fuse to the tip
d. excessive root penetration

3. When the wire electrode speed is increased, the result is usually _____.

a. lower welding heat
b. less penetration
c. higher amperage
d. flatter beads

4. When the gun electrode is in the neutral travel angle, weld penetration is _____.

a. high
b. moderate
c. low
d. negligible

5. The length of the wire that extends beyond the contact tip of the welding gun is the _____.

a. stickout
b. standoff distance
c. electrode extension
d. contact tip extension

6. When a larger ($\frac{3}{32}$" to $\frac{1}{8}$") wire electrode diameter is used, the width of a stringer bead is generally no more than _____ times the wire electrode's diameter.

a. three
b. four
c. five
d. six

7. When weave beads are made, the motion of the electrode is _____.

a. straight
b. slightly side-to-side
c. wide side-to-side
d. up to the welder

8. If a restart has undercut, _____.

a. too much time was spent in the crater
b. not enough time was spent in the crater
c. there was not enough side-to-side motion
d. there were arc strikes outside the weld area

9. A weld termination normally leaves _____.

a. a crack
b. porosity
c. spatter
d. a crater

10. Deposit connective weld beads parallel to one another to make _____ beads.

a. solid
b. equal length
c. congruent
d. overlapping

11. When making flat (1F) fillet welds, increase or decrease the travel speed to control the amount of _____.

a. porosity
b. root penetration
c. base metal fusion
d. weld metal buildup

12. For the first bead in a fillet weld, the work angle is _____ degrees to both plate surfaces.

a. 15
b. 20
c. 45
d. 55

13. Groove welds can be made in _____.

a. all positions
b. the 1G and 2G positions only
c. the 2G and 3G positions only
d. the 1G, 2G, and 3G positions only

14. To make a vertical groove weld with weave beads, use a zigzag pattern across the root bead and pause at each toe _____.

a. for good fusion
b. to prevent excessive spatter
c. for complete penetration
d. to fill in the crater

15. When making a vertical open V-groove weld with weave beads, the uniform weld width should be _____.

a. ±1⁄16"
b. ±1⁄8"
c. ±3⁄8"
d. ±1⁄2"

Trade Terms Introduced in This Module

Axis: The center line of the weld or prepared base metal.

Degassing: Removing the gases formed during the welding process that can cause imperfections in the weld.

Additional Resources

This module is intended to present thorough resources for task training. The following reference works are suggested for further study. These are optional materials for continued education rather than for task training.

AWS B1.10:1999 Guide for the Nondestructive Examination of Welds. Miami, FL: American Welding Society.

AWS B1.11:2000 Guide for the Visual Examination of Welds. Miami, FL: American Welding Society.

AWS C5.6-89 Recommended Practices for Gas Metal Arc Welding. Miami, FL: American Welding Society, 1989.

AWS D3.5-93R Guide for Steel Hull Welding. Miami, FL: American Welding Society, 1993.

E1.10 2009 Product Catalog. Cleveland, OH: The Lincoln Electric Company.

GMAW Welding Guide. Cleveland, OH: The Lincoln Electric Company, 2006.

Lincoln Electric website offers sources for products and training: **http://www.lincolnelectric.com**

OSHA 1910.269, Appendix C, Protection from Step and Touch Potentials. Current edition. Washington, DC: Occupational Safety & Health Administration (OSHA).

OSHA 1926.351, Arc Welding and Cutting. Current edition. Washington, DC: Occupational Safety & Health Administration (OSHA).

The Procedure Handbook of Arc Welding. 14th edition. Cleveland, OH: The James F. Lincoln Arc Welding Foundation, 2000.

Welding Handbook, Volume 1 *Welding Science & Technology*. Miami, FL: American Welding Society, (2001).

Welding Handbook, Volume 2 Part 1: *Welding Processes*. Miami, FL: American Welding Society, (2004).

Appendix

Performance Accreditation Tasks

The Performance Accreditation Tasks (PATs) correspond to and support the learning objectives in *AWS EG2.0, Curriculum Guide for the Training of Welding Personnel: Level–Entry Welder.*

PATs provide specific acceptable criteria for performance and help to ensure a true competency-based welding program for students.

The following tasks are designed to evaluate your ability to run fillet and groove welds with GMAW or FCAW equipment. Perform each task when you are instructed to do so by your instructor. As you complete each task, show it to your instructor for evaluation. Do not proceed to the next task until instructed to do so by your instructor.

MAKE A FILLET WELD IN THE (1F) FLAT POSITION

As directed by the instructor, use GMAW (spray and/or short-circuiting transfer), FCAW, or both processes with carbon steel solid or flux-cored electrode wire sizes, and, if required, appropriate shielding gas to make a six-pass fillet weld using stringer beads on carbon steel plate, as shown.

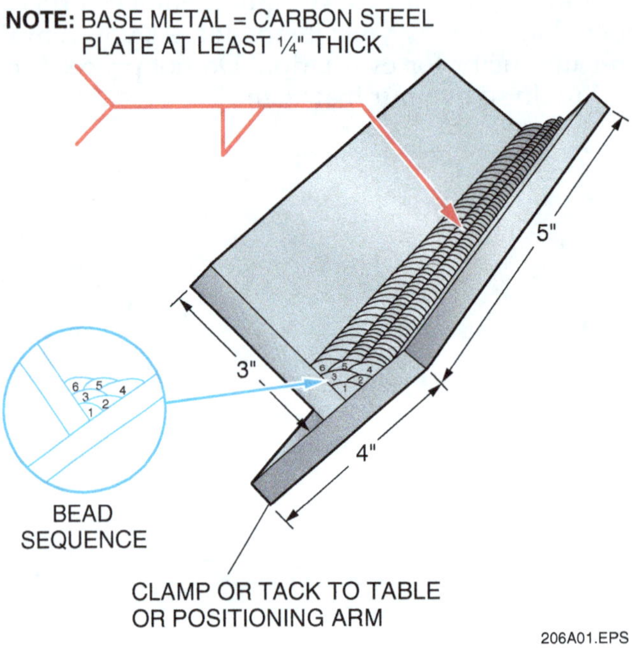

Criteria for Acceptance

- Uniform rippled appearance on the bead face ______
- Craters and restarts filled to the full cross section of the weld ______
- Uniform weld size ±1/16" ______
- Smooth flat transition with complete fusion at the toes of the welds ______
- Acceptable weld profile in accordance with the applicable code or standard ______
- No porosity ______
- No excessive undercut ______
- No overlap ______
- No inclusions ______

MAKE A FILLET WELD IN THE (2F) HORIZONTAL POSITION

As directed by the instructor, use GMAW (spray and/or short-circuiting transfer), FCAW, or both processes with carbon steel solid or flux-cored electrode wire sizes, and, if required, appropriate shielding gas to make a six-pass fillet weld using stringer beads on carbon steel plate, as shown.

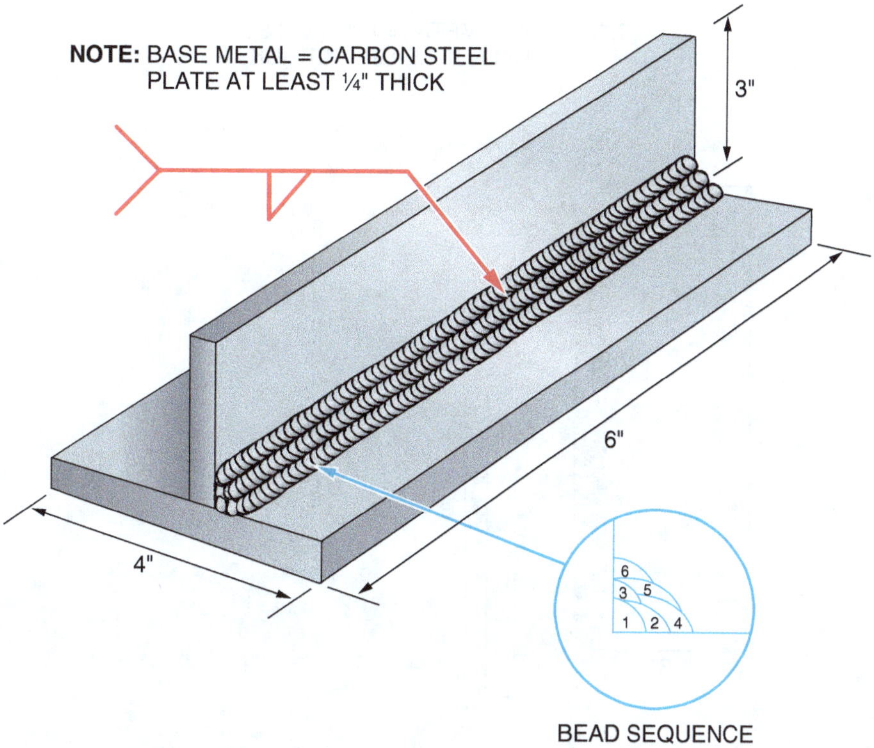

Criteria for Acceptance

- Uniform rippled appearance on the bead face ________
- Craters and restarts filled to the full cross section of the weld ________
- Uniform weld size ±1/16" ________
- Smooth flat transition with complete fusion at the toes of the welds ________
- Acceptable weld profile in accordance with the applicable code or standard ________
- No porosity ________
- No excessive undercut ________
- No overlap ________
- No inclusions ________

MAKE A FILLET WELD IN THE (3F) VERTICAL POSITION

As directed by the instructor, use GMAW, FCAW, or both processes with carbon steel solid or flux-cored electrode wire sizes, and, if required, appropriate shielding gas to make a vertical fillet weld on carbon steel plate, as shown.

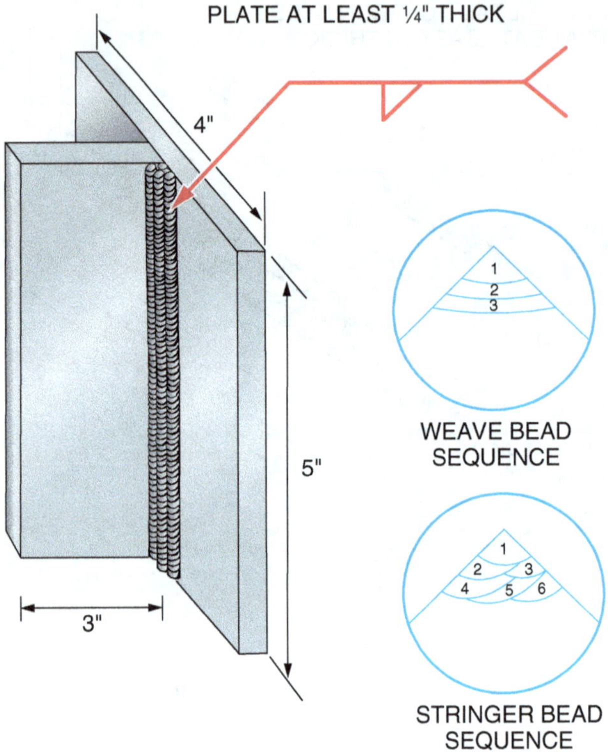

206A03.EPS

Criteria for Acceptance

- Uniform rippled appearance on the bead face ____________
- Craters and restarts filled to the full cross section of the weld ____________
- Uniform weld width ±1⁄16" ____________
- Acceptable weld profile in accordance with the applicable code or standard ____________
- Smooth flat transition with complete fusion at the toes of the weld ____________
- No porosity ____________
- No excessive undercut ____________
- No inclusions ____________
- No cracks ____________

MAKE A FILLET WELD IN THE (4F) OVERHEAD POSITION

As directed by the instructor, use GMAW, FCAW, or both processes with carbon steel solid or flux-cored electrode wire sizes, and, if required, appropriate shielding gas to make a six-pass fillet weld using stringer beads on carbon steel plate, as shown.

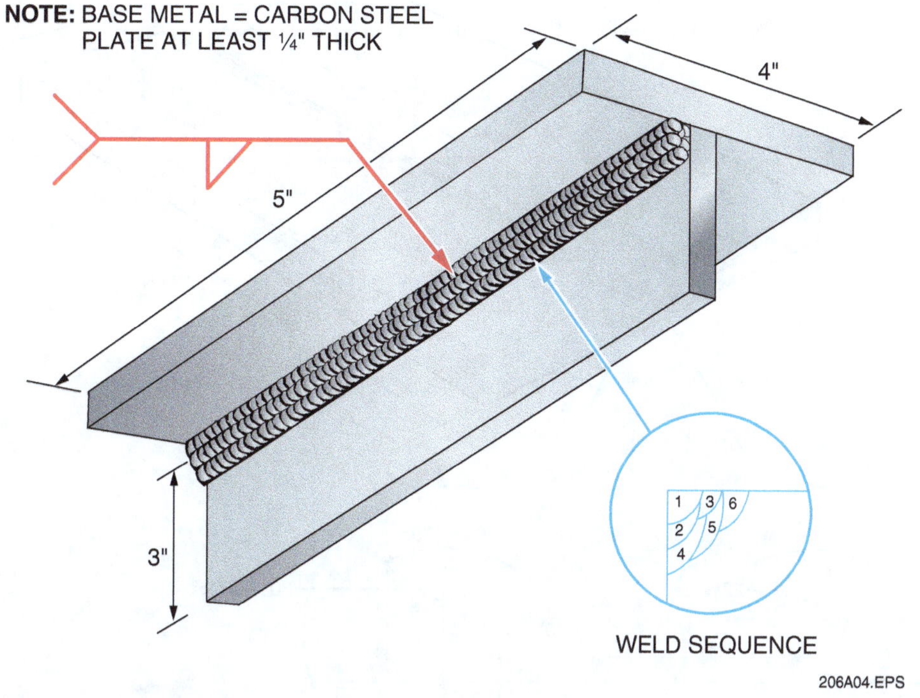

Criteria for Acceptance

- Uniform rippled appearance on the bead face __________
- Craters and restarts filled to the full cross section of the weld __________
- Uniform weld width ±1⁄16" __________
- Acceptable weld profile in accordance with the applicable code or standard __________
- Smooth flat transition with complete fusion at the toes of the weld __________
- No porosity __________
- No excessive undercut __________
- No inclusions __________
- No cracks __________

MAKE A GROOVE WELD, WITH OR WITHOUT BACKING, IN THE (1G) FLAT POSITION

As directed by the instructor, use GMAW (spray and/or short-circuiting transfer), FCAW, or both processes with carbon steel solid or flux-cored electrode wire sizes, and, if required, appropriate shielding gas to make a multiple-pass groove weld using stringer beads on carbon steel plate, with or without backing, as shown.

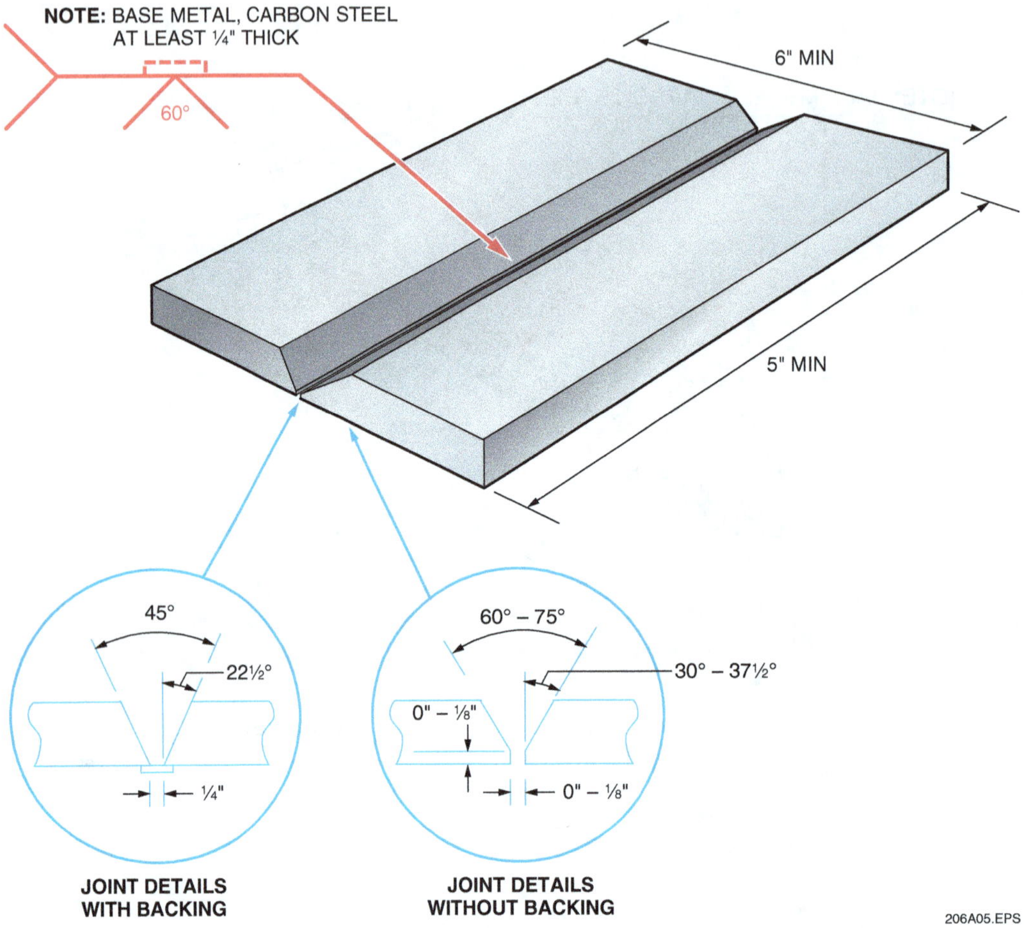

Criteria for Acceptance

- Uniform rippled appearance on the bead face ____________
- Craters and restarts filled to the full cross section of the weld ____________
- Uniform weld width ±1⁄16" ____________
- Acceptable weld profile in accordance with the applicable code or standard ____________
- Complete uniform root reinforcement at least flush with the base metal to a maximum buildup of 1⁄16", if applicable ____________
- Smooth flat transition with complete fusion at the toes of the weld ____________
- No porosity ____________
- No excessive undercut ____________
- No inclusions ____________
- No cracks ____________

MAKE A GROOVE WELD, WITH OR WITHOUT BACKING, IN THE (2G) HORIZONTAL POSITION

As directed by the instructor, use GMAW, FCAW, or both processes with carbon steel solid or flux-cored electrode wire sizes, and, if required, appropriate shielding gas to make a multiple-pass groove weld on carbon steel plate, with or without backing, as shown.

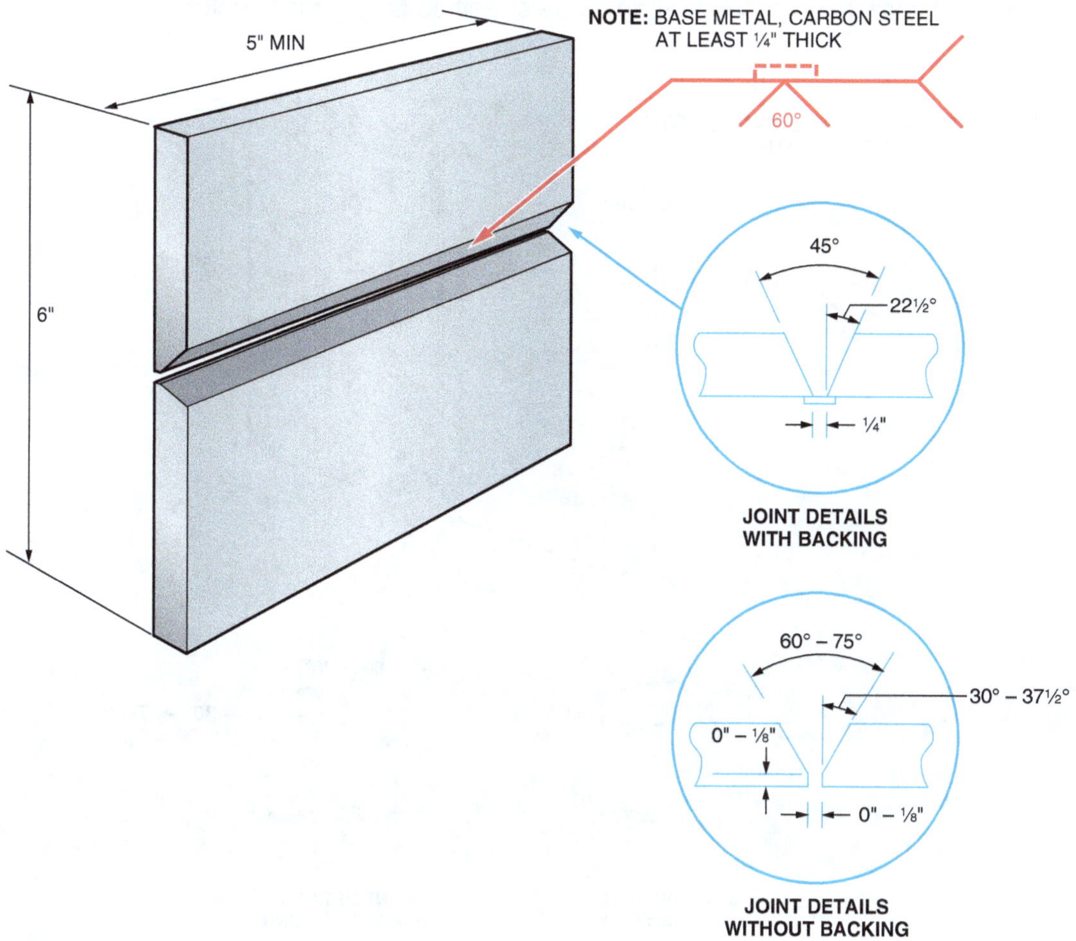

206A06.EPS

Criteria for Acceptance

- Uniform rippled appearance on the bead face ____________
- Craters and restarts filled to the full cross section of the weld ____________
- Uniform weld width ±1⁄16" ____________
- Acceptable weld profile in accordance with the applicable code or standard ____________
- Complete uniform root reinforcement at least flush with the base metal to a maximum buildup of 1⁄16", if applicable ____________
- Smooth flat transition with complete fusion at the toes of the weld ____________
- No porosity ____________
- No excessive undercut ____________
- No inclusions ____________
- No cracks ____________

MAKE A GROOVE WELD, WITH OR WITHOUT BACKING, IN THE (3G) VERTICAL POSITION

As directed by the instructor, use GMAW, FCAW, or both processes with carbon steel solid or flux-cored electrode wire sizes, and, if required, appropriate shielding gas to make a multiple-pass groove weld on carbon steel plate, with or without backing, as shown.

Note: Run the root vertical up or vertical down as specified by your instructor.

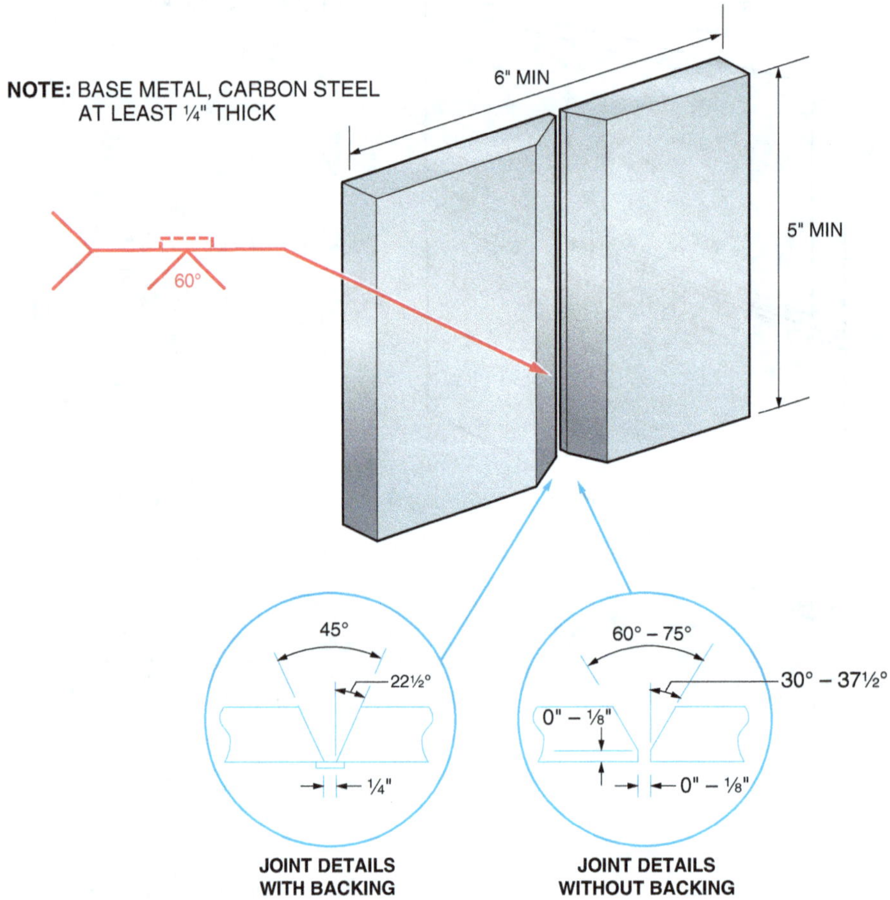

206A07.EPS

Criteria for Acceptance

- Uniform rippled appearance on the bead face ______
- Craters and restarts filled to the full cross section of the weld ______
- Uniform weld width ±1⁄16" ______
- Acceptable weld profile in accordance with the applicable code or standard ______
- Complete uniform root reinforcement at least flush with the base metal to a maximum buildup of 1⁄16", if applicable ______
- Smooth flat transition with complete fusion at the toes of the weld ______
- No porosity ______
- No excessive undercut ______
- No inclusions ______
- No cracks ______

MAKE A GROOVE WELD, WITH OR WITHOUT BACKING, IN THE (4G) OVERHEAD POSITION

As directed by the instructor, use GMAW, FCAW, or both processes with carbon steel solid or flux-cored electrode wire sizes, and, if required, appropriate shielding gas to make a multiple-pass groove weld on carbon steel plate, with or without backing, as shown.

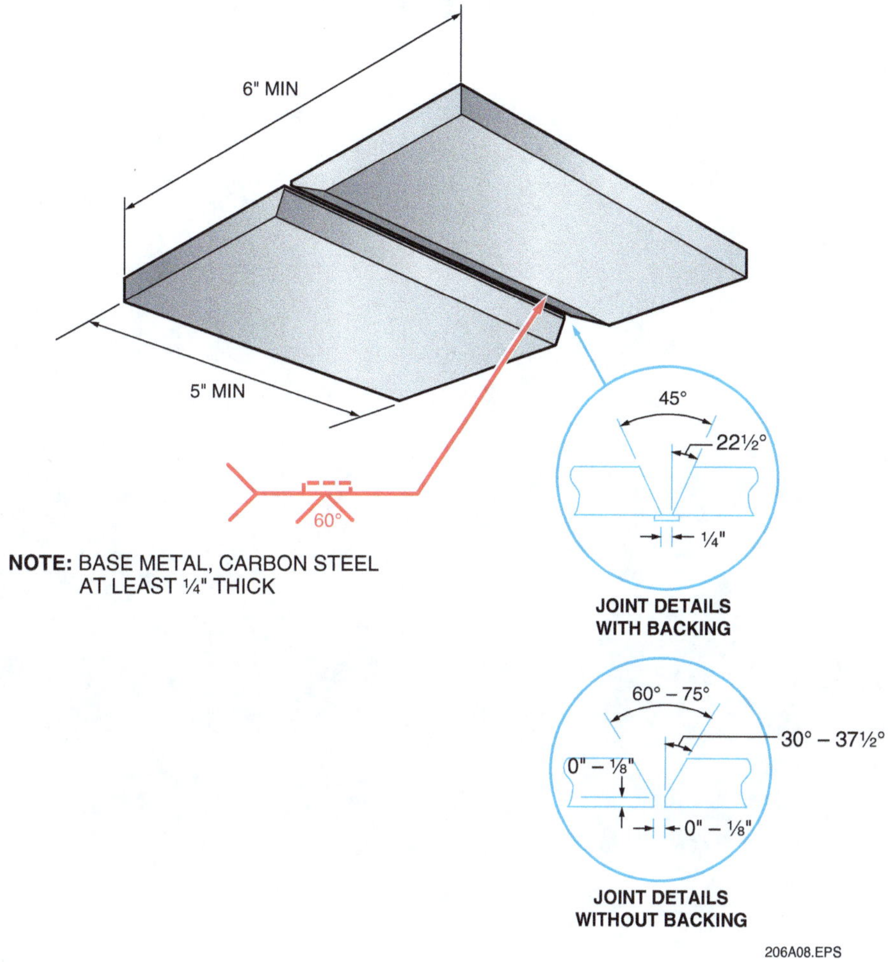

Criteria for Acceptance

- Uniform rippled appearance on the bead face ______
- Craters and restarts filled to the full cross section of the weld ______
- Uniform weld width ±1/16" ______
- Acceptable weld profile in accordance with the applicable code or standard ______
- Complete uniform root reinforcement at least flush with the base metal to a maximum buildup of 1/16", if applicable ______
- Smooth flat transition with complete fusion at the toes of the weld ______
- No porosity ______
- No excessive undercut ______
- No inclusions ______
- No cracks ______

NCCER CURRICULA — USER UPDATE

NCCER makes every effort to keep its textbooks up-to-date and free of technical errors. We appreciate your help in this process. If you find an error, a typographical mistake, or an inaccuracy in NCCER's curricula, please fill out this form (or a photocopy), or complete the online form at **www.nccer.org/olf**. Be sure to include the exact module ID number, page number, a detailed description, and your recommended correction. Your input will be brought to the attention of the Authoring Team. Thank you for your assistance.

Instructors – If you have an idea for improving this textbook, or have found that additional materials were necessary to teach this module effectively, please let us know so that we may present your suggestions to the Authoring Team.

NCCER Product Development and Revision

13614 Progress Blvd., Alachua, FL 32615

Email: curriculum@nccer.org
Online: www.nccer.org/olf

❑ Trainee Guide ❑ Lesson Plans ❑ Exam ❑ PowerPoints Other ________________

Craft / Level: Copyright Date:

Module ID Number / Title:

Section Number(s):

Description:

Recommended Correction:

Your Name:

Address:

Email: Phone:

Physical Characteristics and Mechanical Properties of Metals

29203-09

Trainees with successful module completions may be eligible for credentialing through the NCCER Registry. To learn more, go to **www.nccer.org** or contact us at **1.888.622.3720**. Our website has information on the latest product releases and training, as well as online versions of our *Cornerstone* magazine and Pearson's product catalog.

Your feedback is welcome. You may email your comments to **curriculum@nccer.org**, send general comments and inquiries to **info@nccer.org**, or fill in the User Update form at the back of this module.

This information is general in nature and intended for training purposes only. Actual performance of activities described in this manual requires compliance with all applicable operating, service, maintenance, and safety procedures under the direction of qualified personnel. References in this manual to patented or proprietary devices do not constitute a recommendation of their use.

29203-09
Physical Characteristics and Mechanical Properties of Metals

Objectives

When you have completed this module, you will be able to do the following:

1. Identify and explain the composition and classification of base metals.
2. Explain and demonstrate field identification methods for base metals.
3. Identify and explain the physical characteristics and mechanical properties of metals.
4. Identify and explain forms and shapes of structural metals.
5. Explain metallurgical considerations for welding metals.

Trade Terms

Alloy
American Iron and Steel Institute (AISI)
American Society for Testing Materials International (ASTM)
Anneal
Austenitizing
Casting
Coefficient
Ductile
Ductility
Ferrous
Flange
Gas metal arc welding (GMAW)
Gas tungsten arc welding (GTAW)
Hardenability
Heat-affected zone (HAZ)
Hot shortness
Interpass temperature
Malleable
Mechanical properties
Metallic
Nonferrous metal
Notch toughness
Oxide
Postweld heat treatment (PWHT)
Quench
Residual stress
Shielded metal arc welding (SMAW)
Society of Automotive Engineers (SAE)
Tempering
Tensile strength
Underbead cracking
Unified numbering system (UNS)
Web
Wrought

Required Trainee Materials

1. Pencil and paper
2. Appropriate personal protective equipment, including:
   - Safety glasses with a full face shield or a helmet
   - Flame-retardant clothing
   - Welding gloves
   - 8-inch or taller high-top safety shoes or boots
   - Earmuffs or earplugs

Prerequisites

Before you begin this module, it is recommended that you successfully complete *Core Curriculum*; *Welding Level One*; and *Welding Level Two*, Modules 29201-09 and 29202-09.

Industry Recognized Credentials

If you are training through an NCCER-accredited sponsor, you may be eligible for credentials from NCCER's Registry. The ID number for this module is 29205-09. Note that this module may have been used in other NCCER curricula and may apply to other level completions. Contact NCCER's Registry at 888.622.3720 or go to **www.nccer.org** for more information.

Contents

Topics to be presented in this module include:

Figures and Tables

Figures and Tables (continued)

1.0.0 Introduction

Welders must understand the physical characteristics and mechanical properties of metals to properly identify base metals and select weld filler metals. They must also understand how welding heat affects base material and filler metals.

This module explains the physical characteristics, mechanical properties, classification systems, and weldability of common metals and their alloys. It describes the particular physical characteristics and mechanical properties that influence the selection of a metal for a specific application. It also explains how to identify common base metals and describes the standard commercial structural steel shapes used for fabrication and construction.

2.0.0 Metal Composition and Classifications

All living and nonliving things are made up of chemical elements that can be altered by changes in temperature. A steak left on the grill will eventually burn, and a piece of wood thrown into a fire will burn, too. Metals are no different. When they are heated to a high enough temperature they will melt, and some will burn.

Metals, like other substances, are a collection of chemical elements bound together by natural forces or by man-made production processes. The chemical elements in some metals make them stronger than other metals. Those same chemical elements allow some metals to withstand stress or force better than others.

The composition of common welding base metals varies widely, from metals made of essentially one metallic element, to alloys (mixtures) of metallic and nonmetallic components. Metals are classified into two basic groups: ferrous metals, which are composed mainly of iron, and nonferrous metals, which contain very little or no iron. Ferrous metals include all steel, cast iron, wrought iron, malleable iron, and ductile (nodular) iron. Nonferrous metals and their alloys include the light metals (aluminum, magnesium, titanium); the heavy metals (copper, nickel, lead, tin, zinc); and the precious metals (platinum, gold, silver).

2.1.0 Ferrous Metal Composition and Classification Systems

Ferrous metals contain mostly iron and some carbon. The amount of carbon ranges from almost none in commercially pure iron to about 1.4 percent. Steel contains less than 1.7 percent carbon. Above 1.4 percent carbon, steel develops weak and brittle properties similar to those of cast iron. Carbon steel, the largest group of the ferrous-based metals, contains much smaller amounts of carbon than cast iron. As the carbon content increases in carbon steel, the steel becomes stronger and more wear-resistant. The carbon content of a particular carbon steel is the largest factor in determining its use. Steel castings usually contain a higher percentage of carbon than rolled plate and other rolled shapes. The casting foundry can usually provide the alloy formula.

Steels also contain impurities and trace amounts of elements such as manganese, silicon, phosphorus, copper, and lead. Sometimes additional amounts of these elements are intentionally added to steel to change its characteristics, such as to increase its forgeability (the ability to be heated in a forge and then beaten or hammered into shape) or to decrease a tendency toward forming blowholes (the vents that form when air or gases escape).

Steels, including carbon steels, low-alloy carbon steels, alloy steels, and stainless steels, are classified by various systems, such as specifica-

Hot Tip

Carbon and Alloy Steels

Steel is an alloy of iron and carbon typically containing less than 1 percent carbon. All steels also contain varying amounts of other elements such as manganese, silicon, phosphorus, sulfur, and oxygen. In addition, some standard alloy steels can also contain elements such as nickel, chromium, and molybdenum. There are currently about 3,500 different grades of steel with many different properties. Seventy percent of these steels have been developed in the past 20 years. Most of the steels produced today are carbon and alloy steels. Ninety-five percent of the construction and fabrication metals used worldwide consist of these materials.

203SA01.EPS

Hot Tip

Steel Alloying Elements

Some alloying elements used with steel and the reasons they are added include the following:

- *Manganese* – Promotes forgeability
- *Silicon* – Prevents blowholes
- *Phosphorus* – Improves machinability
- *Copper* – Inhibits corrosion
- *Lead* – Improves machinability

tion number and grade or manufacturer's trade name and number. Steel classification systems include those developed by the American Iron and Steel Institute (AISI), the American Society for Testing Materials International (ASTM), and the Unified Numbering System (UNS), as well as manufacturer's trade names and identification numbers. Stainless steels have their own classification systems. They are sometimes referred to by the percentages of their chromium and nickel content, such as 18/8, 25/20, and 18/10; but this system has largely been replaced by the AISI stainless steel classification system. Common classification systems for steels are explained in the following sections.

2.1.1 Carbon Steel Classification

The principal classification system for carbon steels is the AISI numerical designation of Standard Carbon and Alloy Steels. It was originally a four-digit system developed by the Society of Automotive Engineers (SAE) for carbon steels commonly used in structural shapes, plate, strip, sheet, and welded tubing. Five digits are now used for some alloys. The classification system may be referred to as the AISI, the SAE, or the AISI/SAE system.

The AISI and SAE designations are essentially the same, except that the AISI system sometimes uses a letter prefix to indicate the manufacturing process that produced the steel. The following are examples of letter designations:

- *A* – Open-hearth steel
- *B* – Acid Bessemer carbon steel
- *C* – Basic open-hearth carbon steel
- *D* – Acid open-hearth carbon steel
- *E* – Electric furnace steel

The absence of a letter prefix indicates basic open-hearth or acid-Bessemer carbon steel. The prefixing letter (if any) designates the manufacturing process. The first two numerical digits represent the series (type and class) of steel. The third and fourth (and sometimes fifth) numerical digits specify the approximate percentage of carbon content. *Table 1* shows the AISI/SAE system for designating carbon and alloy steel types and classes and includes the AISI/SAE number.

The following examples explain the AISI/SAE numbers for carbon steels:

- AISI number C1020:
  - C = Indicates basic open-hearth carbon steel
  - 10 = Carbon steel, nonresulfurized
  - 20 = Contains approximately 0.20 percent carbon
- AISI number E2512:
  - E = Indicates electric furnace steel
  - 25 = Designates steel alloyed with approximately 5 percent nickel
  - 12 = Designates steel containing approximately 0.12 percent carbon
- AISI number E52100:
  - E = Indicates electric furnace steel
  - 5 = Contains either approximately 0.50 percent, 1.00 percent, or 1.45 percent chromium, designated by the next digit: 0, 1, or 2, respectively
  - 2 = Designates approximately 1.45 percent chromium
  - 100 = Designates approximately 1 percent carbon

Hot Tip

UNS

The UNS used to identify metals was developed jointly by the ASTM International and SAE as a way of relating the different numbering systems for metals and alloys. Its objective is to eliminate the confusion caused when more than one identification number is used to identify the same material or when the same identification number is used to identify two different materials. It is anticipated that this system will eventually replace the AISI and other systems.

Each UNS number has a letter prefix followed by five digits. Often the prefix letter relates to the family of metals. For example, UNS numbers A00001 through A99999 refer to aluminum and aluminum alloys, and UNS numbers C00001 through C99999 signify copper and copper alloys. When possible, the digits in the UNS sequence contain the same numbering sequences used in the other systems. For example, the UNS number G10200 corresponds to the AISI/SAE number 1020 for carbon steel. Most engineering and materials reference books contain tables that cross-reference AISI and SAE numbers for metals to the corresponding UNS numbers.

Table 1 AISI/SAE System for Designating Carbon and Alloy Steel Types and Classes

AISI/ASE Number	Carbon Steel Types and Classes
1XXX	Carbon steels
10XX	Plain (nonresulfurized) carbon steel grades
11XX	Free machining, resulfurized (screw stock)
12XX	Free machining, resulfurized, rephosphorized
13XX	Manganese 1.75 percent
15XX	High-manganese (1.00 to 1.65 percent) carburizing steels
2XXX	Nickel steels
23XX	3.50 percent nickel
25XX	5.00 percent nickel
3XXX	Nickel-chromium steels
31XX	1.25 percent nickel, 0.65 or 0.80 percent chromium
32XX	1.75 percent nickel, 1.07 percent chromium
33XX	3.50 percent nickel, 1.50 or 1.57 percent chromium
34XX	3.00 percent nickel, 0.77 percent chromium
4XXX	Molybdenum steels
40XX	0.20 or 0.25 percent molybdenum
41XX	0.50, 0.80, or 0.95 percent chromium, 0.12, 0.20, 0.25, or 0.30 percent molybdenum
43XX	1.82 percent nickel, 0.50 or 0.80 percent chromium, 0.25 percent molybdenum
46XX	0.85 or 1.82 percent nickel, 0.20 or 0.25 percent molybdenum
47XX	1.05 percent nickel, 0.45 percent chromium, 0.20 or 0.35 percent molybdenum
48XX	3.50 percent nickel, 0.25 percent molybdenum
5XXX	Chromium steels
50XX	0.27, 0.40, 0.50, or 0.65 percent chromium
51XX	0.80, 0.87, 0.92, 0.95, 1.00, or 1.05 percent chromium
5XXXX	1.00 carbon, 0.50, 1.02, or 1.45 percent chromium
6XXX	Chromium-vanadium steels
61XX	0.60, 0.80, or 0.95 percent chromium, 0.10 or 0.15 percent (minimum) vanadium
86XX	0.55 percent nickel, 0.50 percent chromium, 0.20 percent molybdenum
92XX	0.65, 0.82, or 0.85 percent manganese, 1.40 or 2.00 percent silicon, 0.00 or 0.65 percent chromium
93XX	3.25 percent nickel, 1.20 percent chromium, 0.12 percent molybdenum
94XX	0.45 percent nickel, 0.40 percent chromium, 0.12 percent molybdenum
97XX	0.55 percent nickel, 0.20 percent chromium, 0.20 percent molybdenum
98XX	1.00 percent nickel, 0.80 percent chromium, 0.25 percent molybdenum
XXBXX	Boron steels
XXLXX	Leaded steels

203T01.EPS

The common group classification of carbon steels is primarily based on carbon content. Plain carbon steels (AISI series 10XX through 11XX) are basically iron-carbon alloys. High-strength low-alloy (HSLA) carbon steels (AISI series 13XX through 98XX) have small amounts of alloying elements to improve strength, hardness, and toughness or to increase resistance to corrosion, heat, and environmental damage. The plain carbon steels are commonly grouped as follows:

- *Low-carbon* – 0.10 percent to 0.15 percent carbon, 0.25 percent to 1.50 percent manganese
- *Mild-carbon* – 0.15 percent to 0.30 percent carbon, 0.60 percent to 0.70 percent manganese
- *Medium-carbon* – 0.30 percent to 0.50 percent carbon, 0.60 percent to 1.65 percent manganese
- *High-carbon* – 0.50 percent to 1.00 percent carbon, 0.30 percent to 1.00 percent manganese

The HSLA carbon steels are alloyed with one or more of the elements manganese, nickel, chromium, or molybdenum to provide higher strength, better toughness, weldability, and, in some cases, greater resistance to corrosion. As the percentage of these elements increases, changes in the weldability of the low-alloy steels make electrode selection and welding procedures more critical. HSLA steels include the following:

- Low-nickel steels
- Low-nickel chrome steels
- Low-manganese steels
- Low-alloy chromium steels
- Weathering steels

Another class of steel in the AISI series is sulfur steel. It is not an alloy steel but is still included in the series.

2.1.2 ASTM Steel Specification System

ASTM International develops and publishes specifications for use in material production and testing. Welding codes and procedures typically use the ASTM designation to identify metal types because it includes specifications for mechanical and physical characteristics of the material as well as chemical composition. Piping and tubing are the largest group of materials fabricated by welding covered in ASTM specifications.

Prefix letters are a general part of each specification and provide a general idea of the specification content. For example, *A* is used for ferrous metals, *B* for nonferrous metals, and *E* for miscellaneous subjects, such as examination and testing.

Sometimes an *S* is added before the prefix letter. This indicates that ASME International, publishers of the *Boiler and Pressure Vessel Code*, has adopted that specification.

With the exception of a few mild steel grades, such as SA-36 (mild steel plate) and SA-53 (mild steel pipe), ASTM specifications are three-digit numbers. Most specifications also have various grades, classes, or types. *Table 2* gives examples of some of the ASTM specifications for alloy steels.

NOTE

The complete collection of ASTM standards consists of 15 sections comprising 65 volumes, with a separate index. The following is provided only to illustrate sample ASTM numbers.

When companies buy steel, the manufacturer will provide the customer with a certificate indicating the quality of the steel. This document is often called a mill test report (MTR) or a certified mill test report (CMTR). The report mainly shows the chemical content and testing results of the metal that was purchased (*Figure 1*). Such information is necessary in case the structure being built from the material fails. For liability reasons, this information must be tracked and kept on record.

Included in the mill test report are the ASTM specifications, size, grade, and length of the product, which in this case is wide flange (WF) beams. The report also shows the heat number, chemical analysis, and physical properties of the steel. *Figure 2* shows how the markings look on an actual beam.

The markings show who made the steel, where it was made, its original dimensions, and its heat number or batch number. Heat or batch numbers are tracking numbers that the steel producer assigns to a particular ladle of steel. When a product, such as a beam or plate, is made from that ladle of steel, the tracking number is placed on that product in case of recall or problems with the chemistry. The following are explanations of the letters and numbers shown on the beam in *Figure 2*:

- *SDI* – Typically the name of the manufacturer.
- *USA* – Country of origin where the steel was melted.

Table 2 ASTM Specification Standards

ASTM Number		Tensile	Type of Base Metal
A209	Grade T1	55,000	C-0.05 Mo tubes
	Grade T1a	60,000	C-0.05 Mo tubes
	Grade T1b	53,000	C-0.05 Mo tubes
A217	Grade T17	60,000	1.00 Cr-0.15 V tubes
A334	Grade 5	65,000	5.00 Ni-Steel pipe
A335	Grade P1	55,000	C-0.05 Mo pipe
	Grade P2	55,000	0.60 Cr-0.85 Mo pipe
	Grade P3	60,000	1.57 Cr-0.70 Mo pipe
	Grade P3b	60,000	2.00 Cr-0.55 Mo pipe
	Grade P11	60,000	1.25 Cr-0.55 Mo-Si pipe
	Grade P12	60,000	0.95 Cr-0.55 Mo pipe
	Grade P15	60,000	1.40 Si-0.55 Mo pipe
A517	Grade A	115,000	0.05 Cr-0.025 Mo-Si plates
	Grade B	115,000	0.05 Cr-0.02 Mo-V plates
	Grade P	115,000	0.125 Ni-0.10 Cr-0.05 Mo plates

C – Carbon Cr – Chromium Mo – Molybdenum Ni – Nickel Si – Silicon V – Vanadium

203T02.EPS

Steel Mills, Inc.
First Steel Mill
123 Rebar Way
Narvona, State 904261 USA

Mill Test Report

Chemical and Physical Test Report																			Customer P.O. Number ________
Shape & Size		Grade			Specification														Sales Order ________
32mm Rebar (#10)		420(60)			ASTM A615/A615M-03														
Heat Identification Number	C	Mn	P	S	Si	Cu	Ni	Cr	Mo	V	Nb	B	N	Sn	Al	Ti	Zr	Ca	C Equivalent
123-123	0.32	0.91	0.015	0.067	0.32	0.38	0.13	0.16	0.016	0.000	0.014	0.000	0.000	0.003	0.000	0.000	0.000	0.0000	0.502
Mechanical Test																			
Yield:79800 PSI, 550.2 MPA	Tensile:102000 PSI, 703.27 MA	% EL: 15.0/8 in, 15.0/200MM	Bend: Okay	Def: .073, 1.85MM	%I/h 3.3L														

This material, including the billets, was produced and manufactured in the United States of America.

Signature of Quality Assurance Manager

The above figures are certified extracts from the original chemical and physical test records, as contained in the permanent records of the company.

Signature of Mill Metalurgist

203F01.EPS

Figure 1 Example of a mill test report.

203F02.EPS

Figure 2 Mill markings on a beam.

203F03.EPS

Figure 3 Mill markings on plate steel.

- *A046196* – The heat number or batch number, which can be used to match the material to a material test report. Heat numbers are normally unique to each manufacturer, and they are usually meaningful only to the producer. The number provides a way to trace the product. The material test report lists the mechanical test results and chemistry. These can be checked against the material specification to ensure that the material test report results meet the requirements of the material specification.
- *W 16 × 36* – This beam has a depth of approximately 16 inches (actual depth 15⅞ inches) and weighs 36 pounds per foot of length. If you looked up "W 16 × 36" in the *American Institute of Steel Construction (AISC) Handbook*, you would find the beam's flange width and thickness, web thickness, beam depth, and many other design properties.

> **NOTE**
> One piece of information that is not included on the beam in *Figure 2* is the ASTM specification to which the beam was made (for example, A36, A441, A572, A588, etc.).

Plate steel also has mill markings, such as batch or heat numbers. Refer to *Figure 3*. The information at the top of the plate indicates that it was not heat treated after being rolled. The S7-09438 heat number is the batch number. The size information is self-explanatory. Grade A36 indicates that the plate was manufactured in accordance with the ASTM specifications for A36 steel plate. The other information at the bottom of the plate shows who made it and who bought it.

There is no set format for labeling steel, so steel producers have their own way of labeling their products. The mill markings on the beams and plates shown in *Figures 1, 2,* and *3* were stenciled or sprayed onto the steel. Other manufacturers attach paper labels to their products, as shown in *Figure 4*.

The mill markings on most steel will not be as clearly visible as those shown in these examples. Steel is often stored outside in the weather and may be exposed to other rough conditions or handling that can damage the markings.

When preparations are being made for a welding job, the job planner determines which steel to use. On some jobs, lay-down yard personnel may bring the materials to the work area. However, the fabricators (fitters and welders) must be able to ensure that the delivered steel is the same as that indicated on the welding procedure specification. *Figures 5A* and *5B* show a WPS with standard or English measurements. *Figure 5C* shows a WPS with metric values.

When only a part of the original piece of steel is needed, welding personnel must transfer the heat number of the original piece to the piece that they are cutting from the main piece. The transfer of numbers must be done before the part is cut away from the original piece. Transferring the heat number ensures that the removed piece of steel can be tracked when it is welded into place. Follow site-specific guidelines for transferring numbers.

On high-profile jobs, the process of tracking these numbers can be very involved because of liability concerns. Many jobs today require the welders to closely track the materials being used. The owner representatives (usually the quality control personnel) can use these heat or batch numbers to trace where the raw materials came from, the day and shift on which the material was produced, and who was on duty at the time. Most plants will not allow material to be used if the heat

203F04.EPS

Figure 4 Paper labels attached to steel products.

number has been ground off or rusted over. From a contractor's point of view, material that lacks a heat number should never be used in any critical application. If a failure occurred and the installer could not produce a valid heat number, then the contractor/installer would assume full responsibility, even though the steel was produced by someone else.

2.2.0 Low-Alloy Steel Groups

Low-alloy steels fall into the following general groups:

- High-strength low-alloy steels
- Quench-and-tempered steels
- Heat-treatable low-alloy steels
- Chromium-molybdenum steels

2.2.1 High-Strength Low-Alloy Steels

High-strength low-alloy (HSLA) steels are designed to meet specific mechanical requirements per ASTM specifications. The principal difference between plain-carbon steels and HSLA steels is the addition of alloys such as manganese, chromium, nickel, copper, molybdenum, and vanadium. Typically, these alloys total less than 2 percent. The carbon content of HSLA steels is about the same as that in low-carbon and mild steels (0.06 percent to 0.28 percent). Compared to mild steels, HSLA steels are designed to provide a combination of higher strength, better corrosion resistance, and improved notch toughness. The weldability of HSLA steel is similar to that of mild steel.

XYZ CONSTRUCTION

DATE OF ISSUE	7/20/2005	WELDING PROCEDURE SPECIFICATION IGMA-11-AF	
REVISION	2 - 11/16/06	SUPPORTING PQR NO. AND PAGE: IGMA-11 PAGE B	
PROCESS	GTAW / SMAW		
TYPE	MANUAL		

JOINTS (QW-402)

JOINT DESIGN	V OR U GROOVES AND FILLETS, PER THE PROJECT DRAWINGS AND SPECIFICATIONS
BACKING	NONE
BACKING MATERIAL	COMPATIBLE WITH BASE MATERIALS IF USED
JOINT FIT-UP REQUIREMENTS	1/8" +/- 1/16"
OTHER	NO RETAINERS (QW 402.11) ROOT SPACING PER DRAWINGS AND SPECIFICATIONS

BASE METALS (QW-403)

P-NO	1	GRP NO.		to	P-NO	1	GRP NO.
BASE METAL SPEC.				to			

THICKNESS RANGE QUALIFIED (QW-451)

BASE METAL GROOVE	3/16" TO 8"	FILLET	ALL
DEPOSITED WELD METAL	GTAW - 3/4" MAX (No weld pass > 1/2" thick)		
	SMAW - 8" MAX (No weld pass > 1/2" thick)		
OTHER			

FILLER METALS/ELECTRODES (QW-404); PROGRESSION (QW-405) **POSITIONS: ALL**

LAYER	ROOT	PROCESS	GTAW	AWS CLASS	ER70S-2	PROGRESSION	UPHILL
SFA NO.	5.18	F NO.	6	A NO.	1	SIZES	1/16", 3/32", 1/8"
LAYER	WELDOUT	PROCESS	SMAW	AWS CLASS	E-7018	PROGRESSION	UPHILL
SFA NO.	5.1	F NO.	4	A NO.	1	SIZES	3/32", 1/8", 5/32"
LAYER		PROCESS		AWS CLASS		PROGRESSION	
SFA NO.		F NO.		A NO.		SIZES	
OTHER	NO CONSUMABLE INSERTS (QW 404.22) NO PENETRATION FLUX (QW 404.50)						

PREHEAT (QW - 406)

PREHEAT TEMPERATURE	75° F (175F when t > 1")
MAX INTERPASS TEMPERATURE	600° F
PREHEAT MAINTENANCE	NONE
OTHER	

POSTWELD HEAT TREATMENT (QW-407)

TEMPERATURE	1150° F
TIME RANGE	1HR/INCH ONE HOUR MINIMUM
MAX HEATING & COOLING RATES	PER THE APPLICABLE ASME DESIGN CODE

203F05A.EPS

(A)

Figure 5 Sample WPS with standard measurements and metric values (1 of 3).

XYZ CONSTRUCTION

REVISION 2 - 11/16/06

WELDING PROCEDURE SPECIFICATION
IGMA-11-AF

GAS (QW-408)

SHIELDING GAS	GTAW - 100% ARGON	FLOW RATE	15-35 CFH
PURGING GAS	NONE	FLOW RATE	N/A
TRAILING GAS	NONE	FLOW RATE	N/A
OTHER			

ELECTRICAL CHARACTERISTICS (QW-409)

PROCESS:	GTAW	PROCESS:	SMAW	PROCESS:	
ELECTRODE:	See Note 1	ELECTRODE:	E-7018	ELECTRODE:	
CURRENT:	DC	CURRENT:	DC	CURRENT:	
POLARITY:	SP	POLARITY:	RP	POLARITY:	

CHARACTERISTICS: ER70S-2				CHARACTERISTICS: E7018				CHARACTERISTICS:			
SIZE	AMPS	VOLTS	TRAVEL SPEED IPM	SIZE	AMPS	VOLTS	TRAVEL SPEED IPM	SIZE	AMPS	VOLTS	TRAVEL SPEED IPM
1/16"	50-150	N/A	Manual	3/32"	50-120	N/A	Manual				
3/32"	75-180	N/A	Manual	1/8"	70-150	N/A	Manual				
1/8"	100-250	N/A	Manual	5/32"	90-200	N/A	Manual				

GTAW - PULSED CURRENT = NONE

TECHNIQUE (QW-410)

STRING OR WEAVE BEAD	USE STRINGER BEAD FOR POSITION 2G. USE EITHER STRINGER BEAD OR WEAVE BEAD FOR POSITIONS 1G, 3G, 4G, 5G, AND 6G.
ORIFICE OR GAS CUP SIZE	4 Thru 12
INITIAL AND INTERPASS CLEANING	BRUSHING AND/OR GRINDING
METHOD OF BACK GOUGING	THERMAL OR MECHANICAL IF REQUIRED (If Thermal grind back to bright metal)
OSCILLATION	N/A
CONTACT TUBE TO WORK DISTANCE	N/A
MULTIPLE OR SINGLE PASS PER SIDE	MULTIPLE
MULTIPLE OR SINGLE ELECTRODES	SINGLE
PEENING	NONE
OTHER	

NOTES NO AUTOGENOUS WELDING

1.) A potential health risk exists from inhaling grinding dust from EWTh-2 tungsten electrodes. As an alternative, EWCe-2 and/or EWLa-2 tungsten electrodes are acceptable.

203F05B.EPS

(B)

Figure 5 Sample WPS with standard measurements and metric values (2 of 3).

Weld Procedure Number	30 P1 TIG 01 Issue A
Qualifying Welding Procedure (WPAR)	WP T17/A

Manufacturer:	National Fabs Ltd 25 Lane End Birkenshaw Leeds	Method Of Preparation and Cleaning:	Machine and Degrease
Location:	Workshop	Parent Metal Specification:	Grade 304L Stainless Steel
Welding Process:	Manual TIG	Parent Metal Thickness	3 to 8mm Wall
Joint Type:	Single Sided Butt Weld	Pipe Outside Diameter	25 to 100mm
		Welding Position:	All Positions
		Welding Progression:	Upwards

Joint Design

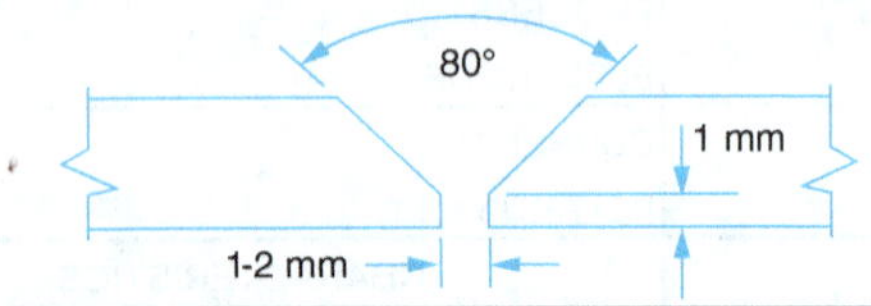

Welding Sequences

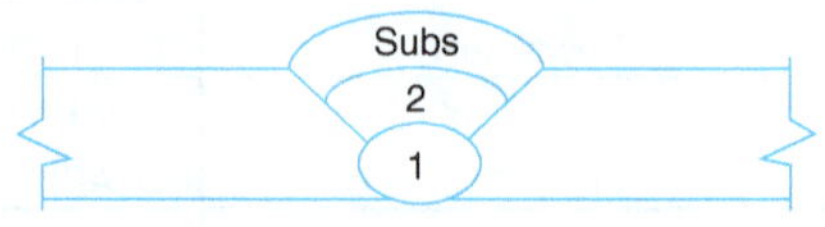

Run	Process	Size of Filler Metal	Current A	Voltage V	Type of Current/Polarity	Wire Feed Speed	Travel Speed	Heat Input
1 2 And Subs	TIG TIG	1.2 mm 1.6 mm	70 - 90 80 - 140	N/A	DC– DC–	N/A	N/A	N/A

Welding Consumables:-	
Type, Designation Trade Name:	BS 2901 Part 2 : 308S92
Any Special Baking or Drying:	No
Gas Flux:	Argon 99.99% Purity
Gas Flow Rate - Shield:	8 - 12 LPM
- Backing:	5 LPM
Tungsten Electrode Type/ Size:	2% Thoriated 2.4mm Dia
Details of Back Gouging/Backing:	Gas Backing
Preheat Temperature:	5°C Min
Interpass temperature:	200°C Max
Post Weld Heat Treatment Time, temperature, method: Heating and Cooling Rates*:	Not Required

Production Sequence

1. Clean weld and 25mm borders to bright metal using approved solvent.
2. Position items to be welded ensuring good fit up and apply purge.
3. Tack weld parts together using TIG, tacks to at least 5mm min length.
4. Deposit root run using 1.2mm dia. wire.
5. Inspect root run internally.
6. Complete weld using 1.6mm dia wire using stringer beads as required.
7. 100% Visual inspection of completed weld

Revision History

Date	Issue	Changes	Authorization
26/11/2000	A	First changes	Jack Straw *Jack Straw*

203F05C.EPS

(C)

Figure 5 Sample WPS with standard measurements and metric values (3 of 3).

HSLA steels can be divided into two general groups: Group A and Group B. Steels in Group A are designed for high strength, while steels in Group B are designed for high strength, improved atmospheric corrosion resistance, and notch toughness. Typical steels in Group A are A441, A572, and A633, which are used for structural applications such as pipelines, buildings, bridges, machinery, and railroad equipment. Other steels in Group A are A225 and A737, which are used for pressure vessels. Typical steels in Group B are A242 and A588, which are used for high-strength structural applications that may include unpainted applications. Group B steels specifically intended for low-temperature applications (the nickel content is as high as 9 percent) are A203 and A353.

CAUTION

The welding recommendations provided are general guidelines only. When welding HSLA steel, the site WPS or site quality standards must be followed. Failure to follow proper standards can result in defective welds.

HSLA steels can be welded using the common welding processes, although successful welding requires consideration of preheating, interpass temperature, and control of hydrogen, including the use of low-hydrogen electrodes when applicable. Typically, only the root pass and possibly the hot pass are made with the gas tungsten arc welding (GTAW) process on HSLA steels. Fill passes and cover passes are often made with faster processes such as shielded metal arc welding (SMAW) or gas metal arc welding (GMAW). GMAW and GTAW are processes that you will learn about in future modules. One advantage of using the GTAW process is that hydrogen is easier to control because fluxes are not used. Minimum preheat and interpass temperatures are determined based on the alloy type and the thickness of the thinner part being welded. Generally, HSLA steels less than ¾ inch thick do not require preheating.

CAUTION

Steels that are wet or below 32°F should always be preheated to at least 70°F and maintained at that temperature to remove moisture and prevent its formation. Failure to preheat metals will result in defective welds.

Table 3 provides minimum preheat and interpass temperatures for typical HSLA steels and also includes the ASTM steel numbers and grades.

Table 3 Minimum Preheat and Interpass Temperatures for Typical High-Strength Low-Alloy Steels Welded With Low-Hydrogen Electrodes

ASTM Steel	Thickness (Inches)	Minimum Temperature (°F)
A242	Up to 0.75	32
A441	0.81 to 1.50	50
A572, Gr 42, 50	1.56 to 2.50	150
A588	Over 2.50	225
A572, Gr 60, 65	Up to 0.75	50
A633, Gr E	0.81 to 1.50	150
	1.56 to 2.50	225
	Over 2.50	300

203T03.EPS

Hot Tip

Hot Tapping

Steel pipes are commonly hot-tapped to install a bypass around a repair area. With hot tapping, taps for the bypass are welded to the pipe while material or liquid is still flowing in the pipe. Low-alloy pipes cannot be hot-tapped because the material flowing through the pipe will cool the pipe rapidly and prevent proper preheat, interpass, and postheat temperature maintenance. This can result in defective welds as well as pipe failure during welding. Always comply with site-specific lockout/tagout procedures when performing hot taps.

Filler metal for welding HSLA steel depends on the alloy and the application. Mild steel filler metal can be used when joint strength is not critical or when alloy steel is being joined to mild steel. Alloy steel should be used when the strength of the joint must equal or approach that of the base metal. High-alloy stainless steel is used for special applications including welding dissimilar metals or welding high-nickel steel, such as A353.

Typically, HSLA steels are not postweld heat-treated; however, this may be required to improve ductility or to maintain the dimensions of the weldment during machining.

2.2.2 Quench-and-Tempered Steels

Quench-and-tempered (Q&T) steels are furnished in the heat-treated condition, which generally consists of austenitizing, quenching, and tempering. Some Q&T steels fall into the carbon steel classification, some into the low-alloy steel classification, and others into the alloy classification of the AISI/SAE system. Q&T steels combine high yield and tensile strength with good notch toughness, ductility, corrosion resistance, or weldability, depending on their intended use. Q&T steels have less than 0.25 percent carbon, with yield strengths of 50,000 to 180,000 psi. Most Q&T steels are furnished as plate, although some are available as seamless pipe.

Q&T steels are used where high strength and minimum weight are essential. Many of the Q&T steels are covered by ASTM classifications. A common group of Q&T steels known as HY steels are covered by military specifications (MIL). The most common HY steel is HY-80, which has a yield stress of 80,000 psi. HY-100, HY-130, HY-150, and

HY-180 are also available. Types and typical uses of Q&T steels include the following:

- *A514 and A517* – Used for earth-moving equipment, pressure vessels, bridges, TV towers, and ships.
- *A533 Grade B* – Used for nuclear pressure vessels.
- *A543 Type B* – Used in nuclear reactor vessels, ships, and submarines. Classes 1 and 2 steels are similar to HY-80 and HY-100.
- *A553* – Used for cryogenic applications. It contains 8 percent to 9 percent nickel for high strength at extremely low temperatures.
- *A678* – Used for high-strength structural applications.

To prevent cracking, most Q&T steels can be welded by common welding processes using preheat and interpass temperatures as shown in *Table 4*. These temperatures vary depending on the type of steel being welded. Because Q&T steels obtain their strength from heat treatment, slow cooling can destroy the benefits of the original heat treatment. Slow cooling is generally caused by the following:

- Excess preheat
- Exceeding the recommended interpass temperature
- High heat input, such as high amperage with slow travel speed and the use of weave beads instead of stringers

On the other hand, cooling too quickly can cause a hard, brittle zone that may crack. Fast cooling is caused by the following:

- Insufficient preheat
- Insufficient interpass temperature
- Insufficient heat input during welding

Table 4 Minimum Preheat and Interpass Temperatures for Q&T Steels

Thickness Range (Inches)	Temperature (°F)				
	A514/ A517	A533	A537	A543	A678
Up to 0.50	50	50	50	100	50
0.56 to 0.75	50	100	50	125	100
0.81 to 1.00	125	100	50	150	100
1.10 to 1.50	125	200	100	200	150
1.60 to 2.00	175	200	150	200	150
2.10 to 2.50	175	300	150	300	150
Over 2.50	225	300	225	300	—

203T04.EPS

CAUTION

When welding Q&T steels, pay close attention to the welding procedures concerning preheat and interpass temperature recommendations to prevent loss of strength or cracking.

When welding Q&T steels, always use low-hydrogen techniques and stringer beads to minimize heat input. Typically, the GTAW process is used only on thin sections or for the first one or two passes.

Most Q&T steels are designed to be used in the as-welded condition, with no postheat. Sometimes thick sections of A533 or A537 steels may require postheat treatment. If postheat treatment is required, use extreme care and follow the procedures exactly to prevent adversely affecting the base metal.

2.2.3 Heat-Treatable Low-Alloy Steels

Heat-treatable low-alloy (HTLA) steels are alloy steels that contain carbon in the range of about 0.25 to 0.45 percent, and small amounts of chromium, nickel, and/or molybdenum to enhance hardenability. They are generally welded in the annealed condition, and then the entire weldment is heat treated. The relatively high carbon content of HTLA steels allows them to be heat treated to a very high strength and hardness. Because HTLA steels must be heat treated after welding, they are typically used for smaller parts that can be placed in tempering ovens.

HTLA steels are produced to contain very low levels of impurities. During welding, low hydrogen techniques are used, and extreme care must be taken to keep the weld joint and filler metal absolutely clean and free of impurities. For this reason, GTAW is the preferred process for welding HTLA steels. Heat input is also very important, and welding procedures that take into account current, voltage, travel speed, and preheat temperatures must be followed.

HTLA steels generally require higher preheat and interpass temperatures than other types of alloy steels to prevent hardening of the weld zone and hot cracks. They also require postweld heat treatment (PWHT). *Table 5* shows minimum preheat and interpass temperatures for HTLA steels and also includes the AISI steel numbers and thickness ranges.

HTLA steels should receive postweld heat treatment immediately after they are welded. The type of postweld heat treatment depends on

Table 5 Minimum Preheat and Interpass Temperatures for HTLA Steels

AISI Steel	Thickness Range (Inches)	Minimum Preheat and Interpass Temperature (°F)
4027	Up to 0.50	50
	0.60 to 1.0	150
	1.1 to 2.0	250
4037	Up to 0.50	100
	0.60 to 1.0	200
	1.1 to 2.0	300
4130/5140	Up to 0.50	300
	0.60 to 1.0	400
	1.1 to 2.0	500
4135/4140	Up to 0.50	350
	0.60 to 1.0	450
	1.1 to 2.0	500
4340/8630	Up to 2.0	550
	Up to 0.5	200
	0.6 to 1.0	250
	1.1 to 2.0	300
8640	Up to 0.50	200
	0.60 to 1.0	300
	1.1 to 2.0	350

203T05.EPS

the type of steel and the preheat and interpass temperature maintained during welding. Allowing the weldment to cool after welding without some type of heat treatment generally results in cracks. Immediately after it is welded, HTLA steel should be stress-relieved following strict guidelines depending on the alloy and the preheat and interpass temperatures used. Finally, to achieve the desired mechanical properties, the weldment should be austenitized, quenched, and tempered.

2.2.4 Chromium-Molybdenum Steels

Chromium-molybdenum (Cr-Mo) steels contain 0.5 to 9 percent chromium and 0.5 to 1 percent molybdenum. The carbon content is generally less than 0.20 percent. Cr-Mo steels, which are sometimes called chrome-moly or heat-resisting alloy steels, are widely used in the petroleum industry and in steam-generating plants for elevated-temperature applications. The chromium provides improved resistance to oxidation and corrosion, and the molybdenum increases strength at elevated temperatures. Cr-Mo steels are also used in the aircraft and racing industries as tubing for highly stressed parts (*Figure 6*). Cr-Mo steels that are in the AISI/SAE 4100 series are available in forging, castings, plate, pipe, and tubing to various ASTM specifications.

203F06.EPS

Figure 6 Steel tubing used in the roll cage of a NASCAR-approved race car.

Cr-Mo steels are hardenable, so a welding procedure that includes preheating, and sometimes postheating, must be used. The preheat temperature depends on the amount of alloy, the carbon content, the weld material, and base metal thickness. Generally, the higher the alloy content, carbon content, or thickness, the higher the preheat temperature. *Table 6* shows minimum preheat and interpass temperatures for Cr-Mo steels including their types and ASTM steel designations.

Once the Cr-Mo base metal has been preheated, welding should proceed without interrupting the heating cycle to avoid hard spots and cracking. If the base metal contains less than 9 percent chromium or is less than 1-inch thick, the heating cycle should be 800°F per half hour per 1 inch of thickness. After being heated, wrap the metal and let it cool until it is at ambient temperature.

CAUTION

Check your welding procedures for any special precautions needed when interrupting a heating cycle. Failure to follow proper welding procedures will result in defective welds.

When the welding consumables are a potential source of hydrogen, such as with SMAW, it may be necessary to raise the preheat temperature and hold the weldment at this elevated temperature for a period of time. This allows the hydrogen to escape before the base metal cools, which reduces the risk of cracking. To avoid potential problems that can occur when the heating cycle of a Cr-Mo base metal is interrupted, always plan to complete the heating cycle once it has started.

Some Cr-Mo steels containing less than 1.25 percent Cr and 0.05 percent Mo can be placed in service as-welded if a high preheat temperature was used and the section is relatively thin. Other Cr-Mo steels should either be stress-relieved, annealed, or normalized and tempered.

Stress-relief heat treatment is used to reduce welding stresses and to increase the ductility and toughness of the weld metal and the **heat-affected zone (HAZ)**. The stress-relief temperature ranges from about 1,150°F to 1,400°F and depends on the alloy being treated. The entire weldment or the weld zone is generally held at the stress-relief temperature for one hour per inch of thickness.

The procedures for annealing or normalizing and tempering are generally the same as those used for HTLA steels discussed earlier in this module.

2.3.0 Common Grade Stainless Steel Classifications

Stainless steels are typically classified as being common grade or specialty grade. This section covers common grade stainless steels. The specialty grades are discussed later in this module. Common grade stainless steels (corrosion-resistant steels) are iron-based alloys that normally contain at least 11 percent chromium. Other alloying elements, including nickel, carbon, manganese, and silicon, may be present in varying quantities, depending on the specific type of stainless steel, to enhance its physical and mechanical properties. The main physical characteristics of all stainless steels are their resistance to corrosion and heat. Some have good low- and high-temperature mechanical properties. When

Table 6 Minimum Preheat and Interpass Temperatures for Cr-Mo Steels*

ASTM Steel	Type	Up to 0.5 inch	0.5 to 1.0 inch	Over 1.0 inch
A335-P2 A335-P12 A335-P11	1/2 Cr - 1/2 Mo 1 Cr - 1/2 Mo 11/4 Cr - 1/2 Mo	300°F	300°F	300°F
A369-FP3b A335-P22	2 Cr - 1/2 Mo 21/4 Cr - 1Mo	300°F to 350°F	300°F to 350°F	350°F
A335-P21	3 Cr - 1 Mo	300°F to 350°F	300°F to 350°F	400°F
A335-P5 A335-P7 A335-P9	5 Cr - 1/2 Mo 7 Cr - 1/2 Mo 9 Cr - 1 Mo	400°F	400°F	500°F

*Maximum carbon content of 0.15 percent. For higher carbon content, preheat should be increased 100°F to 200°F.

203T06.EPS

compared with mild steels, stainless steels have the following characteristics:

- Lower coefficients of thermal conductivity that increase the chances of distortion
- Higher coefficients of thermal expansion that increase the chances of distortion
- Higher electrical resistances that increase the tendency to build up heat from welding current

Common grade stainless steels are classified by their grain structures. The type of grain structure is determined by the specific alloy content of the stainless steel and its heat treatment during manufacture. Based on their microcrystalline structures, the classification of common grade stainless steels is divided into these three groups:

- Austenitic
- Ferritic
- Martensitic

2.3.1 Austenitic Stainless Steels

Austenitic stainless steels are nonmagnetic in the annealed condition and nonhardenable by heat treatment. However, they can be hardened significantly by cold working. Austenitic stainless steels combine excellent corrosion and heat resistance with good mechanical properties over a broad temperature range. Austenitic stainless steels make up the largest of the three common grade stainless steel groups. For this reason, they are the type of stainless steel encountered most often by welders. They include the AISI 200 and 300 series stainless steels (*Table 7*) that all contain significant amounts of both chromium and nickel. Austenitic steels are sometimes further subdivided in two classifications based on their compositions: chromium-nickel (AISI 300 series) and chromium-nickel-magnesium (AISI 200 series). The range of applications for austenitic stainless steels includes housewares, containers, industrial piping and vessels, and architectural facades.

Austenitic stainless steels are also available in low-carbon (L-grade) and high-carbon (H-grade) types. The letter L after a stainless steel type, such as 304L, indicates a low-carbon content of 0.03 percent or under. Similarly, the letter H after a stainless steel type indicates a high-carbon content, ranging between 0.04 and 0.10 percent. L-grades are typically used when annealing after welding is impractical, such as in the field where pipe and fittings are welded. H-grades are used when the steel will be subjected to extreme temperatures; the higher carbon content helps it retain strength.

2.3.2 Ferritic Stainless Steels

Ferritic stainless steels are always magnetic. They are hardened to some extent by cold working, not by heat treatment. Ferritic stainless steels combine corrosion and heat resistance with fair mechanical properties over a narrower temperature range than austenitic steels. Ferritic stainless steels are straight chromium stainless steels containing 11.5 to 27 percent chromium, about 1 percent manganese, and little or no nickel. The carbon content is 0.20 percent or less. Examples of ferritic stainless steels include AISI types 405, 409, 430, 442, 444, and 446. They typically are used for decorative trim, sinks, and automotive applications, particularly exhaust systems.

2.3.3 Martensitic Stainless Steels

Martensitic stainless steels are magnetic and can be hardened by quenching and tempering. They are excellent for use in mild environments such as the atmosphere, freshwater, steam, and with weak acids. However, they are not resistant to severely corrosive solutions. Martensitic stainless steels comprise two groups: chromium-martensitic and martensitic. The chromium-martensitic stainless steels contain from 11.5 to 18 percent chromium, about 1 percent manganese, and in some cases 0 to 2.5 percent nickel. Examples of chromium-martensitic stainless steels are AISI types 403, 410, 414, 420, 422, 431, and 440. Martensitic stainless steels contain only 4 to 6 percent chromium and no nickel. For this reason, they are not considered true stainless steels, although their corrosion resistance is much greater than mild carbon steels, even at elevated temperatures. Examples of martensitic stainless steels are AISI types 501 and 502.

2.4.0 Specialty Grade Stainless Steel Classifications

In addition to the common grades of stainless steel, the following four specialty grades of stainless steel are also available:

- Precipitation-hardening
- Superaustenitic
- Superferritic
- Duplex

2.4.1 Precipitation-Hardening Stainless Steels

Precipitation-hardening (PH) stainless steels contain alloying additions, such as aluminum, that enable them to be hardened by a solution and

Table 7 AISI Classification Numbers and Composition for Selected Stainless Steels

AISI NO.	C%	Mn%	Si%	Cr%	Ni%	Other Elements
Chromium-nickel-magnesium-austenitic (Nonhardenable)						
201	0.15 Max	5.5–7.5	1.0	16.0–18.0	3.0–5.5	N 0.25 Max
202	0.15 Max	7.5–10	1.0	17.0–19.0	4.0–6.0	N 0.25 Max
Chromium-nickel-austenitic (Nonhardenable)						
301	0.15 Max	2.0	1.0	16.0–18.0	6.0–8.0	—
302	0.15 Max	2.0	1.0	17.0–19.0	6.0–10.0	—
303	0.15 Max	2.0	1.0	17.0–19.0	8.0–10.0	Mo 0.0–0.6
304	0.08 Max	2.0	1.0	18.0–20.0	8.0–10.5	—
304L	0.03 Max	2.0	1.0	18.0–20.0	8.0–12.0	—
308	0.08 Max	2.0	1.0	19.0–21.0	10.0–12.0	—
309	0.20 Max	2.0	1.0	22.0–24.0	12.0–15.0	—
310	0.25 Max	2.0	1.5	24.0–26.0	19.0–22.0	—
316	0.08 Max	2.0	1.0	16.0–18.0	10.0–14.0	Mo 2.0–3.0
316L	0.03 Max	2.0	1.0	16.0–18.0	10.0–14.0	Mo 2.0–3.0
321	0.08 Max	2.0	1.0	17.0–19.0	9.0–12.0	Ti, 5xC Min
347	0.08 Max	2.0	1.0	17.0–19.0	9.0–13.0	Cb+Ta10xC Min
348	0.08 Max	2.0	1.0	17.0–19.0	9.0–12.0	Ti 0.10 Max
Chromium-martensitic (Hardenable)						
403	0.15 Max	1.0	0.5	11.5–13.0	—	—
410	0.15 Max	1.0	1.0	11.5–13.5	—	—
414	0.15 Max	1.0	1.0	11.5–13.5	1.25–2.5	—
416	0.15 Max	1.0	1.0	12.0–14.0	—	S 0.15 Min
420	0.15 Min	1.0	1.0	12.0–14.0	—	—
422	0.20–0.25	1.0	0.75	11.0–13.0	0.5–10	Mo 0.75–1.25 W 0.75–1.25 V 0.15–0.3
440A	0.60–0.75	1.0	1.0	16.0–18.0	—	Mo 0.75 Max
440B	0.75–0.95	1.0	1.0	16.0–18.0	—	Mo 0.75 Max
440C	0.95–1.2	1.0	1.0	16.0–18.0	—	Mo 0.75 Max
Chromium-ferritic (Nonhardenable)						
405	0.08 Max	1.0	1.0	11.5–14.5	—	Al 0.10–0.30
409	0.08 Max	1.0	1.0	10.5–11.75	—	Ti, 6xC Min
430	0.12 Max	1.0	1.0	16.0–18.0	—	—
442	0.20 Max	1.0	1.0	18.0–23.0	—	Ti, 0.20 + 4 (%C+%N)
444	0.025 Min	1.0	1.0	17.5–19.5	1.0	Mo 1.75–2.5, N 0.035 Max (Nb+T2), 0.2 + 4 (%C+%N) Min
446	0.20 Max	1.50	1.0	23.0–27.0	1.0	N 0.25 Max
Martensitic						
501	0.11 Max	1.0	1.0	4.0–6.0	—	Mo 0.4–0.65
502	0.10 Max	1.0	1.0	4.0–6.0	—	Mo 0.4–0.65

203T07.EPS

Hot Tip

Why Stainless Steel Doesn't Rust

When steel comes in contact with oxygen and water vapor in the atmosphere, a chemical reaction occurs, and the steel begins to change to its original form of iron oxide. The chromium added to steel prevents it from rusting or staining, thus the name stainless steel. The chromium reacts with the oxygen and water in the atmosphere to form a microscopic and transparent layer of chrome-oxide called a passive film. The sizes of the chromium atoms and their oxides are about the same, so they join to form a tight, stable layer only a few atoms thick on the surface of the steel. If the stainless steel surface is cut or scratched, disrupting the passive film, it is self-healing because more oxide immediately forms and covers the exposed surface, protecting it from rust again. This is why after repeated use and sharpening, stainless steel cutlery remains bright and shiny. However, if the surface of stainless steel is contaminated by a ferrous metal, the ferrous metal will rust.

203SA02.EPS

aging heat treatment. Precipitation-hardening stainless steels were developed after World War II because the design and manufacture of jet aircraft required the use of stainless steels with a better weight-to-strength ratio. Precipitation-hardening stainless steels have designations such as 17-4PH, PH13-8MO, and AM350.

2.4.2 Superaustenitic Stainless Steels

Superaustenitic stainless steels have compositions that fall between 300 series stainless steels and nickel-alloy steels. They are highly resistant to corrosion and stress-corrosion cracking caused by chloride-containing solutions. Many of these alloys contain deliberate amounts of copper and/or nitrogen to increase their strength and corrosion resistance. These stainless steel types are widely used in process industry applications in place of common austenitic stainless steels that had a tendency to fail because of pitting, crevice corrosion, or chloride-induced stress corrosion cracking.

2.4.3 Superferritic Stainless Steels

Superferritic stainless steels differ from common grade ferritic stainless steels in their lower carbon and nitrogen content and higher molybdenum and/or chromium content. This combination provides much higher corrosion resistance in oxidizing and chloride chemical environments and excellent resistance to corrosion in salt water. Examples of superferritic stainless steel designations include E-Brite®, Sea-Cure®, and AL 29-4-2®.

2.4.4 Duplex Stainless Steels

Duplex stainless steels are supplied with approximately equal amounts of ferrite and austenite, thus the name duplex. They also contain about 24 percent chromium and 5 percent nickel. Developed around 1970, duplex steels are among the most recently developed stainless steels. They offer higher yield strength and superior resistance to pitting and stress cracking than the 300 series stainless steels. An example of duplex stainless steel is Type 2205, which accounts for more than 80 percent of duplex use. Duplex stainless steels are used for heat exchangers, tubes, and pipes for applications involving the production and handling of gas and oils, and for desalinization.

Highly alloyed, high-performance duplex stainless steels, called super-duplex stainless steels, are also made. These contain higher levels of chromium, molybdenum, and nitrogen. They were developed to provide outstanding resistance to acids, acid chlorides, and other caustic solutions typical of those encountered in the chemical, petrochemical, and pulp and paper industries.

GOING GREEN

Recycled Steel

Steel is the world's most recycled material. According to the International Iron and Steel Institute, more than 435 million tons of steel were recycled in 2001.

2.5.0 Nonferrous Metals Classification Systems and Groups

Nonferrous metals include all metals except iron. Commonly used nonferrous metals include aluminum, magnesium, titanium, copper, nickel, zinc, tin, lead, and the precious metals. Nonferrous metals with densities lower than that of steel are considered light metals. They include, in order of increasing density, magnesium, beryllium, aluminum, and titanium. The heavier common nonferrous metals include copper, nickel, lead, tin, and zinc, as well as the precious metals platinum, gold, silver, and their alloys. Each of the nonferrous metals and alloys is classified by a different and unrelated classification system.

2.5.1 Aluminum Alloys

Pure aluminum is a soft, lightweight metal that can be used in its nearly pure form. Because aluminum is very soft and ductile and has a low tensile strength, it is often alloyed with elements such as copper, manganese, silicon, magnesium, and zinc to make different alloys with specific properties for very different purposes. When aluminum is exposed to the atmosphere, a microscopic and transparent aluminum-oxide layer forms immediately that protects the metal from further oxidation. This characteristic gives aluminum its high resistance to corrosion. Aluminum is lightweight, has one-third the weight of steel, and can be made very strong, ductile, and malleable. It is a good conductor of heat and electricity and is highly resistive to weathering and corrosion from many acids. However, aluminum can be corroded easily by alkalis because they attack the oxide layer. Aluminum is also noncombustible and nonmagnetic and is widely used in electrical shielding and near inflammable or explosive substances. It is decorative and easily formed, cast, and machined. Polished aluminum has the highest reflectivity of any material, including mirror glass.

Aluminum alloys are broadly classified as either casting alloys or wrought alloys for rolling, forging, or extruding. The Aluminum Association, Inc. (AA) uses a three-digit plus one decimal system to designate each type of casting alloy (*Table 8*). The first digit (1 through 9) shows the main alloying element or elements. For the 1XX.X series this is essentially pure aluminum (99.00 percent aluminum). With the exception of the 1XX.X series, the second and third digits identify the specific alloys. For the 1XX.X series only, the second and third digits give the degree of aluminum purity above 99.00 percent. For example, if the second and third digits are 60, the alloy contains a minimum of 99.60 percent aluminum. The decimal indicates if the alloy composition is for the final casting (0.0) or for ingot (0.1 or 0.2, depending on impurity limits). Modifications to the alloys are indicated by the use of prefix letters such as A, B, C.

Table 8 AA System Code Designations For Basic Cast Aluminum Alloys

Designation	Major Alloying Element
1XX.X	Near pure aluminum
2XX.X	Copper
3XX.X	Silicon plus copper and/or magnesium
4XX.X	Silicon
5XX.X	Magnesium
6XX.X	Unused series
7XX.X	Zinc
8XX.X	Tin
9XX.X	Other element

203T08.EPS

Similarly, the AA uses a four-digit numbering system for wrought aluminum alloy types (*Table 9*) that is not related to the three-digit casting alloy designation system. In this wrought aluminum system, the first digit identifies the alloy group by the major alloying element. For the 1XXX group, the second digit designates modifications and

GOING GREEN

Recycling Aluminum

Aluminum can be recycled repeatedly with no loss of quality or properties.

Table 9 AA System Code Designations for the Basic Wrought Alloy Groups

Designation	Major Alloying Element
1XXX	Non-aluminum elements are 1 percent or less
2XXX	Copper
3XXX	Manganese
4XXX	Silicon
5XXX	Magnesium
6XXX	Magnesium and silicon
7XXX	Zinc
8XXX	Other elements
9XXX	Unused series

203T09.EPS

GOING GREEN

Why Aluminum Must Be Cleaned Before Welding

The aluminum-oxide coating protecting aluminum must be thoroughly cleaned from the metal before welding. This is because, while pure aluminum melts at about 1,200°F, the aluminum oxide coating that protects the metal does not melt until it reaches the higher temperature of 3,700°F.

All cleaning materials used must be contained and disposed of in accordance with local environmental laws. Cleaning something as large as a boat presents a greater challenge when it comes to containing the runoff of cleaning materials.

203SA03.EPS

impurity limits. A zero (0) indicates no special control on individual impurities. A second digit of 1 through 9 indicates special control over one or more impurities and is assigned consecutively. For groups 2XXX through 8XXX, the second digit indicates alloy modifications and is consecutively assigned. A second digit of 0 indicates the original alloy. The last two digits indicate the minimum percentage of aluminum.

The following examples explain AA wrought aluminum numbers:

- AA Number 1075:
  - 1075 = Indicates aluminum content of 99.0 percent or greater
  - 1075 = Indicates no special control on the impurities
  - 1075 = Indicates an aluminum content of 99.75 percent
- AA Number 1180:
  - 1180 = Indicates aluminum content of 99.0 percent or greater
  - 1180 = Indicates impurities are limited
  - 1180 = Indicates an aluminum content of 99.80 percent

In general, aluminum and its alloys can be welded by the following methods: oxyfuel or air-fuel, electric arc, electrical resistance, and brazing.

2.5.2 Magnesium Alloys

Magnesium is mainly used in alloy form. It is the lightest of the structural metals, having less than one-quarter the density of steel. It is mainly used in the aerospace and automotive industries. Magnesium is the eighth most common element in the world and the sixth most abundant metal, comprising about 2.5 percent of the earth's surface. Because seawater contains about 0.13 percent magnesium, some production facilities process seawater to obtain the magnesium. This is done after the precipitation of other salts from the seawater, which leaves a magnesium-rich brine. The uses of magnesium center on three properties of the metal: its ability to form intermetallic compounds with other metals, its high chemical reactivity, and its low density. Magnesium can be extruded, forged, rolled, and machined.

Magnesium alloys are broadly classified as casting alloys or wrought (rolled, forged, or extruded) alloys. Both of these classifications use the ASTM recommendation B275-05, *Standard Practice for Codification of Certain Nonferrous Metals and Alloys, Cast and Wrought*. Designations consist of one or two letters representing the alloying elements, followed by their respective percentages, and rounded off to the nearest whole number. The letters representing the alloying elements are as follows:

- *A* – Aluminum
- *B* – Bismuth
- *C* – Copper
- *D* – Cadmium
- *E* – Rare Earths
- *F* – Iron
- *G* – Magnesium
- *H* – Thorium
- *K* – Zirconium
- *L* – Beryllium
- *M* – Manganese
- *N* – Nickel

Hot Tip

Magnesium

Magnesium is widely used in specialty racing cars and some commercial cars as a lightweight material to minimize weight and fuel consumption because of the 20 to 25 percent weight saving over aluminum. It typically is used in the construction of transmission casings, intake manifolds, and cylinder head covers. Magnesium pistons and various other engine parts are also being used in some race cars.

It is difficult to visually differentiate aluminum and magnesium. Use positive metal identification (PMI) test equipment to verify these materials.

203SA04.EPS

- *P* – Lead
- *Q* – Silver
- *R* – Chromium
- *S* – Silicon
- *T* – Tin
- *Y* – Antimony
- *Z* – Zinc

Magnesium alloy designations also contain information indicating the temper condition. This part of the designation, separated by a hyphen from the main designation of letters and numbers, consists of a letter or a letter and number. The letters and numbers that indicate the temper are as follows:

- *F* – As fabricated
- *O* – Annealed
- *H10, H11* – Strain hardened
- *H23, H24, H26* – Strain hardened and annealed
- *T4* – Solution heat treated
- *T5* – Artificially aged
- *T6* – Solution heat treated and artificially aged
- *T8* – Solution heat treated, cold worked, and artificially aged

In the alloy AZ92A-T6, the first A is aluminum and the Z is zinc. The 9 indicates that the aluminum percentage is between 8.6 and 9.4. The 2 shows that the zinc percentage is between 1.5 and 2.5. The final A denotes that this is the first alloy to receive the AZ92 designation. The letter and number T6 indicate solution heat treatment and artificially aged tempering.

Magnesium and its alloys are best welded by the GTAW process for thin sections and the GMAW process for thick sections. The composition of the filler metals should be compatible with the base metal. However, zinc above 1 percent increases **hot shortness** and can result in cracking. *Table 10* shows examples of magnesium alloys and their principal uses.

2.5.3 Titanium Alloys

Titanium is mainly used in alloy form. Its alloys are high-strength, lightweight (about one-half the density of steel), corrosion-resistant structural metals that retain their strength at elevated temperatures. It is used in aircraft engines and struc-

Table 10 Examples of Magnesium Alloys

ASTM Number	Nominal Alloy Composition (By %)						
	Al	Mn	Zn	Zr	Rare Earths	Th	Mg
Sand and Permanent Mold Castings							
AZ92A-T6	9.0	0.10	2.0	—	—	—	Balance
AZ63A-T6	6.0	0.15	3.0	—	—	—	Balance
AZ81A-T4	7.6	0.13 Min	0.7	—	—	—	Balance
AZ91C-T6	8.7	0.13	0.7	—	—	—	Balance
EZ33A-T5	—	—	2.6	0.8	3.3	—	Balance
HK31A-T6	—	—	0.3	0.7	—	3.3	Balance
HZ32A-T6	—	—	2.1	0.8	—	3.3	Balance
Die Castings							
AZ91A-F	9.0	0.13	0.7	—	—	—	Balance
AM60A-F	6.0	0.13	—	—	—	—	Balance
AZ91D-F	9.0	0.13	0.7	—	—	—	Balance
Extrusions							
AZ61A-F	6.5	0.15	1.0	—	—	—	Balance
MIA-F	—	1.20	—	—	—	—	Balance
AZ80A-T5	8.5	0.12	0.5	—	—	—	Balance
AZ10A-F	1.3	0.20	0.4	—	—	—	Balance
AZ31B-F	3.0	0.20	1.0	—	—	—	Balance
ZK60A-F	—	—	5.5	0.45	—	—	Balance
Sheet and Plate							
AZ31B-H24	3.0	0.20	1.0	—	—	—	Balance
HK31A-H24	—	—	—	0.7	—	3.3	Balance

203T10.EPS

tures, chemical processing, surgical implants, sporting goods, and marine and other applications where corrosion resistance is required.

Titanium is a reactive metal. At high temperatures, it readily combines with oxygen, hydrogen, and nitrogen to form stable compounds. Similar to aluminum and magnesium, titanium forms a tight, protective, microscopic oxide film on clean surfaces at room temperature, when in contact with oxygen. This makes it highly resistant to corrosion at low temperatures. Titanium alloys are classified into three groups according to the crystal structures that are stable at room temperature.

- *Alpha alloys* – Are the usual crystalline structures of pure titanium. Aluminum is added to stabilize the crystalline structure. These alloys have excellent strength and are oxidation-resistant from 600°F to 1,100°F. They have good ductility and weldability and can be hardened by working, but not by heat treatment.
- *Beta alloys* – Can contain vanadium, chromium, and aluminum. Iron and chromium are added to make them heat-treatable, high-strength alloys. They have fair weldability.
- *Alpha-beta alloys* – Contain one or more of the elements manganese, molybdenum, tin, vanadium, iron, and aluminum. They have a wide range of mechanical characteristics, but have poor weldability.

Titanium is considered difficult to weld because the weld must be kept in an inert (nonreactive) gas atmosphere until it cools to below the reactive temperature. Small amounts of hydrogen or large amounts of oxygen or nitrogen can cause embrittlement in the HAZ, which indicates hardness that leads to breakability. Titanium is usually welded by the GTAW process.

About 34 different titanium alloys are currently being produced, including the ones shown in *Table 11*.

2.5.4 Beryllium Alloys

Beryllium (Be) is a high-strength, lightweight metal. It is only about two-thirds as heavy as aluminum, but is six times stiffer than steel. Beryllium's heat capacity is five times that of copper, meaning that one pound of it will absorb as much

Table 11 Examples of Titanium Alloys

Designation	Nominal Alloy Composition (By %) Al	Sn	Zr	Mo	Other
Unalloyed Grades					
ASTM Grade 1	—	—	—	—	—
ASTM Grade 7	—	—	—	—	0.2 Pd
ASTM Grade 11	—	—	—	—	0.12–0.25 Pd
ASTM Grade 12	—	—	—	0.3	0.6–0.9 Ni
Alpha and Near Alpha Alloys					
Ti-5Al-2.5Sn	5.0	2.5	—	—	—
Ti-5Al-2.5Sn-ELI	5.0	2.5	—	—	—
Ti-6Al-2Sn-4Zr-2Mo	6.0	2.0	4.0	2.0	—
Ti-6Al-2Nb01Ta-0.8Mo	6.0	—	—	1.0	2 Nb, 1 Ta
Alpha-Beta Alloys					
Ti-6Al-4V	6.0	—	—	—	4.0 V
Ti-6Al-4V-ELI	6.0	—	—	—	6.0 V
Ti-6Al-6V-2Sn	6.0	2.0	—	—	0.75 Cu, 6.0 V
Ti-8Mn	—	—	—	—	8.0 Mn
Ti-7Al-4Mo	7.0	—	—	4.0	—
Ti-6Al-2Sn-4Z-6Mo	6.0	2.0	4.0	6.0	—
Beta Alloys					
Ti-13V-11Cr-3Al	3.0	—	—	—	11.0 Cr, 13.0 V
Ti-8Mo-8V-2Fe-3Al	3.0	—	—	8.0	8.0
Ti-11.5Mo-6Zr-4.5Sn	—	4.5	6.0	11.5	—

203T11.EPS

Hot Tip

Titanium

Commercial production of titanium did not begin until the 1950s. At that time, titanium was recognized for its importance and was developed as a unique, lightweight, high-strength alloy for use in engine and airframe components of high-performance jet aircraft. Today, titanium alloys are widely used in many applications that previously used metals such as stainless and specialty steels, copper alloys, and nickel-based alloys.

heat as five pounds of copper. For this reason, it is widely used in aerospace and nuclear reactor applications. Beryllium is a reactive metal, and at high temperatures it readily combines with oxygen, hydrogen, and nitrogen to form stable compounds. At low temperatures, it is highly resistant to corrosion. Small amounts of nitrogen and hydrogen or large amounts of oxygen cause embrittlement. *Table 12* lists beryllium alloys and some of their mechanical properties, such as tensile and yield strength.

Beryllium is considered a difficult metal to weld because it must be kept in an inert gas atmosphere while it is being welded.

WARNING!

Dust, fumes, vapors, and fine machining particles of beryllium are cancer-causing agents. If inhaled, they can cause chronic beryllium disease. An independent air supply is required when working with beryllium.

Hot Tip

Beryllium

Beryllium allows the transmission of X-rays better than other metals or glass. Because of this property, along with its high melting point, it is widely used as the window portion of high-intensity X-ray tubes, such as those used in high-resolution X-ray machines and imaging equipment.

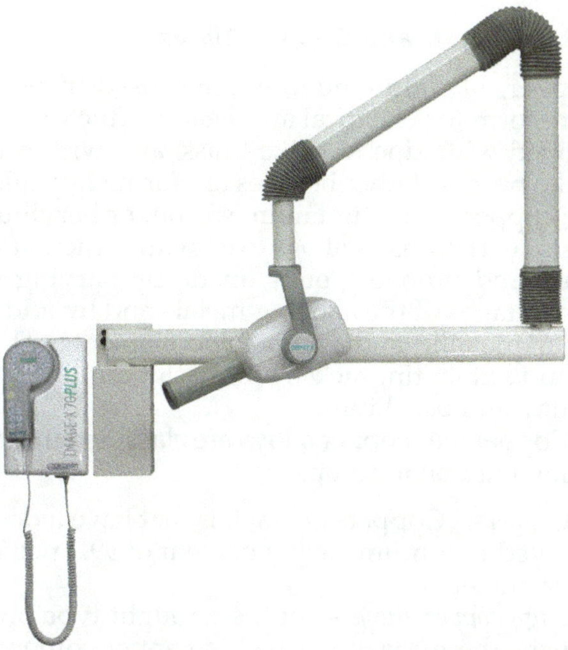

203SA05.EPS

Table 12 Examples of Beryllium Alloys and Mechanical Properties

Designation and Percent Be	Tensile Strength (psi)	Yield Strength (psi)	Commercial Form	Description or Properties
HP-20 (98.0 percent)	48,000	35,000	Block	Pressed powder
HP-8 (99.0 percent)	37,000	27,000	Block	Pressed powder
HP-12 (98.5 percent)	38,000	30,000	Block	Pressed powder
HP-40 (92.0 percent)	65,000	60,000	Block	Pressed powder
C-10 (99.0 percent)	20,000	—	Block	Pressed powder
PR-2 (98.0 percent)	77,000	55,000	Sheet/plate	Some form at 1,350°F
IS-2 (99.0 percent)	45,000	31,000	Sheet/plate	High form at 700°F
XT-20 (98.0 percent)	77,000	40,000	Extrusion	Good mechanical properties
XT-40 (92.0 percent)	90,000	50,000	Extrusion	High strength
Lockalloy	60,000	50,000	Extrusion	Service at 600°F to 800°F

203T12.EPS

GOING GREEN

Beryllium Disposal

To protect both yourself and the environment, make sure that all dust, fumes, vapors, and flying particles resulting from any work done on beryllium are contained and disposed of in accordance with local environmental guidelines.

2.5.5 Copper and Copper Alloys

Copper, like aluminum, is often used in nearly pure form for electrical and heat conductors. It is alloyed with zinc to make brass, and with tin to make bronze. Other bronzes are formed by alloying copper with aluminum, silicon, or beryllium. Many variations of alloys exist within each of the brass and bronze groups, made by varying the percentages of the alloying metals and by adding other metallic and nonmetallic elements. These include zinc, tin, nickel, silicon, aluminum, cadmium, and beryllium.

Copper and copper alloys are classified into the following major groups:

- *Coppers* – Coppers are metals that have a designated minimum copper content of 99.3 percent or higher.
- *High copper alloys* – For the wrought-type products, these are alloys with a copper content of less than 99.3 percent but more than 96 percent that do not fall into any other copper alloy group. Cast-type high copper alloys have a copper content in excess of 94 percent, to which silver may be added for special properties.
- *Brasses* – These alloys contain zinc as the principal alloying element, with or without other alloying elements. The wrought-type alloys are grouped into three categories of brasses: copper-zinc alloys, copper-zinc-lead alloys (leaded brasses), and copper-zinc-tin alloys (tin brasses). Cast alloys are grouped into five categories of brasses: copper-tin-zinc alloys (red, semi-red and yellow brasses); manganese bronze alloys (high-strength yellow brasses); leaded manganese bronze alloys (leaded high-strength yellow brasses); copper-zinc-silicon alloys (silicon brasses and bronzes); and cast copper-bismuth and copper-bismuth-selenium alloys.
- *Bronzes* – Bronzes are copper alloys in which the major alloying element is not zinc or nickel. For wrought-type alloys, there are four groups of bronzes: copper-tin-phosphorus alloys (phosphor bronzes); copper-tin-lead-phosphorus alloys (leaded phosphor bronzes); copper-aluminum alloys (aluminum bronzes); and copper-silicon alloys (silicon bronzes). Cast-type alloys are divided into four groups of bronzes: copper-tin alloys (tin bronzes), copper-tin-lead alloys (leaded and high leaded tin bronzes), copper-tin-nickel alloys (nickel-tin bronzes), and copper-aluminum alloys (aluminum bronzes). Alloys known as manganese bronzes, in which zinc is the major alloying element, are included in the brasses.
- *Copper-nickels* – Copper-nickel alloys have nickel as the principal alloying element, with or without other alloying elements.
- *Copper-nickel-zinc alloys* – These alloys contain zinc and nickel as the principal and secondary alloying elements, with or without other elements. They are commonly called nickel silvers.
- *Leaded coppers* – Leaded coppers comprise a series of cast alloys of copper with 20 percent or more lead, sometimes with a small amount of silver, but without tin or zinc.
- *Special alloys* – These are alloys whose chemical compositions do not fall into any of the above categories.

Two classification systems for copper alloys are the ASTM designation system (B224) (*Table 13*) and the Copper Development Association, Inc. designation system (*Table 14*). The Copper Development Association Inc. system has been updated to fit the UNS, and it includes all commercially available metals and alloys. The format for the UNS designation is CNNN00, in which NNN designates numbers. As shown in *Table 14*, the alloys are divided into wrought and cast alloy categories. For this reason, an alloy made in both a wrought and cast form can have different UNS numbers, depending on the method of manufacture.

Tough-pitch copper is oxygen bearing, and most of the impurities have been oxidized during refining to increase conductivity. Continuous casting methods have allowed the oxygen content to be reduced to 0.03 percent or less to achieve the desired results. During fire refining, the oxygen that comes from the air injected to cause the oxidation of impurities also combines with hydrogen to form steam. The steam creates gas pockets (porosity) in the metal during solidification. (In the past, high porosity was a characteristic of tough-pitch copper.) When tough-pitch copper is reheated by welding or annealing operations, the hydrogen trapped in the copper is diffused into the metal and reacts with cuprous oxide to form insoluble water vapor. The expansion associated

Table 13 ASTM Classification System for Copper Alloys

ASTM Designation	Alloy Type	Typical Uses
	Tough-Pitch Coppers	
ETP	Electrolytic, tough-pitch	Bus bars, brazing rods, wire anodes, forgings
FRHC	Fire-refined, high-conductivity, tough-pitch	Mechanical applications
FRTP	Fire-refined, tough-pitch	Sheets, strips, plate
ATP	Arsenical, tough-pitch	Roofing, radiator cores
STP	Silver-bearing, tough-pitch	Pans, printing rolls, fasteners
SATP	Silver-bearing, arsenical, tough-pitch	Pans, printing rolls, fasteners
	Oxygen-Free Coppers	
OFHC	Oxygen-free (without residual deoxidants)	Tubing wave guides, starting anodes, wire
OFP	Oxygen-free, phosphorus-bearing	Welding rods, forging
OFTPE	Oxygen-free, tellurium- and phosphorus- bearing	Welding rods, forging
OFS	Oxygen-free, silver-bearing	Plate, sheets, rods, forging
OFTE	Oxygen-free, tellurium-bearing	Bars, free-machining
	Deoxidized Coppers	
DHP	Phosphorized, high-residual phosphorus	Tubes, pipes, anodes, projectile rotating bands
DLP	Phosphorized, low-residual phosphorus	Tubes, wave guides, general use
DPS	Phosphorized, silver-bearing	Heat exchangers, steam lines, condenser tubes, tubes for general use
DPA	Phosphorized, arsenical	Heat exchangers, steam lines, condenser tubes, tubes for general use
DPTE	Phosphorized, tellurium-bearing	Free-machining

203T13.EPS

Table 14 UNS Designations for Classifying Copper Alloys

Copper Designation	Composition
	Wrought Alloy Groups
C11X00	99.95% oxygen-free, high-conductivity copper
C12X00	99.88+% copper (tough-pitch copper)
C19X00	96+% copper (high-copper alloys)
C2XX00	Copper-zinc alloys (brasses)
C3XX00	Copper-zinc-lead alloys (leaded brasses)
C4XX00	Copper-zinc-tin alloys (tin brasses)
C51X00	Copper-tin alloys (phosphor bronzes)
C54X00	Copper-tin-lead alloys (leaded phosphor bronzes)
C62X00	Copper-aluminum alloys (aluminum bronzes)
C65X00	Copper-silicon alloys (silicon bronzes)
C70X00	Copper-nickel alloys
	Cast Alloy Groups
C80X00	99+% copper, copper alloys
C81X00	High-copper alloys (beryllium copper)
C83X00	Red brasses and leaded red brasses
C84X00	Yellow brasses and leaded yellow brasses
C86X00	Manganese and leaded manganese-bronze alloys
C87X00	Silicon bronzes and brasses
C90X00	Tin bronzes

203T14.EPS

Hot Tip

Statue of Liberty

The Statue of Liberty contains 179,000 pounds of copper. When copper oxidizes, it forms a blue-green self-protective coating, which is why the copper skin of the Statue of Liberty remained virtually intact after 100 years of being subjected to high sea winds, driving rains, and scorching sun. Close analysis showed that weathering and oxidation of the copper skin had caused only 0.005 inch of wear in a century. For this reason, the copper skin did not need to be significantly rebuilt when the statue was renovated for its centennial. However, high-alloy copper saddles and rivets were used during the restoration to fasten the copper skin to the framing underneath. This was done to ensure the structural integrity of the statue and guard against any galvanic reaction problems. Galvanic action is corrosion caused by an electrical current produced as the result of a chemical reaction between two dissimilar metals.

203SA06.EPS

with this reaction forces the grains apart, which causes embrittlement of the metal. For applications where copper must be resistant to gassing, the oxygen is sometimes eliminated by adding a deoxidizer, such as phosphorus, to the molten metal before casting. Such copper is specified as deoxidized copper.

For welding copper alloys, GMAW and GTAW are most commonly used.

2.5.6 Nickel and Nickel Alloys

Nickel, a corrosion-resistant metal, is used as an alloying element as well as for plating other metals. Nickel-clad steel and nickel-based alloys containing over 50 percent nickel are used extensively in industry. They have good corrosion resistance and elasticity, and unique magnetic thermal expansion and thermal conductivity. Many high-strength, high-temperature nickel super alloys have been developed for aerospace, automotive, and hot-die applications. Alloys of nickel and chromium (Nichrome®) are used to make heating elements of high electrical resistance.

UNS and international standard designations have been assigned to many nickel-alloys. However, trade names, such as Monel 400®, Inconel 600®, and Hastelloy 22® are more commonly used to identify them. *Table 15* lists some common nickel alloys and some of their uses and properties.

In general, nickel and nickel-based alloys can be welded by SMAW, GMAW, and GTAW processes. Welding nickel alloys is very similar to welding austenitic stainless steel, with the following exceptions:

- When welding nickel alloys, the surface oxide melts at a much higher temperature than the base metal, so surfaces should be thoroughly cleaned.
- Embrittlement can be caused by lead, sulfur, phosphorus, or other low-temperature metals. The surface should be free of all contaminants.
- Weld penetration in nickel-based alloys is shallower than with other metals and requires the use of very wide bevel and groove angles.

2.5.7 Lead, Tin, and Zinc

Lead, tin, zinc, and their alloys have very low melting points. For this reason, they are usually soldered and not welded. Zinc-plated, or galvanized, steels can be welded, but the fumes are toxic and must not be inhaled.

2.5.8 Platinum, Gold, Silver, and Other Precious Metals

The precious metals are osmium, iridium, rhodium, platinum, ruthenium, gold, palladium, silver, and their alloys. They are used industrially for electrical contacts, mirror coatings, catalysts, electrodes, high-temperature crucibles, and dental structures and fillings. They are usually brazed, soldered, or welded with special equipment.

Table 15 Common Nickel Alloys and Some of Their Uses

Alloy	Ni	C	Mn	Fe	S	Si	Cu	Cr	Al	Ti	Other	Typical Properties and Uses
Commercially Pure Nickel												
A-Nickel	99.4	0.1	0.2	0.15	0.005	0.05	0.1					Chemical industry, electroplating
D-Nickel	95.2	0.1	4.5	0.15	0.005	0.05	0.05					Resistance to sulfur attack
E-Nickel	97.7	0.1	2.0	0.10	—	0.05	0.05					Similar to d-nickel
L-Nickel	99.4	0.02	0.2	0.15	0.005	0.05	0.1					Severe plastic forming operations
Z-Nickel	94.0	0.16	0.25	0.25	0.005	0.04	0.05					Age hardenable, springs, pump rods, shafts
Nickel-Copper Alloys												
Monel	66.0	0.12	0.90	1.35	0.005	0.15	31.50					Corrosion resistance, toughness, high strength
R-Monel	66.0	0.18	0.90	1.35	0.005	0.15	31.50					Free-machining
K-Monel	65.3	0.15	0.60	1.00	0.005	0.15	29.50		2.80	0.50		Age-hardenable, nonmagnetic
H-Monel	65.0	0.1	0.9	1.5	0.015	3.0	29.50					Age-hardenable, machinable
S-Monel	63.0	0.1	0.9	2.0	0.015	4.0	30.0					Age-hardenable, anti-galling, impellers pump liners
Nickel-Iron Alloys												
Invar	36.0	*	*	63.0	*							Very low expansion coefficient length standards
Platinite	46.0			54.0								Glass seal
Permalloy	78.5			21.5								High-magnetic permeability, submarine telegraph cables
*Combined amounts of silicon, manganese and carbon total less than 1 percent												
Nickel-Chromium and Nickel-Iron-Chromium Alloys												
Inconel	76.4	0.04	0.20	7.20	0.007	0.2	0.10	15.85				Heat and corrosion resisting, nonmagnetic
Inconel 702	78.0	0.02	0.05	0.30	0.007	0.15	0.05	15.85	3.0	0.60		Age-hardenable, operates up to 2,400°F
S-Inconel	68.0	0.20	1.0	8.5	0.008	5.0	0.5	15.5				Age-hardenable, anti-galling
Inconel-X-750	72.9	0.04	0.70	6.8	0.007	0.30	0.05	15.0	0.80	2.5	CB-0.9	Age-hardenable, used up to 1,500°F
Nimonic 80A	74.5	0.05	0.55	0.55	0.007	0.20	0.05	20.0	1.30	2.50		Age-hardenable, similar to Inconel X
Nimonic 90	57.0	0.05	0.50	9.50	0.007	0.20	0.05	20.45	1.65	2.60	CO-17.00	Similar to Nimonic 80A
Nichrome	80.0	0.05	0.01	0.5	—	0.20	—	20.0				Heating elements
Chromel	90.0							10.0				Heating elements, thermocouples
Nicoloy 901	37.35	0.04	0.45	33.75	0.007	0.30	0.10	13.50	0.25	2.50	MO-5.90	Age-hardenable, gas turbine wheels
Ni-O-NEL	41.35	0.03	0.65	31.65	0.007	0.35	1.80	20.2	0.40	0.50	MO-310	Resistant to certain hot acids
60-12 Ni-Cr	62.45	0.55		25.0				12.0				Resistance heating element
65-15 Ni-Cr	65.0			24.00	0.007			15.00				Resistance heating element
67-17 Ni-Cr	72.45	0.55		10.0				17.00				Heat-resisting alloy

203T15.EPS

3.0.0 Field Identification of Base Metals

Many of the common base metals and alloys can be generally identified by simple tests and observation of their characteristic metallic features. This is especially true of the ferrous metals, which can be identified by magnetic attraction, surface appearance, structural form or shape, and grinding sparks.

3.1.0 Metal Labeling

Because it is difficult to identify base metal types, most sites have a system to label or color-code the various types of metals being used. Labeling is performed by printing the ASTM number on the material (shape or plate) with an indelible marker or paintstick. Color-coding involves assigning a color to a base metal type and then spraying that color on the material. When color-coding is used, a master chart is often posted near the storage area. The master chart identifies the color assigned to each base metal.

For stainless steel and nickel alloys, special care should be taken to mark only with low-chloride markers. Using markers other than the low-chloride markers will leave a residue on the metal that can contaminate the weld. Specialty markers, which usually contain white or yellow ink, are available for marking metals such as stainless steel (*Figure 7*).

> **NOTE**
> Check your site quality standards for the type of material labeling that is used at your location. Follow these standards to ensure material traceability.

3.2.0 Identification by Magnet

All carbon steels, cast steels, and cast irons are attracted to a magnet. Of the stainless steels, only ferritic and martensitic stainless steels can be magnetized, and they have the bright, corrosion-free surface typical of stainless steel. The largest group of stainless steels—austenitic stainless steels—are nonmagnetic. Nonferrous metals and manganese steels are also nonmagnetic.

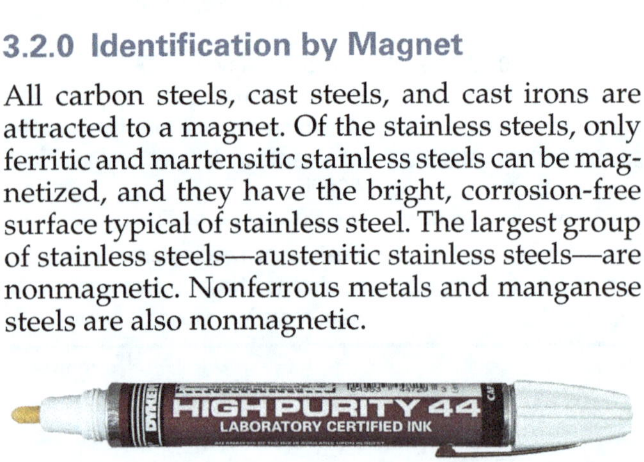

203F07.EPS

Figure 7 A typical low-chloride marker for specialty metals.

> **CAUTION**
> Magnetic testing is sometimes unreliable because of alloys or changes that can occur during welding. Do not rely solely on magnetic testing.

3.3.0 Identification by Appearance

Both hot-rolled and cold-finished carbon steel are milled in standard structural shapes, such as rods, bars, beams, channels, and angles. Unweathered, hot-rolled carbon steel has a blue-gray, smooth-to-scaly surface, but it oxidizes to a rough, red-brown rust when exposed to the weather. The corners of hot-rolled shapes are slightly rounded or chamfered.

Unweathered, cold-rolled carbon steel has a smooth, gray finish that oxidizes to a fine red-brown rust when weathered. Except for round shapes, the sides and edges are very flat and smooth, and the corners are well formed, not rounded or chamfered.

Cast steel and cast iron look similar on the surface. They both exhibit the shape characteristics typical of castings: curved surfaces and rounded edges, thin and thick regions, reinforcing ribs or raised areas, spokes, and irregular shapes. The cast surfaces are usually slightly textured and can contain mold seams. When broken, cast iron has a grainy appearance, while cast steel has a smooth texture. Cast iron is used to make everything from water-meter covers, to engine crankshafts, to fan blades (*Figure 8*).

There are three types of cast iron: gray, white, and malleable. The exterior surface of all the cast irons has a slightly irregular texture that comes from the use of sand molds. Malleable iron is

Hot Tip

Labeling Material

When cutting a piece of material that is labeled or color-coded, it is very important to make sure that the label or color code is visible on the material that is being left behind. If possible, make the cut in such a way as to leave the label or color code still visible. If this is not possible, be sure to mark the material being left behind to ensure that others can correctly identify the needed material. It is also a good idea to mark the material being removed so that everyone involved can identify it as the material specified for the job.

203F08.EPS

Figure 8 A compressor's cooling fan blade made of cast iron.

sometimes forged. If it has been forged, it will have a smoother appearance and die marks.

Gray cast iron contains a high percentage of carbon and silicon. Gray cast iron is only formed by slow cooling, which forms large crystals and a soft, easily machined metal. During the slow cooling process, the high silicon content forces the carbon to separate out in the form of graphite flakes (free carbon). This gives a fresh fracture its characteristic grainy and gray appearance. If you rub a clean finger or white cloth against a freshly fractured gray cast-iron surface, the graphite will leave a dark smear. Gray cast iron can be welded by following special procedures.

White cast iron is chilled rapidly, which forms small crystals and a very hard and brittle metal. The carbon is united with the iron and is not free carbon. Freshly fractured cast iron has an irregular, fine, silvery-white, crystalline surface. Because the weldability of white cast iron is poor, it is generally not welded.

Malleable cast iron is white cast iron that has been annealed. The annealing process results in a casting with an outer shell of white cast iron that grades into a core of gray cast iron. The resulting casting is tougher and stronger than gray cast iron and less brittle than white cast iron. Malleable cast iron can be welded, but special procedures must be used or it will revert back to white cast iron.

Hot Tip

Identification of Nonferrous Metals

The electrical conductivity measuring instrument shown can be used to determine the composition of nonferrous metals, including titanium alloys and silver. By using eddy currents to determine the electrical conductivity of a nonferrous metal under test, conclusions can be made about the composition, structure, and condition of the material.

203SA07.EPS

3.4.0 Identification by X-Ray Fluorescence Spectrometry

Portable metal analyzers that use X-ray fluorescence (XRF) spectrometry provide a fast and accurate analysis for metal and metal alloy identification. A fluorescent X-ray is created when an X-ray of sufficient energy, emitted from a miniaturized X-ray tube or from a small, sealed radioactive isotope, strikes an atom in the sample, dislodging an electron from one of the atom's inner orbital shells. The atom regains stability, filling the vacancy left in the inner orbital shell with an electron from one of the atom's higher energy orbital shells. The electron drops to the lower

Hot Tip

Spark Tests

In the past, spark tests were sometimes used by metal workers to identify various types of iron, steel, and steel alloys. The test consisted of observing the characteristics of the spark shower given off when iron or steel was ground by a power grinder. This method required a great deal of skill and training on the part of the metal worker to interpret the spark shower. Because of the highly accurate metal analyzers available today, the spark test is used only on rare occasions to quickly distinguish between low and high carbon (tool) steels.

energy state by releasing a fluorescent X-ray, and the energy of this X-ray is equal to the specific difference in energy between two quantum states of the electron.

When a sample is measured using X-ray fluorescence, each element present in the sample emits its own unique fluorescent X-ray energy spectrum. By simultaneously measuring the fluorescent X-rays emitted by the different elements in the sample, handheld XRF analyzers can rapidly determine those elements that are present in the sample and their relative concentrations—in other words, the elemental chemistry of the sample. For samples with known ranges of chemical composition, such as common grades of metal alloys, the XRF analyzer also identifies most sample types by grade name, through comparing the analyzed concentrations with known values contained in an alloy library programmed within the analyzer. This entire process is typically completed in less than five seconds, and is completely nondestructive.

Thirty or more elements may be analyzed simultaneously by measuring the characteristic fluorescence X-rays emitted by a sample. *Figure 9* shows a Thermo Fisher Scientific Niton XL3 Series handheld XRF analyzer, its screen, and the analyzer in use.

4.0.0 Physical Characteristics of Metals

Physical characteristics are distinctive for every metal. They include the following variable traits:

- Density (specific gravity)
- Electrical conductivity
- Thermal conductivity
- Thermal expansion
- Melting point
- Corrosion resistance

Hot Tip

Determining the Weight and Density of Materials

One cubic inch of pure water at 62°F weighs 0.0361 pound. If you know the density (specific gravity) of a material, you can determine the weight of a cubic inch of the material by multiplying its density by 0.0361. For example, the density of aluminum is 2.71. To calculate the weight for one cubic inch of aluminum, multiply $2.71 \times 0.0361 =$ 0.0978 lb.

To obtain the weight in pounds per cubic foot (lb/ft^3) of a material, multiply its density by 62.355. For example, one cubic foot of aluminum weighs 168.98 lbs ($2.71 \times 62.355 = 168.98$).

If you know the weight per cubic foot of a material, you can calculate the density of the material by multiplying the weight of one cubic foot of the material by 0.01604. For example, one cubic foot of aluminum weighs 168.98 lbs. The density of aluminum is 2.71 (168.98 lb/ft^3 3 0.01604 = 2.71).

4.1.0 Density

Density is the mass (weight) of a specific volume of metal. Density may also be expressed as specific gravity or specific density. Specific gravity is the weight of a metal compared to the weight of an equal volume of water. The higher the density of a material, the heavier it will be.

Light metals, such as aluminum and magnesium, have low densities (low weights for a given volume) while ferrous metals, such as carbon steel and stainless steel, have much higher densities. An iron or steel item weighs nearly three times as much as an aluminum item of the same size. *Table 16* lists the densities of some common metals in pounds per cubic inch.

4.2.0 Electrical Conductivity

Electrical conductivity is a metal's ability to conduct electricity. All metals can conduct electricity to some degree. Silver is one of the best conductors, but it is too expensive to use for commercial power transmission. Copper is not quite as good a conductor as silver, but for the same diameter conductor or wire, it is a better conductor than aluminum. However, for conductors weighing the same per unit of length, aluminum will carry more current than copper because it has a larger cross section.

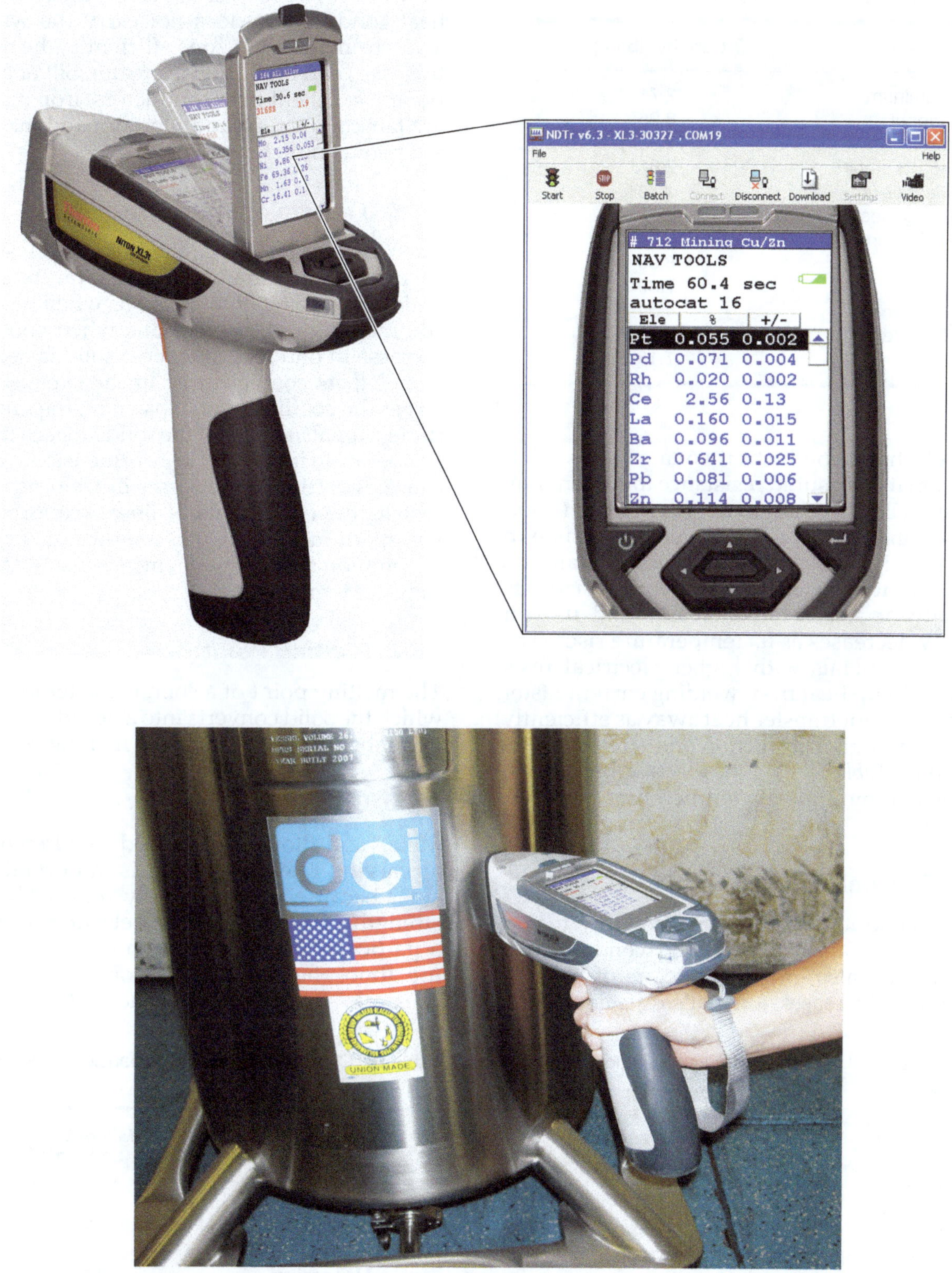

Photos courtesy of Thermo Fisher Scientific.

203F09.EPS

Figure 9 NITON XL3 handheld analyzer.

Table 16 Densities of Some Common Metals

Metal	Density (lb/in³)
Aluminum	2.71
Copper	8.96
Gold	19.32
Iron or steel	7.87
Magnesium	1.74
Manganese	7.43
Molybdenum	10.20
Nickel	8.90
Tin	7.30
Titanium	4.54
Tungsten	19.30
Zinc	7.13

203T16.EPS

The electrical conductivity of a metal is calculated from its measured resistance to current flow (its resistivity). Resistivity is the inverse (opposite) of conductivity. The lower the resistivity value, the higher the conductivity value and the more current a metal can carry for its size. The resistivity of any metal increases and its conductivity decreases as its temperature rises. This means that metals with higher electrical resistances build up heat from welding current faster. They also do not transfer heat away as efficiently, resulting in higher distortion and stresses in the weld zone. *Table 17* lists the electrical resistivity of some common metals in microhms per cubic centimeter.

4.3.0 Thermal Conductivity

Thermal conductivity is a metal's ability to conduct heat. Metals that are good conductors of electricity are also good conductors of heat. As temperature increases, a metal's resistance to the flow of heat increases. Because a metal with poor heat conductivity does not carry the weld heat away from the weld zone efficiently, the temperature rises and distortion is greater. Silver and copper are excellent heat conductors, iron and nickel are fair conductors, and titanium and manganese are poor conductors.

4.4.0 Thermal Expansion

Thermal expansion is the change in size that occurs in a material as its temperature changes. Solids expand in all dimensions when heated and contract from all dimensions when cooled. The increase in unit length when a solid is heated 1°F is called its coefficient of linear expansion. The larger the coefficient of expansion, the greater the dimensional increase of the solid for each degree of temperature increase. Large dimensional changes cause distortion and stresses in weldments. *Table 18* lists the coefficients of linear expansion for a variety of metals. *Figure 10* shows the expansion of common metal pipes in inches per 100 feet for a range of temperatures.

4.5.0 Melting Point

The melting point of a solid is the temperature at which the solid converts into a liquid. *Table 19* lists a number of metals and their melting points.

4.6.0 Corrosion Resistance

Metals and alloys are corroded by different agents at varying rates. Corrosion is important to consider because it can severely reduce the tensile strength of a metal. When selecting a metal for a particular use, its ability to resist corrosion by the atmosphere and by gaseous or liquid chemicals may be an important factor. Certain metals

Table 17 Electrical Resistivity of Some Common Metals

Metal	Electrical Resistivity (Microhms/cm³)
Aluminum	2.92
Copper	1.67
Gold	2.19
Iron or steel	9.71
Magnesium	4.46
Manganese	185.00
Molybdenum	5.17
Nickel	6.84
Titanium	80.00
Zinc	5.91

203T17.EPS

Table 18 Coefficients of Linear Expansion for Some Common Metals

Metal	Millionths of an Inch/°F
Aluminum	13.3
Copper	9.2
Iron	6.5
Magnesium	14.0
Manganese	12.0
Molybdenum	2.7
Nickel	7.4
Titanium	4.7
Zinc	22.1

203T18.EPS

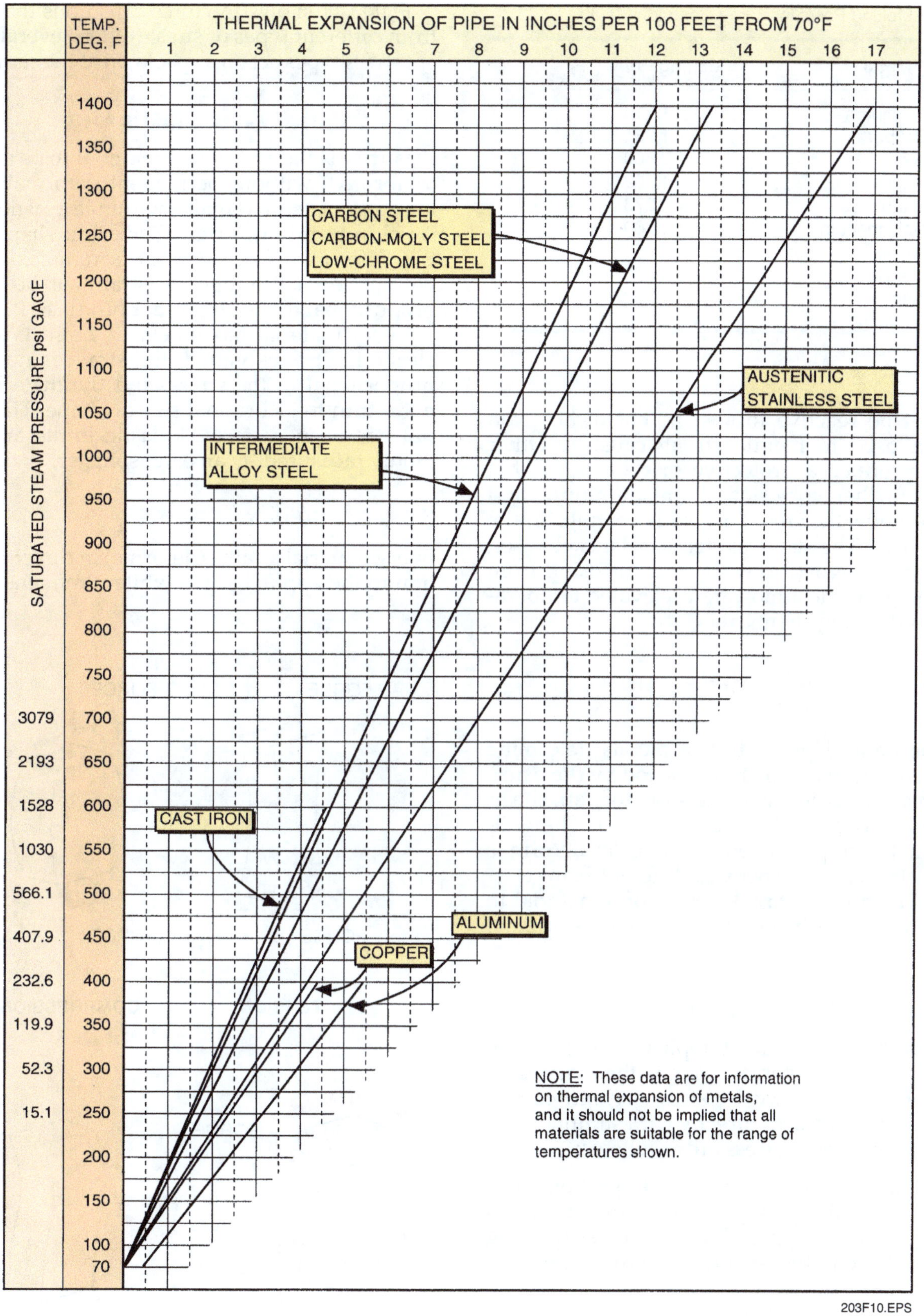

Figure 10 Expansion of different types of pipe.

Table 19 Melting Points for Some Common Metals

Metal	Melting Point (F)
Aluminum	1,215
Copper	1,981
Iron	2,800
Magnesium	1,202
Manganese	2,273
Molybdenum	4,760
Nickel	2,651
Titanium	3,300
Zinc	787

203T19.EPS

may not be reactive to specific chemicals. Iron is often used to handle strong caustics. Other metals produce a protective oxide coating that retards further corrosion. Both aluminum and lead form weather-resistant oxide coatings in the atmosphere. The stainless steels and nickel-based alloys are designed to withstand corrosion from such things as the atmosphere, elevated temperatures, and strong chemicals.

5.0.0 Mechanical Properties of Metals

The mechanical properties of metals determine how they will react to or be affected by the external forces of tension, compression, torsion, shear, impacts, and cold shaping (working). The ability of metals to permanently change or deform is called plasticity. Plasticity is different from elasticity, which refers to the ability of a material to change temporarily and then revert to its original form.

5.1.0 Stress-Strain Relationship

Stress is the force or load applied to a specimen that causes it to deform. Strain is the magnitude of deformation caused by stress. Within certain limits, stress increases directly with strain.

The four types of stress are as follows:

- *Tension* – Opposed, in-line forces pulling away from each other at opposite sides of a specimen
- *Compression* – Opposed, in-line forces pushing toward each other against opposite sides of a specimen
- *Torsion* – Opposed, separated, twisting forces acting on the same axis of a specimen
- *Shear* – Opposed, nonintersecting forces acting opposite each other and perpendicular to a specimen

Figure 11 shows the force directions that result from different types of stress. Often several different stress types act on a specimen at the same time.

5.2.0 Elasticity and Elastic Limit

Elasticity is the ability of a material to be strained (deformed) without permanent deformation. As long as a material remains within its elastic range, it will return to its original size and shape when the deforming force (stress) is removed. While within the elastic range, the strain will be directly proportional to the stress that produces it.

Every material has its elastic limit. When the elastic limit is exceeded, the strain is no longer proportional to the stress, and the material will not return to its original size or shape. The material has been distorted. Elastic limits are commonly measured by an extensometer.

5.3.0 Modulus of Elasticity

Every material strains (deforms) in direct proportion to the applied stress while it remains within

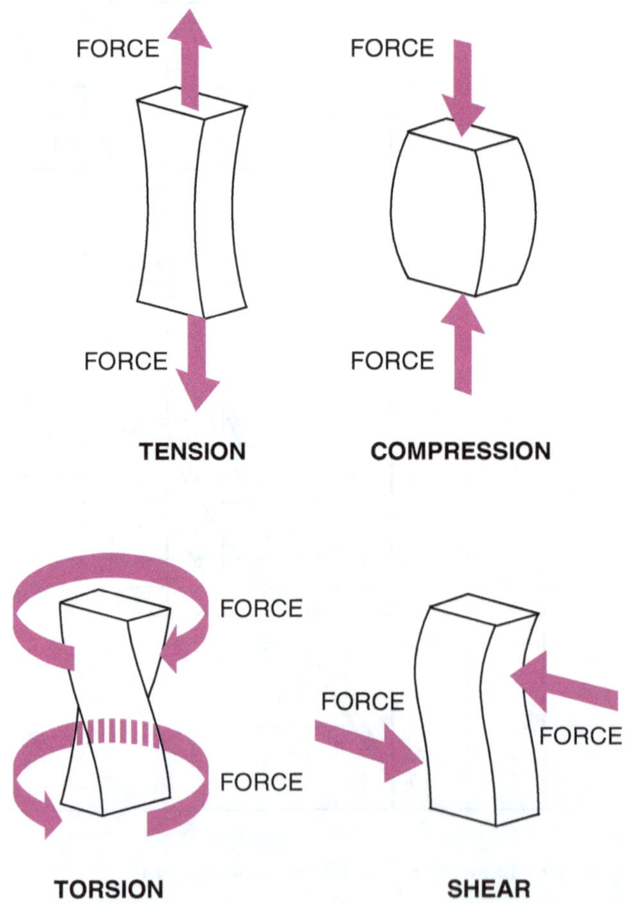

Figure 11 Force directions that result in each stress type.

Hot Tip

Machines Used to Test the Properties of Metal

A wide variety of specialized machines is used to test mechanical properties of metals.

IMPACT TEST MACHINE

ROCKWELL HARDNESS TESTER

PHOTOS COURTESY OF INSTRON®

DESTRUCTIVE TESTING MACHINE FOR TENSILE OR COMPRESSION TESTING

203SA08.EPS

its elastic range; but the amount each material strains for a given stress is not the same. The ratio of stress to strain is called the modulus of elasticity. It remains the same for any material under the same proportional conditions. For example, a material that will stretch 1 inch under a 10-pound tension will stretch 2 inches under a 20-pound tension as long as the elastic limit is not exceeded. Similarly, a material that will stretch ¼ inch under a 10-pound tension will stretch ½ inch under a 20-pound tension as long as the elastic limit is not exceeded. The modulus of elasticity of a metal can be used to predict how much the metal will deform under a specified load, providing the elastic limit is not exceeded.

5.4.0 Tensile Strength

Tensile strength (yield strength) is the maximum tensile stress a material can withstand without exceeding its elastic limit. It is expressed in force per cross-sectional unit area. The maximum load on a tensile material when it fractures is called the ultimate load. Many engineering designs are based on tensile yield strength.

5.5.0 Ductility

Ductility is the plasticity exhibited by a material under tension loading. Ductility is measured by dividing the change in length of a material by the original length at the point of failure. Ductility is the characteristic that allows metals to be drawn into thinner sections without breaking. Copper and aluminum are examples of common ductile metals.

5.6.0 Hardness

The hardness of a material can be defined as its resistance to indentation. It is related to its elastic and plastic properties. The greater the hardness, the greater the resistance to indentation, penetration, and wear. Hardness is measured by pressing a steel ball or diamond point against a surface with a calibrated force. The diameter or depth of the impression is then measured to determine the hardness. Standard hardness testers include the Brinell, Rockwell, and Vickers instruments. *Figure 12* shows a portable hardness tester suitable for testing on the job site.

The Brinell hardness test is performed by pressing a hardened 10-millimeter steel ball into the metal being tested. Force is applied by standard weights: 500 kilograms for aluminum, copper, and other soft materials; and 3,000 kilograms for iron, steel, and other hard metals. The diameter of the indentation is measured under a microscope and then converted to the Brinell hardness number.

The Rockwell hardness tester uses a 1⁄16-inch hardened steel ball with a 100-kilogram weight on soft metals, and a diamond cone and 150-kilogram weight on hard metals. Depth of penetration is read directly from hardness scales on the tester.

The Vickers hardness tester uses a pyramid-shaped diamond penetrator with a variable 1- to 120-kilogram load. The hardness is read from standard tables by comparing the width of the dent with the pressing load.

203F12.EPS

Figure 12 Portable hardness tester.

6.0.0 Structural Steel and Common Milled Shapes

Metals are commercially available in many shapes and sizes. They can be shaped by casting, rolling, drawing, forging, cutting, machining, welding, and spinning. Structural steel and other common milled shapes are formed by rolling hot or cold metal between a succession of rollers. The rollers are specially configured for the shapes they will produce.

Structural steel includes a wide range of steel types, classes, and shapes. To ensure uniform standards, the ASTM specifies the properties of strength, weight, corrosion resistance, and weldability for various steel classifications. *Table 20* lists some typical structural steel types and classifications. Note that in the table the minimum yield stress is expressed as ksi. This is one of the standard notations used for structural steel. The term ksi is defined as kips per square inch, where kips means kilo-pounds (1,000 pounds) of dead load. Hence, a yield stress of 36 ksi means 36,000 pounds per square inch of dead load.

The shape or type of structural steel is identified on drawings by a symbol or abbreviation. Sizes and dimensions are always given in a specified order. The specification format for designating shape and size is shown with the relevant shape in the following sections.

6.1.0 Plate

Plate is rolled metal with a uniform thickness equal to or greater than 3⁄16 inch, and its identifying symbol is PL. Its thickness is given in inches, and its length and width are given in feet.

Example specification:

PL 1⁄4" × 4' × 10'

6.2.0 Sheet Metal

Sheet metal is rolled metal with a uniform thickness less than 3⁄16 inch. Thickness is given in inches or indicated by an American Wire Gauge (AWG) number. Length and width are given in feet. Sheet metal is often rolled into coils.

Example specification:

Sheet No. 12 AWG × 4' × 8'

Table 20 Common Types and Classifications of Structural Steel

Steel Type	ASTM Class	Minimum Yield Stress (ksi)	Form		Remarks
Carbon	A 36	36	Plate Shapes Bars	Sheets, strips Rivets Bolts, nuts	For buildings and general structures; available in high-toughness grades
	A 529	42	Plate Shapes	Bars	For buildings and similar construction
High-strength	A 440	42 to 50	Plate Shapes	Bars	Lightweight and superior corrosion resistance
High-strength, low-alloy	A 441	40 to 50	Plate Shapes	Bars	Primarily for lightweight welded buildings and bridges
	A 572	42 to 65	Several types Plate	Shapes Bars	Lightweight high-toughness for buildings, bridges, and similar structures
Corrosion-resistant, high-strength, low-alloy	A 242	42 to 50	Plate Shapes	Bars	Lightweight and added durability weathering grades available
	A 588	42 to 50	Plate Shapes	Bars	Lightweight, durable in high thickness; weathering grades available
Quenched and tempered alloy	A 514	90 to 100	Several types Plate	Shapes Bars	Strength varies with thickness and type

203T20.EPS

6.3.0 Bars

Bars are rolled to a variety of cross-sectional shapes and sizes (*Figure 13*). Standard bar shapes include the following:

- *Round* – Round bar is specified by BAR, followed by the bar diameter, a circle with a slash through it, and then the length.
- *Square* – Square bar is specified by BAR, followed by the face width, a square with a slash through it, and then the length.
- *Bar (flat)* – Flat bar is specified by BAR, followed by the wide dimension, the narrow dimension, and then the length.
- *Z-bar* – Z-bar is specified by Z, followed by the flange width, the web depth, the flange and web thickness, and then the length. The flange and the web are always the same thickness.
- *Hexagonal* – Hexagonal bar is specified by HEX, followed by the bar thickness (measured across the flats), and then the length.
- *Octagonal* – Octagonal bar is specified by OCT, followed by the bar thickness (measured across the flats), and then the length.

6.4.0 Angles

Angles are L-shaped bars that have either equal-sized or unequal-sized legs that are always at 90 degrees to each other. *Figure 14* shows both types of angle bar stock and the specification formats for each type. The specification format represents the size and thickness of the angle bar stock legs in inches. Thus, L2 × 2 × ⅜ indicates a 2"× 2" angle of ⅜" thickness. For unequal leg stock, the longest leg is listed first, as shown in *Figure 14*.

6.5.0 Channels

Channels are U-shaped forms made of two flanges connected by a common web. The flanges extend from the same side of the web. The flanges can be of uniform thickness or tapered toward the outer

Hot Tip

Miscellaneous Channel Stock

In addition to channels specified in the American Standard Channel format, some channels are specified in the Miscellaneous Channels designation format. Miscellaneous Channels are usually lighter channels than American Standard Channels. The only difference in the specification format is that the format for the Miscellaneous Channels starts with the letters MC instead of the C that is used with the American Standard Channels.

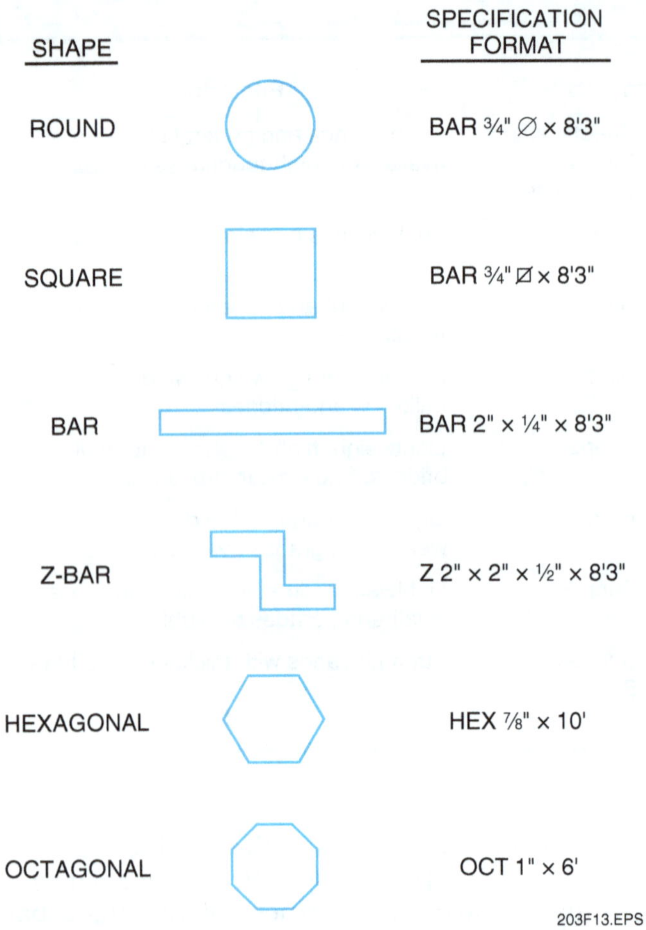

Figure 13 Bar shapes and specification formats.

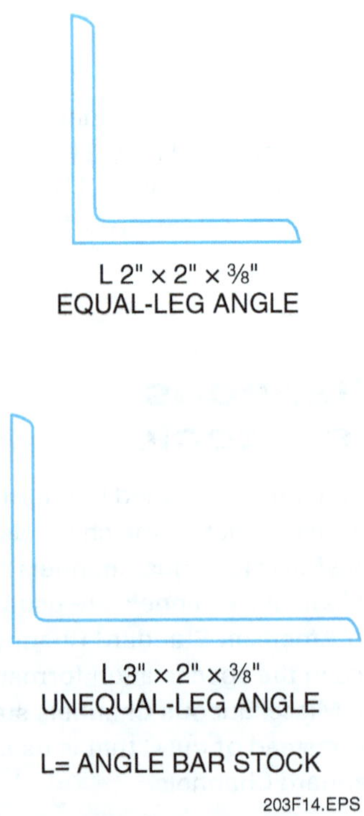

Figure 14 Angle bar stock and specification formats.

edges. There are several channel variations, each having a unique specification format designation. *Figure 15* shows two different channels and their related American Standard Channels specification format designations. The channel C8 × 11.5 is a channel having a nominal depth of 8 inches and a weight of 11.5 pounds per foot of length. Similarly, the channel C8 × 18.75 has a nominal depth of 8 inches and a weight of 18.75 pounds per foot of length. As shown in *Figure 15*, the channel C8 × 11.5 has a flange width of 2¼ inches and a web thickness of ¼ inch; while the C8 × 18.75 channel has a flange width of 2¼ inches and a web thickness of ½ inch. The specific dimensions for the beam depth, flange width, and web thickness for all common channel designations can be found in related tables of dimensions and properties in most structural engineering and architectural standards reference books.

6.6.0 Beams and Shapes from Beams

Beams (*Figure 16*) are made in I-shaped, H-shaped, and T-shaped cross sections, and they are made with flat or tapered flanges. In addition to different flange widths and web depths, the thickness of the flanges and webs varies with beam size. Also, beam weight (pounds per linear foot) for a beam of a given dimension can be increased by adding thickness to the web and flanges with very little change in beam width or depth. The specification

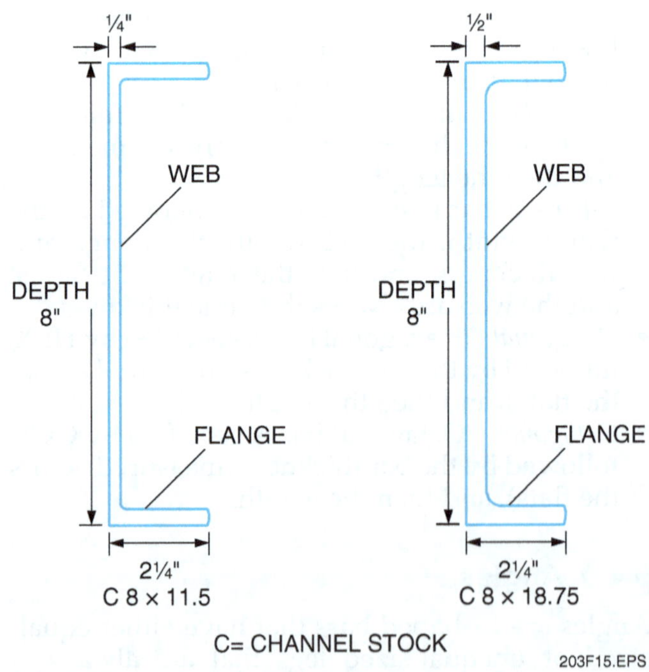

Figure 15 Channels and specification formats.

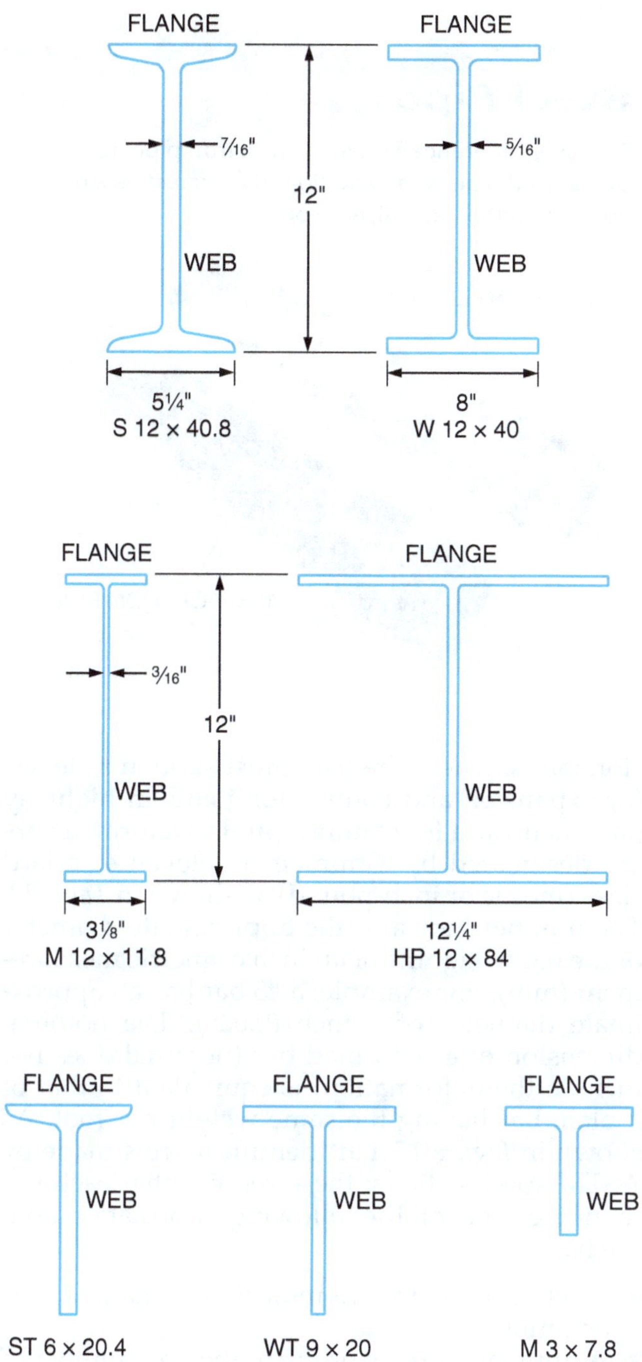

203F16.EPS

Figure 16 Structural beam and T-beam shapes.

designator describes the beam type and its nominal depth in inches and the weight of the beam per foot of length. An example specification designator for a beam is S24 × 120, where S24 represents the beam type (S) and the nominal depth in inches (24). The 120 is the weight in pounds per foot of length. The specific dimensions for the depth, web thickness, and flange thickness for the various designations of beams are given in related tables

Hot Tip

I-Beams

The older American Standard I-beams have been replaced by the American Institute of Steel Construction (AISC) wide flange (W) beams. These wide flange shapes are now used for most beams and columns. The wider flange provides a stronger beam than the older I-beam with the same web width.

of dimensions and properties in most structural engineering and architectural standards books.

Beams are identified with a standard set of symbols. The I-beam and H-beam letter symbols include the following:

- *S (American Standard)* – I-shaped beams with tapered flanges
- *W (Wide flange)* – Wider flanges than S beams with thinner webs and nontapered flanges
- *M* – Similar to W beams but with short flanges
- *HP* – H-shaped beams with nontapered flanges

T-beams include structural tees and tee shapes. Structural tees are made by cutting S, W, and M beams down the center, usually by shearing. The conventional symbol for a structural tee is the original beam letter followed by a T. For example, a tee cut from an S8 × 18.4 beam would be identified as ST4 × 9.2.

T-shapes are rolled into their final tee shape. The symbol for tee shapes is a T, without any other letter. The specification gives the nominal depth, flange width, thickness, and length. An example specification is T3 × 7.8. T-shapes are short pieces, commonly used as connections or supports.

6.7.0 Pipe

Pipe size is specified as a nominal ID (approximate inside diameter) up through 12 inches. Pipe 14 inches and larger is specified by the actual OD (outside diameter). Pipe weight (determined by the wall thickness) is specified by a schedule number or strength. Schedule numbers are 5, 10, 20, 40, 80, and 160. Strengths are specified as standard (STD), extra strong (XS), and extra-extra strong (XXS). Standard falls between Schedules 20 and 40. Extra strong falls between Schedules 40 and 80. Extra-extra strong is Schedule 160 or higher. Because the OD of a given pipe size is always the same, the ID becomes smaller as the pipe wall thickness increases with higher schedule pipe.

Hot Tip

Why Doesn't the Pipe Size Match Pipe ID or OD?

When pipe was first made, it was made from lead, which required a thick wall to hold the specific pressure for a given ID. When the switch was made from lead pipe to steel pipe, a decision had to be made whether to maintain the OD so the steel pipe would fit existing fittings, or to change the OD and keep the ID size.

The decision was made to maintain the OD to provide direct replacement with the existing fittings. At that point, neither nominal ID nor OD corresponded to size.

Table 21 lists standard pipe sizes and schedules with nominal wall thicknesses.

Hot Tip

Steel Pipe

The only difference between black iron pipe and galvanized pipe is the addition of a protective zinc coating on the galvanized pipe.

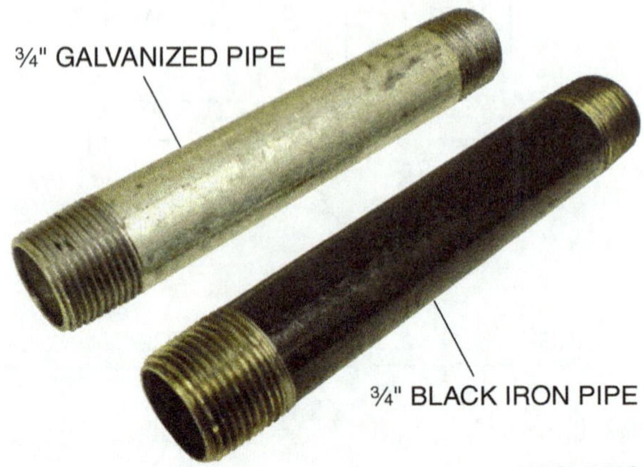

203SA09.EPS

6.8.0 Reinforcing Bars

Reinforcing bars (rebars), sometimes called rerods, are used for concrete reinforcement. They are available in several grades. These grades vary in yield strength, ultimate strength, percentage of elongation, bend-test requirements, and chemical composition. Reinforcing bars can be coated with different compounds, such as epoxy, for use in concrete where corrosion can be a problem. The ASTM International has established standard specifications for reinforcing bars. These grades appear on bar-bundle tags, in color coding, in rolled-on markings on the bars, and/or on bills of materials. The ASTM specifications are as follows:

- *A615, Standard Specification for Deformed and Plain Billet-Steel Bars for Concrete Reinforcement*
- *A706, Standard Specification for Low-Alloy Steel Deformed and Plain Bars for Concrete Reinforcement*
- *A996, Standard Specification for Rail-Steel and Axle-Steel Deformed Bars for Concrete Reinforcement*

The standard configuration for reinforcing bars is the deformed bar. Different patterns may be impressed upon the bars depending on the manufacturer, but all are rolled to conform to ASTM specifications. The purpose of the deformation is to improve the bond between the concrete and the bar and to prevent the bar from moving within the concrete. Plain bars are smooth and round without deformations on them. Plain bars are used for special purposes: dowels at expansion joints, where the bars must slide in a sleeve; for expansion and contraction joints in highway pavement; and for column spirals. Deformed bars are designated by a number in eleven standard sizes (metric or inch-pound) as shown in *Table 22*. The number indicates the approximate diameter of the bar in eighths of an inch or metric measurement (mm). For example, a #5 bar has an approximate diameter of ⅝ inch (0.625). The nominal dimension of a deformed bar (nominal does not include the deformation) is equivalent to that of a plain bar having the same weight per foot. As shown in *Figure 17*, bar identification is made by ASTM specifications; these require that each bar manufacturer roll the following information onto the bar:

- A letter or symbol to indicate the manufacturer's mill
- A number corresponding to the size number of the bar (*Table 22*)
- A symbol or marking to indicate the type of steel (*Table 23*)
- A marking to designate the grade (*Table 24*)

The grade represents the minimum yield or tension strength, measured in kips per square inch (ksi) or megapascals (MPa), that the type of steel will withstand before it permanently stretches (elongates). Today, Grade 420 is the most commonly used rebar. Rebars are normally supplied from the mill bundled in 60-foot lengths (*Figure 18*) that are cut in the field to the required length.

Table 21 Standard Pipe Sizes and Schedules with Nominal Wall Thicknesses

		Nominal Wall Thickness (in)												
Nominal Pipe Size	**Outside Diam.**	**Sched. 10**	**Sched. 20**	**Sched. 30**	**STD**	**Sched. 40**	**Sched. 60**	**XS**	**Sched. 80**	**Sched. 100**	**Sched. 120**	**Sched. 140**	**Sched. 160**	**XXS**
⅛	0.405	–	–	–	0.068	0.068	–	0.095	0.095	–	–	–	–	–
¼	0.540	–	–	–	0.088	0.088	–	0.119	0.119	–	–	–	–	–
⅜	0.675	–	–	–	0.091	0.091	–	0.126	0.126	–	–	–	–	–
½	0.840	–	–	–	0.109	0.109	–	0.147	0.147	–	–	–	0.188	0.294
¾	1.050	–	–	–	0.113	0.113	–	0.154	0.154	–	–	–	0.219	0.308
1	1.315	–	–	–	0.133	0.133	–	0.179	0.179	–	–	–	0.250	0.358
1¼	1.660	–	–	–	0.140	0.140	–	0.191	0.191	–	–	–	0.250	0.382
1½	1.900	–	–	–	0.145	0.145	–	0.200	0.200	–	–	–	0.281	0.400
2	2.375	–	–	–	0.154	0.154	–	0.218	0.218	–	–	–	0.344	0.436
2½	2.875	–	–	–	0.203	0.203	–	0.276	0.276	–	–	–	0.375	0.552
3	3.500	–	–	–	0.216	0.216	–	0.300	0.300	–	–	–	0.438	0.600
3½	4.000	–	–	–	0.226	0.226	–	0.318	0.318	–	–	–	–	–
4	4.500	–	–	–	0.237	0.237	–	0.337	0.337	–	0.438	–	0.531	0.674
5	5.563	–	–	–	0.258	0.258	–	0.375	0.375	–	0.500	–	0.625	0.750
6	6.625	–	–	–	0.280	0.280	–	0.432	0.432	–	0.562	–	0.719	0.864
8	8.625	–	0.250	0.277	0.322	0.322	0.406	0.500	0.500	0.594	0.719	0.812	0.906	0.875
10	10.750	–	0.250	0.307	0.365	0.365	0.500	0.500	0.594	0.719	0.844	1.000	1.125	1.000
12	12.750	–	0.250	0.330	0.375	0.406	0.562	0.500	0.688	0.844	1.000	1.125	1.312	1.000
14 OD	14.000	0.250	0.312	0.375	0.375	0.438	0.594	0.500	0.750	0.938	1.094	1.250	1.406	–
16 OD	16.000	0.250	0.312	0.375	0.375	0.500	0.656	0.500	0.844	1.031	1.219	1.438	1.594	–
18 OD	18.000	0.250	0.312	0.438	0.375	0.562	0.750	0.500	0.938	1.156	1.375	1.562	1.781	–
20 OD	20.000	0.250	0.375	0.500	0.375	0.594	0.812	0.500	1.031	1.281	1.500	1.750	1.969	–
22 OD	22.000	0.250	0.375	0.500	0.375	–	0.875	0.500	1.125	1.375	1.625	1.875	2.125	–
24 OD	24.000	0.250	0.375	0.562	0.375	0.688	0.969	0.500	1.218	1.531	1.812	2.062	2.344	–
26 OD	26.000	0.312	0.500	–	0.375	–	–	0.500	–	–	–	–	–	–
28 OD	28.000	0.312	0.500	0.625	0.375	–	–	0.500	–	–	–	–	–	–
30 OD	30.000	0.312	0.500	0.625	0.375	–	–	0.500	–	–	–	–	–	–
32 OD	32.000	0.312	0.500	0.625	0.375	0.688	–	0.500	–	–	–	–	–	–
34 OD	34.000	0.312	0.500	0.625	0.375	0.688	–	0.500	–	–	–	–	–	–
36 OD	36.000	0.312	0.500	0.625	0.375	0.750	–	0.500	–	–	–	–	–	–
42 OD	42.000	–	–	–	0.375	–	–	0.500	–	–	–	–	–	–

203T21.EPS

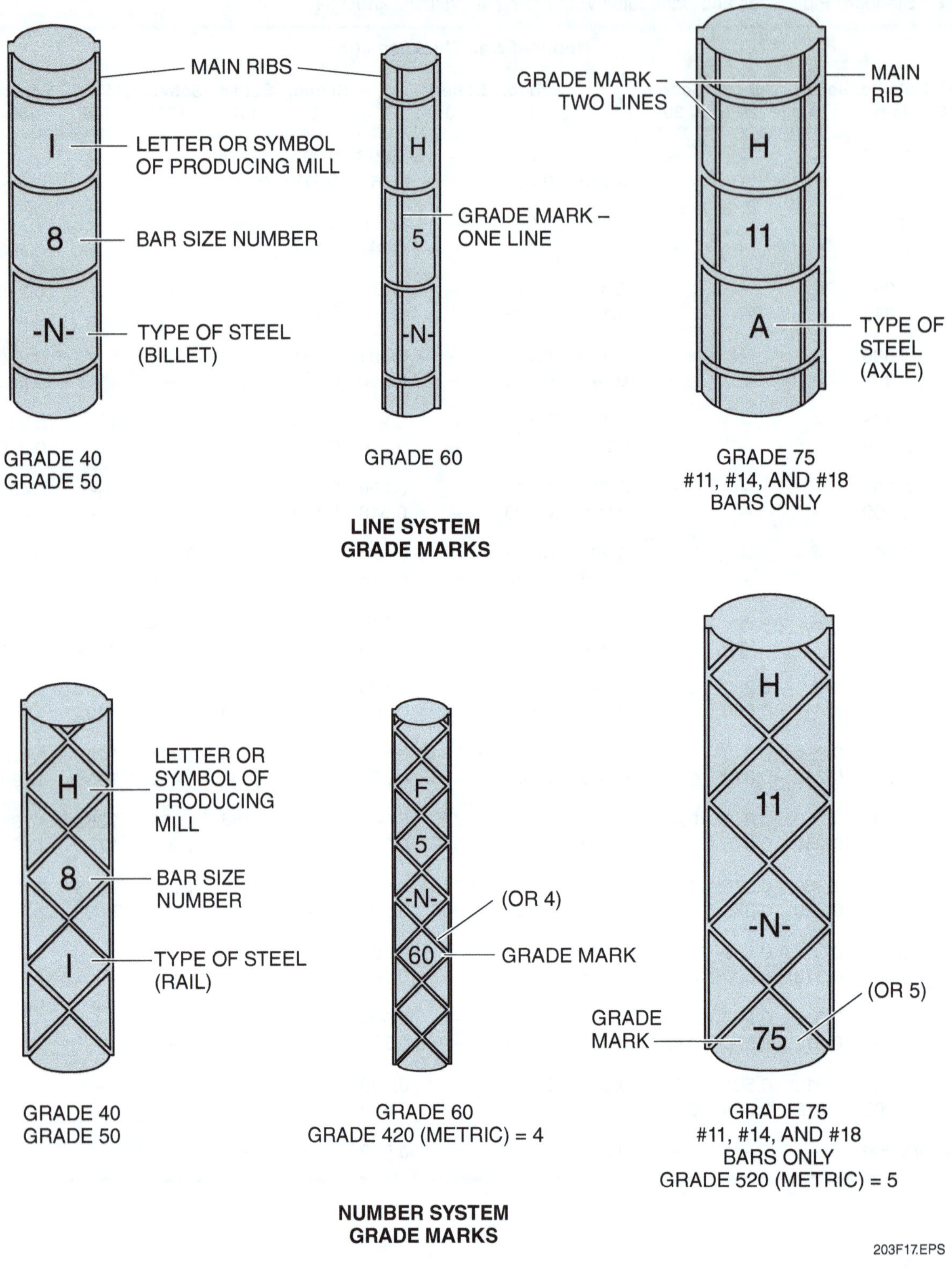

Figure 17 Reinforcement bar identification.

Table 22 ASTM Standard Metric and Inch-Pound Reinforcing Bars

Bar Size		Nominal Characteristics: Diameter		Cross-Sectional Area		Weight	
Metric	[in-lb]	mm	[in]	mm	[in]	kg/m	[lbs/ft]
#10	[#3]	9.5	[0.375]	71	[0.11]	0.560	[0.376]
#13	[#4]	12.7	[0.500]	129	[0.20]	0.944	[0.668]
#16	[#5]	15.9	[0.625]	199	[0.31]	1.552	[1.043]
#19	[#6]	19.1	[0.750]	284	[0.44]	2.235	[1.502]
#22	[#7]	22.2	[0.875]	387	[0.60]	3.042	[2.044]
#25	[#8]	25.4	[1.000]	510	[0.79]	3.973	[2.670]
#29	[#9]	28.7	[1.128]	645	[1.00]	5.060	[3.400]
#32	[#10]	32.3	[1.270]	819	[1.27]	6.404	[4.303]
#36	[#11]	35.8	[1.410]	1006	[1.56]	7.907	[5.313]
#43	[#14]	43.0	[1.693]	1452	[2.25]	11.38	[7.65]
#57	[#18]	57.3	[2.257]	2581	[4.00]	20.24	[13.60]

203T22.EPS

Table 23 Reinforcement Bar Steel Types

Symbol/Marking	Type of Steel
A	Axle (*ASTM A996*)
S or N	Billet (*ASTM A615*)
I or IR	Rail (*ASTM A996*)
W	Low-alloy (*ASTM A706*)

203T23.EPS

6.9.0 Tubing

Tubing is manufactured in square, rectangular, and round cross sections. Round tubing can be distinguished from pipe by its dimensions. Standard round tubing is measured by a nominal OD.

The cross sections of square and rectangular tubing have even outside dimensions and slightly rounded corners. *Figure 19* shows three standard tubing shapes and their specification formats.

203F18.EPS

Figure 18 Reinforcing bars.

Table 24 Reinforcement Bar Grades

Grade	Identification	Minimum Yield Strength
40 and 50	None	40,000 to 50,000 psi (40 to 50 ksi)
60	One line or the number 60	60,000 psi (60 ksi)
75	Two lines or the number 75	75,000 psi (75 ksi)
420	The number 4	420 MPa (60,000 psi or 60 ksi)
520	The number 5	520 MPa (75,000 psi or 75 ksi)

203T24.EPS

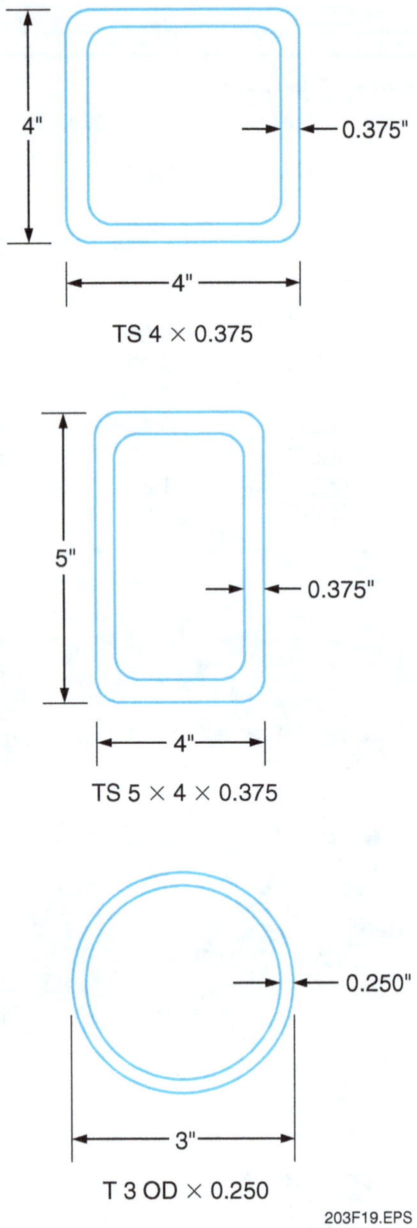

Figure 19 Standard tubing shapes and specification formats.

6.10.0 Seamed and Seamless Tubing and Pipe

Both standard and nonstandard shapes can also be produced by other processes. For example, tubing can be produced by extrusion or by rolling and welding. Extrusions can be produced in an almost limitless variety of shapes. When tubing is produced by rolling and welding, the most common welding process used is electronic resistance welding (ERW).

Hot Tip

ACR Copper Pipe

Copper pipe (tubing) used in air conditioning and refrigeration systems and equipment is different from the copper pipe used for general plumbing applications. Air conditioning and refrigeration (ACR) copper pipe and fittings are manufactured specifically for use in refrigeration systems. ACR pipe is thoroughly cleaned, dried, capped, and sometimes charged with nitrogen to help prevent contamination of air conditioning and refrigeration systems. Another important difference is that the nominal size of ACR pipe is expressed by its OD, while the copper pipe used in general plumbing applications is expressed in terms of its ID.

6.11.0 Forged Shapes

Forging is used to produce specialized shapes where high strength is required. An example is forged high-pressure pipe fittings such as flanges and elbows.

6.12.0 Cast Shapes

Casting is another method of producing shapes. Cast iron and steel are used for machine parts, automobile components, and large castings.

6.13.0 Powdered Metals

Commercially prepared powdered metals are used for welding fillers, surface coatings, and mold castings. When cast to make parts, the molded powder is usually sintered (partially welded without melting) to fuse it into a solid metal. This is a common technique for manufacturing complex-shaped parts or porous parts. It is also used with high-temperature metals that cannot be liquid cast.

7.0.0 Metallurgical Considerations for Welding

Many metallurgical variables must be considered when determining the effects of welding on both the base metal and the weld. Base metal composition and thickness, filler metal, and joint configuration all affect the weld quality. There are also several processes and design considerations used to change the metal characteristics in order to produce a more desirable weld. The following sections explain the metallurgical factors that must be considered to make acceptable welds:

- Base metal preparations
- Joint design considerations
- Filler metal selection
- Filler metal and electrode selection considerations
- Preheating and interpass temperature control
- Postweld heat treatment

7.1.0 Base Metal Preparations

Because of the defects caused by surface contaminants and the possibility of toxic fumes, all welding codes require surface cleaning. In general, the codes state that the surface to be thermally cut or welded must be clean and free from paint, oil, rust, scale, and other material that would be detrimental to either the weld or the base metal when heat is applied. Cleaning is performed by mechanical and/or chemical means.

The most common defect caused by surface contamination is porosity, which occurs when gas pockets or voids appear in the weld metal. When the porosity has a length greater than its width and is approximately perpendicular to the weld face, it is called piping porosity. Piping porosity is formed as the gas pocket floats toward the surface of the weld, leaving a void. The gas pocket is trapped as the weld metal solidifies, but as the next layer of weld is deposited, it continues to float up. Porosity and piping porosity may not be visible on the surface of the weld.

7.1.1 Chemical Cleaning

Chemicals are used to clean oil and grease from metal surfaces that require high-quality welds. These chemicals are very strong, which enables them to do their job quickly. Because they are so strong, they can be very hazardous if mishandled.

There are several different types of chemicals used to clean metal. The type of chemical used depends on the base metal to be cleaned. Using the wrong chemical can cause a reaction with the base metal that results in burning, pitting, or discoloration of the surface. The cleaning chemical should be selected specifically for the type of base metal and the application.

WARNING!

Never use a chemical unless you have been specifically and recently trained in its use. The mishandling of cleaning chemicals can cause severe burns to the skin, lungs, and eyes. Some chemicals can also cause death or blindness. Always use approved solvents. Procedures and methods for using, storing, and disposing of most chemicals are given in the MSDS for the specific chemical being used. These are available at your shop or job site.

7.1.2 Mechanical Cleaning

Mechanical cleaning is the most common method of removing surface contamination. Tools used for mechanical cleaning include hand tools, power tools, and special sandblasting equipment.

WARNING!

When performing mechanical cleaning, be sure to wear safety glasses and a face shield for protection from the flying particles produced during the cleaning operation. In addition, special clothing is required for sandblasting.

7.2.0 Joint Design Considerations

Welded joints are selected primarily for the safety and strength required for the service conditions. When selecting the joint, groove, and weld to use, there are many considerations to take into account. The following sections explain those considerations.

7.2.1 Codes and Welding Procedure Specifications

A welding code is a detailed listing of the rules and principles that apply to specific welded products. Codes ensure that safe and reliable welded products will be produced and that the weldments will be reasonably safe. Clients specify which codes to follow when placing orders or letting contracts. These orders or contracts impose severe penalties for not conforming to the code(s) specified. In addition, when codes are specified, the use of these codes is mandated with the force of law by one or more government jurisdictions. Always check the contract, order, or project specification for the code(s) specified.

Codes require that a WPS be written for all critical welds. A WPS is a written set of instructions for producing sound welds. It includes the type of joint to use as well as the groove preparation, if required. Each WPS is written and tested in accordance with a particular welding code or specification. All welding requires that acceptable industry standards are followed, but not all welds require a WPS. If a weld does require a WPS, it must be followed. The requirement to use a WPS is often listed on job blueprints as a note or in the tail of the welding symbol. If you are unsure whether the welding being performed requires a WPS, do not proceed until you check with your supervisor.

7.2.2 Joint Preparation

One of the easiest joint preparations is the square groove joint because it only requires butting the edges of the plate together. Butt joints are welded with a square groove weld that can have a partial or complete joint penetration weld. A partial joint penetration weld has much less strength than a complete penetration weld. *AWS D1.1* imposes restrictions for complete penetration welds. With SMAW, the maximum base metal thickness is ¼ inch and welding from both sides is required. In addition, a root opening of half the thickness of the base metal is required, and you must gouge the root of the first weld before making the second weld. For GMAW and FCAW, the maximum base metal thickness is ⅜ inch with the same requirements for welding from both sides and for back-gouging.

Fillet welds also require very little joint preparation. When using a fillet weld on outside corner welds, two types of fit-up can be used: corner to corner or half lap. The corner-to-corner joint is difficult to assemble because neither plate can support the other, and you must be careful not to burn through the corner when welding. The half-lap joint is easier to assemble, requires less weld metal, and has less danger of corner burn-through. The half-lap requires a second weld on the inside of the corner. If a half-lap fit-up is used, you must make allowances in the plate dimensions for the lap. Several methods of joint preparation are explained in the following items:

- *Single- and double-groove welds (Figure 20)* – When possible, you should use a double-groove joint in place of a single groove. The double groove requires half the weld metal of the single groove. Also, welding from both sides reduces distortion because the forces of distortion work against each other.

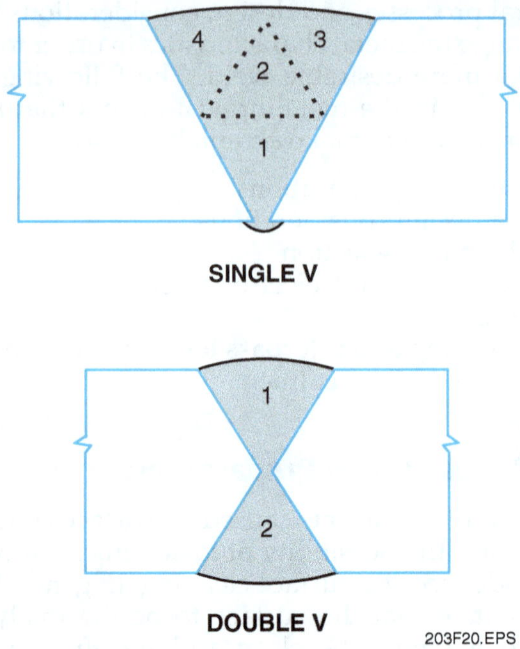

Figure 20 Single- and double-groove welds.

- *U- and J-groove preparation* – On thick joints, U- and J-grooves often require less weld metal than V or bevel joints. The major disadvantage of U- and J-grooves is that they require more preparation time. For this reason, V and bevel joints are used much more frequently than J- or U-grooves.
- *Groove angles and root openings (Figure 21)* – The purpose of the groove angle is to allow electrode access to the root of the weld. If the groove angle is larger than necessary, it requires additional weld metal to fill. This increases the time and cost to complete the weld and increases the chance of distortion. If the groove angle is too small, it can result in weld discontinuities. The root preparation is sized to control melt-through. Increasing or decreasing the root preparation (root opening and root face) results in excess melt-through or insufficient root penetration. As a general rule, as the groove angle decreases, the root opening increases to compensate. Root faces are used with open-root joints, but not when metal backing strips are used (except for aluminum).
- *Groove welds with root openings* – For groove welds on plate, the groove angle is 60 degrees, and the maximum size of the root opening and root face is ⅛ inch. For open root welds on pipe, the groove angle is 60 or 75 degrees, depending on the code or specifications used. The maximum size of the root opening and root faces is referred to in the WPS. *Figure 22* shows open-root joint preparation, including the groove angles.

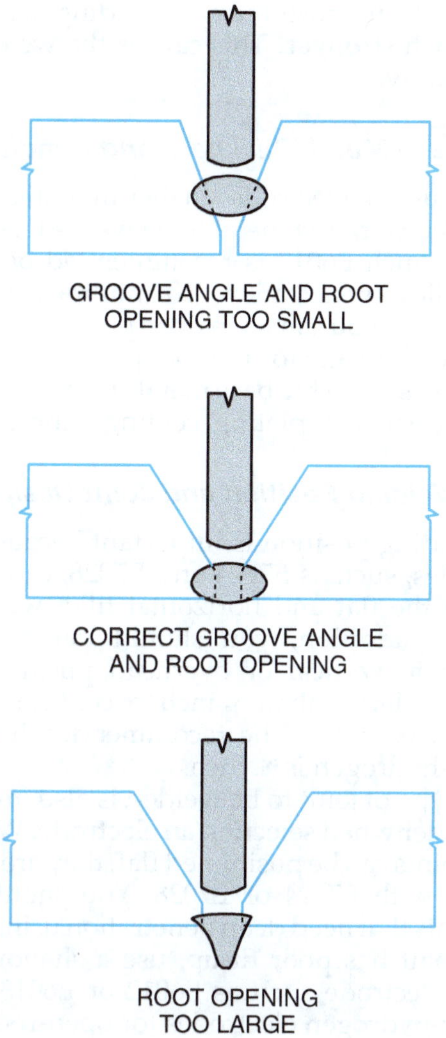

Figure 21 Groove angles and root openings.

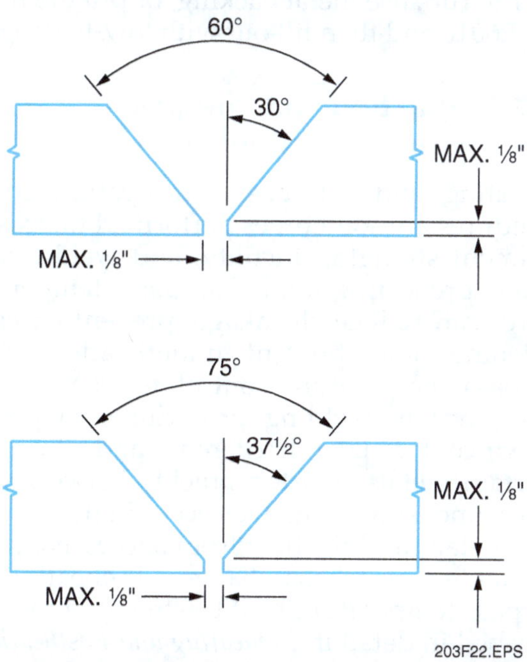

Figure 22 Open-root joint preparation.

Hot Tip

Joint and Surface Preparation

The methods to use for joint and surface preparation and the required degree of surface cleaning should be designated in the welding procedure specification. The methods may include oxyfuel gas, air-carbon arc, or plasma arc cutting, with or without surface cleaning. Surface preparation may involve blasting, machining, or grinding, followed by vapor, ultrasonic, dip, or lint-free cloth cleaning. All methods and practices should be compatible with the application and the metal.

7.3.0 Filler Metal Selection

Because so many different electrodes and filler metals are available, you must be able to distinguish between them and select the correct ones for the job. Electrodes and filler metals are often classified into one of four groups, based on their general characteristics. These groups are as follows:

- Fast-freeze
- Fast-fill
- Fill-freeze
- Low-hydrogen

7.3.1 Fast-Freeze

Fast-freeze electrodes are all-position electrodes that provide deep penetration. However, the arc has a lot of fine weld spatter. The weld bead is flat with distinct ripples and a light slag coating that can be difficult to remove. The arc is easy to control, making fast-freeze electrodes good for vertical and overhead welding.

7.3.2 Fast-Fill

Fast-fill electrodes have iron powder in the flux, which gives them high deposition rates, but they can only be used for flat welds and horizontal fillet welds. These electrodes have shallow penetration with excellent weld appearance and almost no spatter. The heavy slag covering is easy to remove.

7.3.3 Fill-Freeze

Fill-freeze electrodes, also called fast-follow electrodes, are all-position electrodes, but they are most commonly used for downhill or flat welding. They have medium deposition rates and penetration and are excellent for sheet metal. The weld bead ranges from smooth and ripple-free to even with distinct ripples. The medium slag covering is easy to remove.

7.3.4 Low-Hydrogen

Low-hydrogen electrodes are designed for welding high-sulphur, phosphorus, and medium- to high-carbon steels, which have a tendency to develop porosity and underbead cracking. These problems are caused by hydrogen, which is absorbed during welding. Low-hydrogen electrodes are available in fast-fill and fill-freeze types. Low-hydrogen electrodes produce dense, X-ray-quality welds with excellent notch toughness and ductility.

7.4.0 Filler Metal and Electrode Selection Considerations

If there is a WPS for the weld to be made, the filler metal will be specified. It will be given as an AWS classification and/or manufacturer's standard and will sometimes include a manufacturer's name.

> **NOTE**
> The filler metal specified in the WPS must be used.

When selecting filler metals and electrodes not specified in a WPS, you must consider the factors explained in the following sections.

7.4.1 Base Metal Type

The filler metal should be compatible with the base metal's chemical composition and mechanical properties. You can obtain the chemical composition and mechanical properties of the base metal from the base metal mill specifications; then compare the mill specifications to various tables that list base metal and filler metal acceptability.

A weld made on mild steel with a high-tensile electrode, such as E12018, is weaker than one made with E7018. The high-tensile weld is so strong that when stress is placed on the weld, it does not give. It transfers all stress to the heat-affected zone (HAZ) along the weld. This can cause cracking. A weld made with E7018 will give along with the base metal and distribute the stress across the weld, which makes a stronger joint. The opposite occurs when mild-steel electrodes are used to weld high-strength steel. All the stress occurs in the weld because the surrounding base metal is so much stronger. This causes the weld to fail prematurely.

7.4.2 Base Metal Thickness and Condition

Always use an electrode smaller in diameter than the thickness of the metal to be welded. For sheet metal, 3⁄32-inch E6013 is recommended because it has shallow penetration. To minimize the joint preparation, you can use a deep penetration E6010 or E7010 electrode for the root pass. Before welding, always clean the base metal surface to remove surface corrosion, plating, coatings, and paint.

7.4.3 Welding Position and Joint Design

The welding position is important because some electrodes, such as E7024 and E7028, can only be used in the flat and horizontal fillet weld positions. In addition, you should never attempt to weld in vertical or overhead positions with electrodes larger than 3⁄16 inch, except for low-hydrogen electrodes. The recommended limitation for low-hydrogen is 5⁄32 inch.

The type of joint to be welded is also important to consider when selecting an electrode. When lap and T joints can be positioned flat, they are usually welded with E7024 or E7028. You should weld butt joints that need deep penetration using E6010. If the joint has poor fit-up, use a shallow-penetrating electrode such as E6012 or E6013. If you use low-hydrogen electrodes for open-root joints, back-gouging may be required. To eliminate the need for back-gouging and to make the root pass easier to run, use metal backing, or put the root in with E6010 and then fill out with low-hydrogen.

7.5.0 Preheating and Interpass Temperature Control

Preheating and interpass temperature control procedures are sometimes performed to preserve weldment strength, ductility, and weld quality. Proper preheating and interpass temperature control can reduce shrinkage, prevent excessive hardening, and prevent underbead cracking. Interpass temperature control is also required during certain welding procedures to prevent or reduce localized hardening and shrinkage. If certain metals cool too quickly between weld passes, undesirable changes occur in the base and weld metals, and the heat from successive passes will not reverse these changes. Preheating and interpass temperature heat treatment of metals is described in detail in *Preheating and Postheating of Metals*.

7.6.0 Postweld Heat Treatment

Postweld heat treatment (PWHT) is the heating of a weldment or assembly after welding has been completed. It can be used on most types of base metals. Postweld heat treatment may be necessary to reduce residual stresses and to improve corrosion resistance, dimensional stability, fatigue resistance, and impact and low-temperature strength. Postweld heat treatment equipment and techniques include furnaces, banding, and localized heat treatment. Postweld heat treatments include stress relieving, annealing, normalizing, and tempering. Postweld heat treatment of metals is described in detail in *Preheating and Postheating of Metals*.

SUMMARY

Making proper welds depends on your ability to identify base metals and understand the systems used to classify them. Knowing how to identify metals in the field is an important part of making good welds. You must also understand the characteristics of metals and their physical and mechanical properties. Structural metals come in many different forms and shapes, and you must be familiar with the requirements of each.

Metallurgical considerations for welding include cleaning, joint design, codes and specifications, filler metals, preweld and postweld heat treatments, and tempering and finishing processes. An understanding of the characteristics and properties of metals is essential as you advance in your welding career.

Review Questions

1. The largest group of the ferrous-based metals includes the _____.
   a. cast irons
   b. alloy steels
   c. heavy metals
   d. carbon steels

2. Quench-and-tempered steels and chromium-molybdenum steels are examples of _____ steels.
   a. common grade stainless
   b. specialty grade stainless
   c. weathering
   d. low-alloy

3. A lower coefficient of thermal conductivity, higher coefficient of thermal expansion, and higher electrical resistance are characteristics of _____ steel.
   a. nickel
   b. carbon
   c. stainless
   d. chromium

4. The principal alloying element in brasses is _____.
   a. zinc
   b. silicon
   c. beryllium
   d. Muntz metal

5. The three types of cast iron are malleable, white, and _____.
   a. ductile
   b. green
   c. blue
   d. gray

6. Malleable cast iron is white cast iron that has been _____.
   a. chilled rapidly
   b. chilled slowly
   c. annealed
   d. forged

7. A metal's ability to conduct heat is known as its _____.
   a. density
   b. melting point
   c. thermal conductivity
   d. electrical conductivity

8. Corrosion is an important consideration because it can severely reduce a metal's _____.
   a. ductility
   b. tensile strength
   c. thermal expansion
   d. thermal conductivity

9. The ability of a material to be strained (deformed) without permanent deformation is known as _____.
   a. elasticity
   b. modulus
   c. ductility
   d. tensile strength

10. The resistance to indentation of a material is called _____.
    a. ductility
    b. hardness
    c. tensile strength
    d. modulus of elasticity

11. A channel has a designation of C12 × 30. This means it is a(n) _____.
    a. American Standard Channel with a depth of 12 inches
    b. Miscellaneous Channel with a depth of 30 inches
    c. Miscellaneous Channel with a weight of 12 pounds per foot
    d. American Standard Channel with a weight of 12 pounds per foot

12. The most common defect caused by surface contamination of a metal is _____.
    a. rust
    b. porosity
    c. corrosion
    d. toxic fumes

13. The type of chemical used to clean metal depends on the _____.

  a. base metal
  b. filler metal
  c. thickness of the metal
  d. amount of cleaning required

14. Fast-fill electrodes have shallow penetration with excellent weld appearance and _____.

  a. almost no spatter
  b. a light slag coating
  c. a medium slag covering
  d. excellent notch toughness and ductility

15. When compared to the thickness of the metal to be welded, the electrode should always be _____.

  a. less than 3/32 inch in diameter
  b. smaller in diameter
  c. larger in diameter
  d. the same size

Trade Terms Introduced in This Module

Alloy: A metal that has had other elements added to it that substantially change its mechanical properties.

American Iron and Steel Institute (AISI): An industry organization responsible for preparing standards for steel and steel alloys that are based upon the Society of Automotive Engineers (SAE) code system.

American Society for Testing Materials (ASTM): An organization that developed a code system for identifying and labeling steels.

Anneal: To subject an alloy, metal, or glass to a process of heating and slow cooling to make it tougher and less brittle.

Austenitizing: Heating a steel up to or above the transformation temperature range to achieve partial or complete transformation to austenite grain structure.

Casting: The metal object produced by pouring molten metal into a mold.

Coefficient: A numerical measure of a physical or chemical property that is constant for a system under specified conditions, such as the coefficient of friction.

Ductile: The term used to describe a material that can be bent or shaped without breaking.

Ductility: The characteristic of metal that allows it to be stretched, drawn, or hammered without breaking.

Ferrous: Relating to iron or an alloy that is mostly iron; a metal containing iron.

Flange: The outer members of a structural beam, joist, or rail.

Gas metal arc welding (GMAW): In gas metal arc welding, sometimes called MIG, a consumable wire fed through the welding gun is used to make the weld; commonly used on steel and other metals.

Gas tungsten arc welding (GTAW): In gas tungsten arc welding, sometimes called TIG, a nonconsumable tungsten electrode is used to produce the weld; commonly used on stainless steel and light metals, such as aluminum and magnesium.

Hardenability: A characteristic of a metal that enables it to become hard, usually through heat treatment.

Heat-affected zone (HAZ): The part of the base metal that has not been melted but has been altered structurally by the weld heat.

Hot shortness: The condition of metal when it proves to be very brittle and unbendable at red heat but can be bent without showing signs of brittleness when cold or at white heat. This condition is often a result of high sulfur and phosphorus content.

Interpass temperature: In a multiple-pass weld, the temperature of the base metal at the time the next weld pass is started; it is usually the same temperature that is used for preheating.

Malleable: Able to be hammered or pressed into another shape without breaking.

Mechanical properties: The characteristics or traits that indicate how flexible and strong a metal will be under stress.

Metallic: Having characteristics of metal, such as ductility, malleability, luster, and heat and electrical conductivity.

Nonferrous metal: A metal, such as aluminum, copper, or brass, that lacks sufficient quantities of iron to be affected by its properties.

Notch toughness: The impact load required to break a test piece of weld metal measured by a notch toughness test (Charpy V-notch test). This test may be performed on metal below room temperature, at which point it may be more brittle.

Oxide: The scale that forms on the surfaces of metals when they are exposed to air.

Postweld heat treatment (PWHT): The process of heat treating the base metal and weld after the weld has been made.

Quench: To rapidly cool a hot metal using air, water, or oil.

Residual stress: The strain that occurs in a welded joint after the welding has been completed, as a result of mechanical action, thermal action, or both.

Society of Automotive Engineers (SAE): An automotive society responsible for creating an early system for classifying carbon steels.

Shielded metal arc welding (SMAW): The most common type of welding used on iron and steel, but can also be used on stainless steel; uses a flux-coated consumable electrode.

Tempering: The process of reheating quench hardened or normalized steel to a temperature below the transformation range and then cooling it at any rate desired.

Tensile strength: The maximum stress or force that a material can withstand without breaking.

Underbead cracking: Subsurface cracking in the base metal under or near the weld.

Unified numbering system (UNS): An industry-accepted numbering system for identifying ferrous metals and alloys.

Web: The plate joining the flanges of a girder, joist, or rail.

Wrought: Formed or shaped by hammering or rolling.

Additional Resources

This module is intended to present thorough resources for task training. The following reference works are suggested for further study. These are optional materials for continued education rather than for task training.

AWS B1.10: 1999 Guide for the Nondestructive Examination of Welds. Miami, FL: American Welding Society.

AWS B1.11: 2000 Guide for the Visual Examination of Welds. Miami, FL: American Welding Society.

AWS D3.5-93R: Guide for Steel Hull Welding, 1993. Miami, FL: American Welding Society.

AWS D3.6M: 1999 Specification for Underwater Welding. Miami, FL: American Welding Society.

AWS D3.7: 2004 Guide for Aluminum Hull Welding. Miami, FL: American Welding Society.

Lincoln Electric website offers sources for products and training: http://www.lincolnelectric.com

Machinery's Handbook. Erik Oberg, Franklin D. Jones, and Christopher J. McCauley. New York, NY: Industrial Press, Inc. 2004.

Metals and How to Weld Them. T. B. Jefferson. Cleveland, OH: The James F. Lincoln Arc Welding Foundation.

OSHA Standard 1926.351, Arc Welding and Cutting.

Stick Electrode Product Catalog. Cleveland, OH: The Lincoln Electric Company, 2008.

Stick Electrode Welding Guide. Cleveland, OH: The Lincoln Electric Company, 2004.

The Procedure Handbook of Arc Welding, 2000. Cleveland, OH: The Lincoln Electric Company.

The Procedure Handbook of Arc Welding. 14th ed. Cleveland, OH: The James F. Lincoln Welding Foundation, 2000.

Welding Handbook, 4 vols. Miami, FL: American Welding Society.

NCCER CURRICULA — USER UPDATE

NCCER makes every effort to keep its textbooks up-to-date and free of technical errors. We appreciate your help in this process. If you find an error, a typographical mistake, or an inaccuracy in NCCER's curricula, please fill out this form (or a photocopy), or complete the online form at **www.nccer.org/olf**. Be sure to include the exact module ID number, page number, a detailed description, and your recommended correction. Your input will be brought to the attention of the Authoring Team. Thank you for your assistance.

Instructors – If you have an idea for improving this textbook, or have found that additional materials were necessary to teach this module effectively, please let us know so that we may present your suggestions to the Authoring Team.

NCCER Product Development and Revision

13614 Progress Blvd., Alachua, FL 32615

Email: curriculum@nccer.org
Online: www.nccer.org/olf

❑ Trainee Guide ❑ Lesson Plans ❑ Exam ❑ PowerPoints Other ____________________

Craft / Level: Copyright Date:

Module ID Number / Title:

Section Number(s):

Description:

Recommended Correction:

Your Name:

Address:

Email: Phone:

Fundamentals of Crew Leadership

46101-11

NCCER
President: Don Whyte
Director of Product Development: Daniele Dixon
Fundamentals of Crew Leadership Project Manager: Patty Bird
Production Manager: Tim Davis
Quality Assurance Coordinator: Debie Hicks
Editor: Chris Wilson
Desktop Publishing Coordinator: James McKay
Production Specialist: Megan Casey

Editorial and production services provided by Topaz Publications, Liverpool, NY
Lead Writer/Project Manager: Tom Burke
Desktop Publisher: Joanne Hart
Art Director: Megan Paye
Permissions Editors: Andrea LaBarge, Alison Richmond

46101-11
Fundamentals of Crew Leadership

Foreword

Work gets done most efficiently if workers are divided into crews with a common purpose. When a crew is formed to tackle a particular job, one person is appointed the leader. This person is usually an experienced craftworker who has demonstrated leadership qualities. To become an effective leader, it helps if you have natural leadership qualities, but there are specific job skills that you must learn in order to do the job well.

This module will teach you the skills you need to be an effective leader, including the ability to communicate effectively; provide direction to your crew; and effectively plan and schedule the work of your crew.

As a crew member, you weren't required to think much about project cost. However, as a crew leader, you need to understand how to manage materials, equipment, and labor in order to work in a cost-effective manner. You will also begin to view safety from a different perspective. The crew leader takes on the responsibility for the safety of the crew, making sure that workers follow company safety polices and have the latest information on job safety issues.

As a crew leader, you become part of the chain of command in your company, the link between your crew and those who supervise and manage projects. As such, you need to know how the company is organized and how you fit into the organization. You will also focus more on company policies than a crew member, because it is up to you to enforce them within your crew. You will represent your team at daily project briefings and then communicate relevant information to your crew. This means learning how to be an effective listener and an effective communicator.

Whether you are currently a crew leader wanting to learn more about the requirements, or a crew member preparing to move up the ladder, this module will help you reach your goal.

This program consists of a Participant's Manual and Lesson Plans for the Instructor. The Participant's Manual contains the material that the participant will study, along with self-check exercises and activities, to help in evaluating whether the participant has mastered the knowledge needed to become an effective crew leader. The Lesson Plans include instructional outlines, suggested classroom activities, and homework based on the material in the Participant's Manual.

For the participant to gain the most from this program, it is recommended that the material be presented in a formal classroom setting, using a trained and experienced instructor. If the student is so motivated, he or she can study the material on a self-learning basis by using the material in both the Participant's Manual and the Lesson Plans. Recognition through the National Registry is available for the participants provided the program is delivered through an Accredited Sponsor by a Master Trainer or ICTP instructor. More details on this program can be received by contacting NCCER at **www.nccer.org**.

Participants in this program should note that some examples provided to reinforce the material may not apply to the participant's exact work, although the process will. Every company has its own mode of operation. Therefore, some topics may not apply to every participant's company. Such topics have been included because they are important considerations for prospective crew leaders throughout the industries supported by NCCER.

Industry-Recognized Credentials

If you are training through an NCCER-accredited sponsor, you may be eligible for credentials from NCCER's Registry. The ID number for this module is 46101-11. Note that this module may have been used in other NCCER curricula and may apply to other level completions. Contact NCCER's Registry at 888.622.3720 or go to **www.nccer.org** for more information.

Contents

Topics to be presented in this module include:

Contents (continued)

Contents (continued)

Contents (continued)

Figures and Tables

Acknowledgments

This curriculum was revised as a result of the farsightedness
and leadership of the following sponsors:

ABC South Texas Chapter
HB Training & Consulting
Turner Industries Group, LLC
University of Georgia
Vision Quest Academy
Willmar Electric Service

This curriculum would not exist were it not for the dedication and unselfish energy of those volunteers who served on the Authoring Team. A sincere thanks is extended to the following:

John Ambrosia
Harold (Hal) Heintz
Mark Hornbuckle
Jonathan Liston
Jay Tornquist
Wayne Tyson
Antonio "Tony" Vazquez

NCCER Partners

American Fire Sprinkler Association
Associated Builders and Contractors, Inc.
Associated General Contractors of America
Association for Career and Technical Education
Association for Skilled and Technical Sciences
Construction Industry Institute
Construction Users Roundtable
Design Build Institute of America
Gulf States Shipbuilders Consortium
ISN
Manufacturing Institute
Mason Contractors Association of America
Merit Contractors Association of Canada
NACE International
National Association of Women in Construction
National Insulation Association
National Technical Honor Society
NAWIC Education Foundation
North American Crane Bureau
North American Technician Excellence
Pearson
Prov
SkillsUSA
Steel Erectors Association of America
U.S. Army Corps of Engineers
University of Florida, M.E. Rinker, Sr. School of Construction Management
Women Construction Owners & Executives, USA

NCCER Business Partners

Section One
The Basics

Objectives

Upon completion of this section, you should be able to:

1. Describe the opportunities in the construction and power industries.
2. Describe how workers' values change over time.
3. Explain the importance of training and safety for the leaders in the construction and power industries.
4. Describe how new technologies are beneficial to the construction and power industries.
5. Identify the gender and minority issues associated with a changing workforce.
6. Describe what employers can do to prevent workplace discrimination.
7. Differentiate between formal and informal organizations.
8. Describe the difference between authority, responsibility, and accountability.
9. Explain the purpose of job descriptions and what they should include.
10. Distinguish between company policies and procedures.

1.0.0 Industry Today

Today's managers, supervisors, and crew leaders face challenges different from those of previous generations of leaders. To be a leader in industry today, it is essential to be well prepared. Today's crew leaders must understand how to use various types of new technology. In addition, they must have the knowledge and skills needed to manage, train, and communicate with a culturally diverse workforce whose attitudes toward work may differ from those of earlier generations and cultures. These needs are driven by changes in the workforce itself and in the work environment, and include the following:

- A shrinking workforce
- The growth of technology
- Changes in employee attitudes and values
- The emphasis on bringing women and minorities into the workforce
- The growing number of foreign-born workers
- Increased emphasis on workplace health and safety
- Greater focus on education and training

1.1.0 The Need for Training

Effective craft training programs are necessary if the industry is to meet the forecasted worker demands. Many of the skilled, knowledgeable craftworkers, crew leaders, and managers—the so-called baby boomers—have reached retirement age. In 2010, these workers who were born between 1946 and 1964, represented 38 percent of the workforce. Their departure leaves a huge vacuum across the industry spectrum. The Department of Labor (DOL) concludes that the best way for industry to reduce shortages of skilled workers is to create more education and training opportunities. The DOL suggests that companies and community groups form partnerships and create apprenticeship programs. Such programs could provide younger workers, including women and minorities, with the opportunity to develop job skills by giving them hands-on experience.

When training workers, it is important to understand that people learn in different ways. Some people learn by doing, some people learn by watching or reading, and others need step-by-step instructions as they are shown the process. Most people learn best by a combination of styles. It is important to understand what kind of a learner you are teaching, because if you learn one way, you tend to teach the way you learn. Have you ever tried to teach somebody and failed, and then another person successfully teaches the same thing in a different way? A person who acts as a mentor or trainer needs to be able to determine what kind of learner they are addressing and teach according to those needs.

The need for training is not limited to craftworkers. There must be supervisory training to ensure there are qualified leaders in the industry to supervise the craftworkers.

1.1.1 Motivation

As a supervisor or crew leader, it is important to understand what motivates your crew. Money is often thought to be a good motivator. Although that may be true to some extent, it has been proven to be a temporary solution. Once a person has reached a level of financial security, other factors come into play. Studies show that many people tend to be motivated by environment and conditions. For those people, a great workplace may provide more satisfaction than pay. If you give someone a raise, they tend to work harder for a period of time. Then the satisfaction dissipates and they may want another raise. People are often motivated by feeling a sense of accomplishment. That is why setting and working toward recognizable goals tends to make employees more productive. A person with a feeling of involvement or a sense of achievement is likely to be better motivated and help to motivate others.

1.1.2 Understanding Workers

Many older workers grew up in an environment in which they were taught to work hard and stay with the job until retirement. They expected to stay with a company for a long time, and companies were structured to create a family-type environment.

Times have changed. Younger workers have grown up in a highly mobile society and are used to rapid rewards. This generation of workers can sometimes be perceived as lazy and unmotivated, but in reality, they simply have a different perspective. For such workers, it may be better to give them small projects or break up large projects into smaller pieces so that they feel repetitively rewarded, thus enhancing their perception of success.

- *Goal setting* – Set short-term and long-term goals, including tasks to be done and expected time frames. Help the trainees understand that things can happen to offset the short-term goals. This is one reason to set long-term goals as well. Don't set them up for failure, as this leads to frustration, and frustration can lead to reduced productivity.
- *Feedback* – Timely feedback is important. For example, telling someone they did a good job last year, or criticizing them for a job they did a month ago, is meaningless. Simple recognition isn't always enough. Some type of reward should accompany positive feedback, even if it is simply recognizing the employee in a public way. Constructive feedback should be given in private and be accompanied by some positive action, such as providing one-on-one training to correct a problem.

1.1.3 Craft Training

Craft training is often informal, taking place on the job site, outside of a traditional training classroom. According to the American Society for Training and Development (ASTD), craft training is generally handled through on-the-job instruction by a qualified co-worker or conducted by a supervisor.

The Society of Human Resources Management (SHRM) offers the following tips to supervisors in charge of training their employees:

- *Help crew members establish career goals.* Once the goals are established, the training required to meet the goals can be readily identified.
- *Determine what kind of training to give.* Training can be on the job under the supervision of a co-worker. It can be one-on-one with the supervisor. It can involve cross-training to teach a new trade or skill, or it can involve delegating new or additional responsibilities.
- *Determine the trainee's preferred method of learning.* Some people learn best by watching, others from verbal instructions, and others by doing. When training more than one person at a time, try to use all three methods.

Communication is a critical component of training employees. The SHRM advises that supervisors do the following when training their employees:

- *Explain the task, why it needs to be done, and how it should be done.* Confirm that the trainees understand these three areas by asking questions. Allow them to ask questions as well.
- *Demonstrate the task.* Break the task down into manageable parts and cover one part at a time.
- *Ask trainees to do the task while you observe them.* Try not to interrupt them while they are doing the task unless they are doing something that is unsafe and potentially harmful.
- *Give the trainees feedback.* Be specific about what they did and mention any areas where they need to improve.

1.1.4 Supervisory Training

Given the need for skilled craftworkers and qualified supervisory personnel, it seems logical that companies would offer training to their employees through in-house classes, or by subsidizing outside training programs. While some contractors have their own in-house training programs or participate in training offered by associations and other organizations, many contractors do not offer training at all.

There are a number of reasons that companies do not develop or provide training programs, including the following:

- Lack of money to train
- Lack of time to train
- Lack of knowledge about the benefits of training programs
- High rate of employee turnover
- Workforce too small

Did you know?

There is a lot more to training than simply telling a group of people what you know about a subject or talking your way through a series of slides. Studies have shown that what trainees retain depends on several factors. In general, people retain:

10 percent of what they read
20 percent of what they hear
30 percent of what they see
50 percent of what they see and hear
70 percent of what they see, hear, and respond to
90 percent of what they see, hear, and do

- Past training involvement was ineffective
- The company hires only trained workers
- Lack of interest from workers
- Lack of company interest in training

For craftworkers to move up into supervisory and managerial positions, it will be necessary for them to continue their education and training. Those who are willing to acquire and develop new skills have the best chance of finding stable employment. It is therefore critical that employees take advantage of training opportunities, and that companies employ training as part of their business culture.

Your company has recognized the need for training. Your participation in a leadership training program such as this will begin to fill the gap between craft and supervisory training.

1.2.0 Impact of Technology

Many industries, including the construction industry, have made the move to technology as a means of remaining competitive. Benefits include increased productivity and speed, improved quality of documents, greater access to common data, and better financial controls and communication. As technology becomes a greater part of supervision, crew leaders need to be able to use it properly. One important concern with electronic communication is to keep it brief, factual, and legal. Because the receiver has no visual or auditory clues as to the sender's intent, the sender can be easily misunderstood. In other words, it is more difficult to tell if someone is just joking via e-mail because you can't see their face or hear the tone of their voice.

Cellular telephones, voicemail, and handheld communication devices have made it easy to keep in touch. They are particularly useful communication sources for contractors or crew leaders who are on a job site, away from their offices, or constantly on the go.

Cellular telephones allow the users to receive incoming calls as well as make outgoing calls. Unless the owner is out of the cellular provider's service area, the cell phone may be used any time to answer calls, make calls, and send and receive voicemail or email. Always check the company's policy with regard to the use of cell phones on the job.

Handheld communication devices known as smart phones allow supervisors to plan their calendars, schedule meetings, manage projects, and access their email from remote locations. These computers are small enough to fit in the palm of the hand, yet powerful enough to hold years of information from various projects. Information can be transmitted electronically to others on the project team or transferred to a computer.

2.0.0 Gender and Cultural Issues

During the past several years, the construction industry in the United States has experienced a shift in worker expectations and diversity. These two issues are converging at a rapid pace. At no time has there been such a generational merge in the workforce, ranging from The Silent Generation (1925–1945), Baby Boomers (1946–1964), Gen X (1965–1979), and the Millennials, also known as Generation Y (1980–2000).

This trend, combined with industry diversity initiatives, has created a climate in which companies recognize the need to embrace a diverse workforce that crosses generational, gender, and ethnic boundaries. To do this effectively, they are using their own resources, as well as relying on associations with the government and trade organizations. All current research indicates that industry will be more dependent on the critical skills of a diverse workforce—a workforce that is both culturally and ethnically fused. Across the United States, construction and other industries are aggressively seeking to bring new workers into their ranks, including women and racial and ethnic minorities. Diversity is no longer solely driven by social and political issues, but by consumers who need hospitals, malls, bridges, power plants, refineries, and many other commercial and residential structures.

Some issues relating to a diverse workforce will need to be addressed on the job site. These issues include different communication styles of men and women, language barriers associated with cultural differences, sexual harassment, and gender or racial discrimination.

2.1.0 Communication Styles of Men and Women

As more and more women move into construction, it becomes increasingly important that communication barriers between men and women are broken down and that differences in behaviors are understood so that men and women can work together more effectively. The Jamestown, New York Area Labor Management Committee (JALMC) offers the following explanations and tips:

- *Women tend to ask more questions than men do.* Men are more likely to proceed with a job and figure it out as they go along, while women are more likely to ask questions first.
- *Men tend to offer solutions before empathy; women tend to do the opposite.* Both men and women should say what they want up front, whether it's the solution to a problem, or simply a sympathetic ear. That way, both genders will feel understood and supported.
- *Women are more likely to ask for help when they need it.* Women are generally more pragmatic when it comes to completing a task. If they need help, they will ask for it. Men are more likely to attempt to complete a task by themselves, even when assistance is needed.
- *Men tend to communicate more competitively, and women tend to communicate more cooperatively.* Both parties need to hear one another out without interruption.

This does not mean that one method is more effective than the other. It simply means that men and women use different approaches to achieve the same result.

2.2.0 Language Barriers

Language barriers are a real workplace challenge for crew leaders. Millions of workers speak languages other than English. Spanish is commonly spoken in the United States. As the makeup of the immigrant population continues to change, the number of non-English speakers will rise dramatically, and the languages being spoken will also change. Bilingual job sites are increasingly common.

Companies have the following options to overcome this challenge:

- Offer English classes either at the work site or through school districts and community colleges.
- Offer incentives for workers to learn English.

As the workforce becomes more diverse, communicating with people for whom English is a second language will be even more critical. The following tips will help when communicating across language barriers:

- Be patient. Give workers time to process the information in a way that they can comprehend.
- Avoid humor. Humor is easily misunderstood and may be misinterpreted as a joke at the worker's expense.
- Don't assume that people are unintelligent simply because they don't understand what you are saying.
- Speak slowly and clearly, and avoid the tendency to raise your voice.
- Use face-to-face communication whenever possible. Over-the-phone communication is often more difficult when a language barrier is involved.
- Use pictures or drawings to get your point across.
- If a worker speaks English poorly but understands reasonably well, ask the worker to demonstrate his or her understanding through other means.

2.3.0 Cultural Differences

As workers from a multitude of backgrounds and cultures are brought together, there are bound to be differences and conflicts in the workplace.

To overcome cultural conflicts, the SHRM suggests the following approach to resolving cultural conflicts between individuals:

- *Define the problem from both points of view.* How does each person involved view the conflict? What does each person think is wrong? This involves moving beyond traditional thought processes to consider alternate ways of thinking.
- *Uncover cultural interpretations.* What assumptions are being made based on cultural programming? By doing this, the supervisor may realize what motivated an employee to act in a particular manner.

- *Create cultural synergy.* Devise a solution that works for both parties involved. The purpose is to recognize and respect other's cultural values, and work out mutually acceptable alternatives.

2.4.0 Sexual Harassment

In today's business world, men and women are working side-by-side in careers of all kinds. As women make the transition into traditionally male industries, such as construction, the likelihood of sexual harassment increases. Sexual harassment is defined as unwelcome behavior of a sexual nature that makes someone feel uncomfortable in the workplace by focusing attention on their gender instead of on their professional qualifications. Sexual harassment can range from telling an offensive joke or hanging a poster of a swimsuit-clad man or woman, to making sexual comments or physical advances.

Historically, sexual harassment was thought to be an act performed by men of power within an organization against women in subordinate positions. However, the number of sexual harassment cases over the years, have shown that this is no longer the case.

Sexual harassment can occur in a variety of circumstances, including but not limited to the following:

- The victim as well as the harasser may be a woman or a man. The victim does not have to be of the opposite sex.
- The harasser can be the victim's supervisor, an agent of the employer, a supervisor in another area, a co-worker, or a non-employee.
- The victim does not have to be the person harassed, but could be anyone affected by the offensive conduct.
- Unlawful sexual harassment may occur without economic injury to or discharge of the victim.
- The harasser's conduct must be unwelcome.

The Equal Employment Opportunity Commission (EEOC) enforces sexual harassment laws within industries. When investigating allegations of sexual harassment, the EEOC looks at the whole record, including the circumstances and the context in which the alleged incidents occurred. A decision on the allegations is made from the facts on a case-by-case basis. A crew leader who is aware of sexual harassment and does nothing to stop it can be held responsible. The crew leader therefore should not only take action to stop sexual harassment, but should serve as a good example for the rest of the crew.

Did you know?

Some companies employ what is known as sensitivity training in cases where individuals or groups have trouble adapting to a multicultural, multi-gender workforce. Sensitivity training is a psychological technique using group discussion, role playing, and other methods to allow participants to develop an awareness of themselves and how they interact with others.

Prevention is the best tool to eliminate sexual harassment in the workplace. The EEOC encourages employers to take steps to prevent sexual harassment from occurring. Employers should clearly communicate to employees that sexual harassment will not be tolerated. They do so by developing a policy on sexual harassment, establishing an effective complaint or grievance process, and taking immediate and appropriate action when an employee complains.

Both swearing and off-color remarks and jokes are not only offensive to co-workers, but also tarnish a worker's image. Crew leaders need to emphasize that abrasive or crude behavior may affect opportunities for advancement. If disciplinary action becomes necessary, it should be covered by company policy. A typical approach is a three-step process in which the perpetrator is first given a verbal reprimand. In the event of further violations, a written reprimand and warning are given. Dismissal typically accompanies subsequent violations.

2.5.0 Gender and Minority Discrimination

More attention is being placed on fair recruitment, equal pay for equal work, and promotions for women and minorities in the workplace. Consequently, many business practices, including the way employees are treated, the organization's hiring and promotional practices, and the way people are compensated, are being analyzed for equity.

Once a male-dominated industry, construction companies are moving away from this image and are actively recruiting and training women, younger workers, people from other cultures, and workers with disabilities. This means that organizations hire the best person for the job, without regard for race, sex, religion, age, etc.

To prevent discrimination cases, employers must have valid job-related criteria for hiring, compensation, and promotion. These measures must be used consistently for every applicant interview, employee performance appraisal, and hiring or promotion decision. Therefore, all workers responsible for recruitment, selection, and supervision of employees, and evaluating job performance, must be trained on how to use the job-related criteria legally and effectively.

3.0.0 Business Organizations

An organization is the relationship among the people within the company or project. The crew leader needs to be aware of two types of organizations. These are formal organizations and informal organizations.

A formal organization exists when the activities of the people within the work group are directed toward achieving a common goal. An example of a formal organization is a work crew consisting of four carpenters and two laborers led by a crew leader, all working together toward a common goal.

A formal organization is typically documented on an organizational chart, which outlines all the positions that make up an organization and shows how those positions are related. Some organizational charts even depict the people within each position and the person to whom they report, as well as the people that the person supervises. *Figures 1* and *2* show examples of organization charts for fictitious companies. Note that each of these positions represents an opportunity for advancement in the construction industry that a crew leader can eventually achieve.

An informal organization allows for communication among its members so they can perform as a group. It also establishes patterns of behavior that help them to work as a group, such as agreeing to use a specific training program.

An example of an informal organization is a trade association such as Associated Builders and Contractors (ABC), Associated General Contractors (AGC), and the National Association of Women in Construction (NAWIC). Those, along with the thousands of other trade associations in the U.S., provide a forum in which members with common concerns can share information, work on issues, and develop standards for their industry.

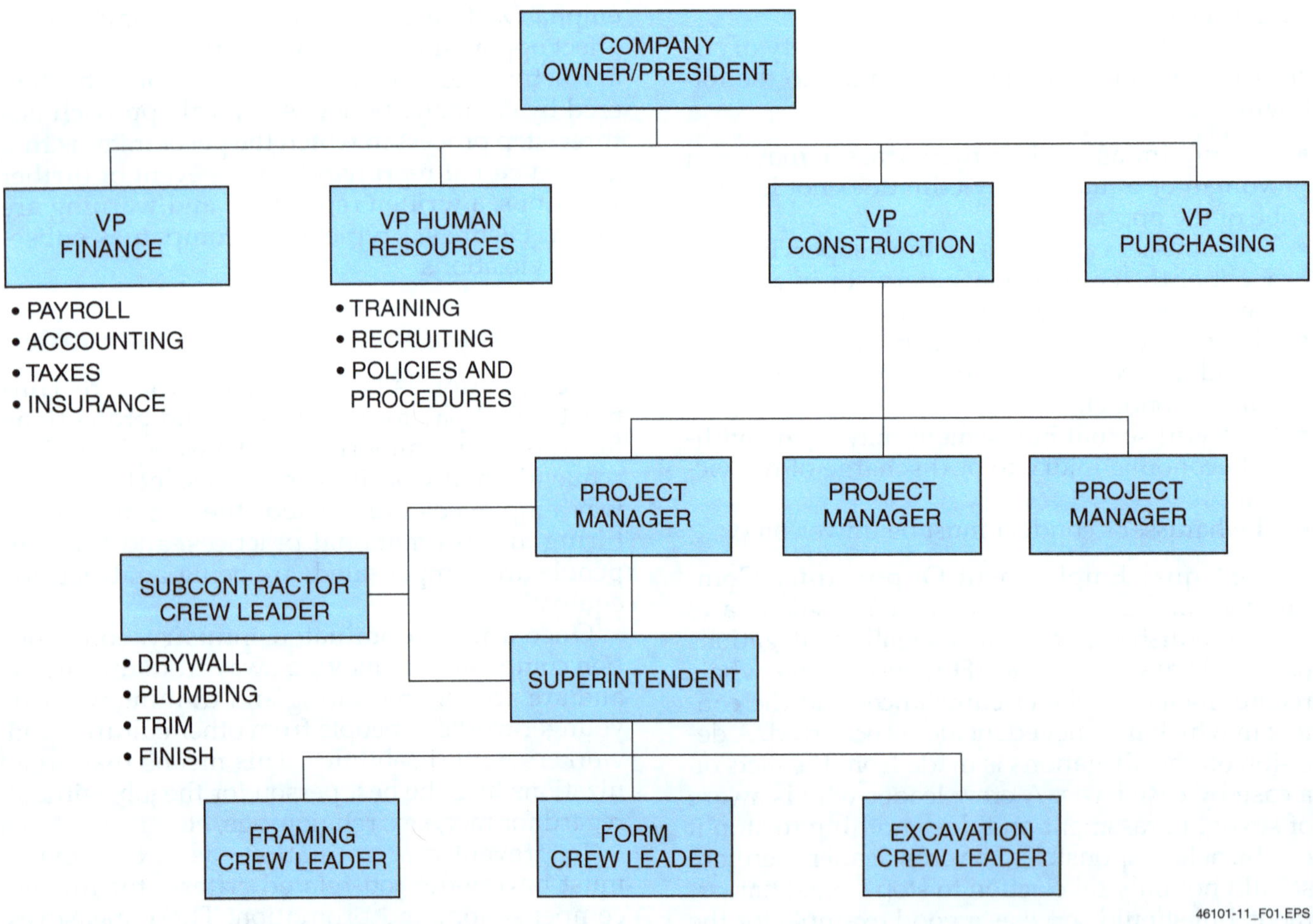

Figure 1 Sample organization chart for a construction company.

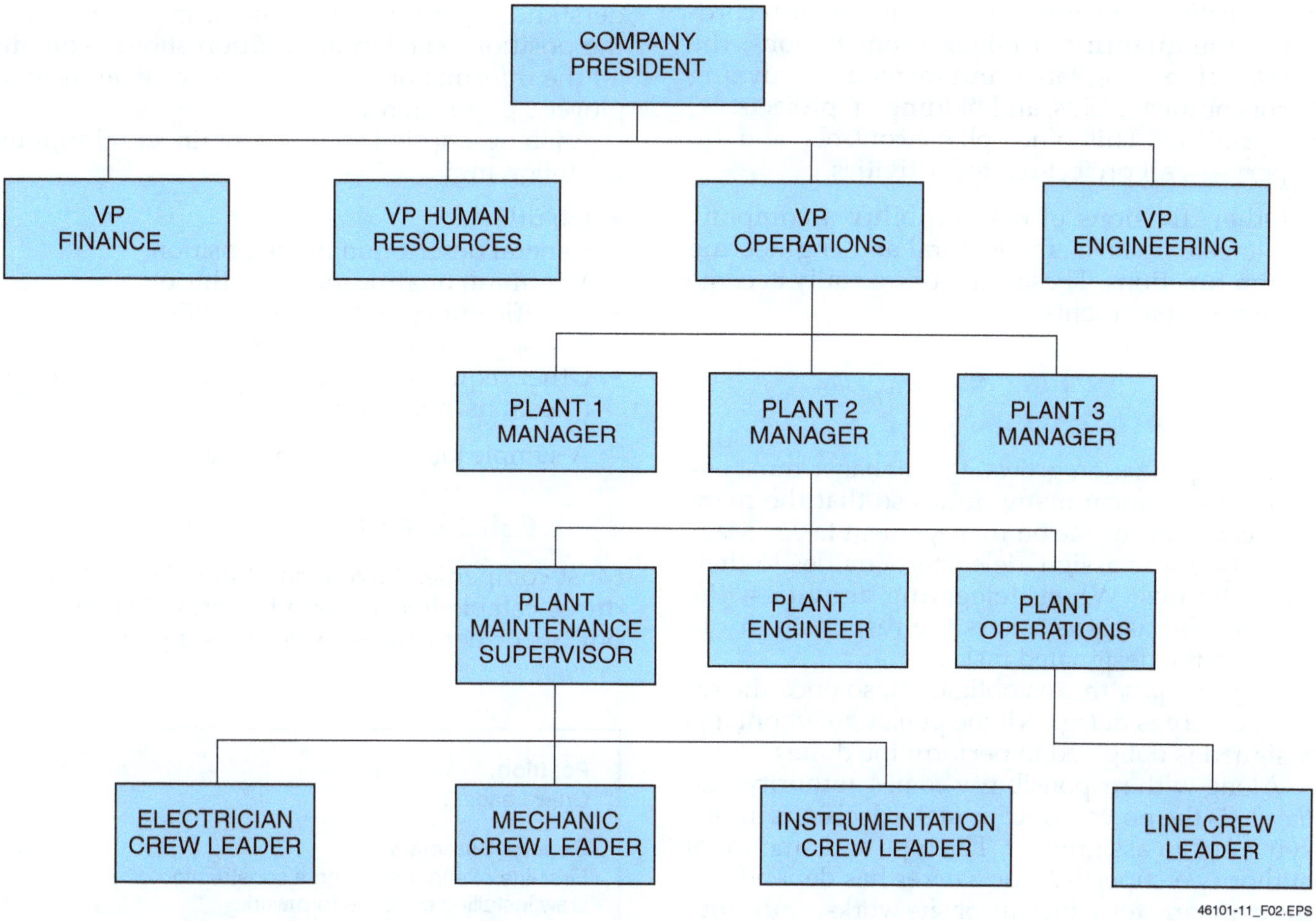

Figure 2 Sample organization chart for an industrial company.

Both types of organizations establish the foundation for how communication flows. The formal structure is the means used to delegate authority and responsibility and to exchange information. The informal structure is used to exchange information.

Members in an organization perform best when each member:

- Knows the job and how it will be done
- Communicates effectively with others in the group
- Understands his or her role in the group
- Recognizes who has the authority and responsibility

3.1.0 Division of Responsibility

The conduct of a business involves certain functions. In a small organization, responsibilities may be divided between one or two people. However, in a larger organization with many different and complex activities, responsibilities may be grouped into similar activity groups, and the responsibility for each group assigned to department managers. In either case, the following major departments exist in most companies:

- *Executive* – This office represents top management. It is responsible for the success of the company through short-range and long-range planning.
- *Human Resources* – This office is responsible for recruiting and screening prospective employees; managing employee benefits programs; advising management on pay and benefits; and developing and enforcing procedures related to hiring practices.
- *Accounting* – This office is responsible for all record keeping and financial transactions, including payroll, taxes, insurance, and audits.
- *Contract Administration* – This office prepares and executes contractual documents with owners, subcontractors, and suppliers.
- *Purchasing* – This office obtains material prices and then issues purchase orders. The purchasing office also obtains rental and leasing rates on equipment and tools.

- *Estimating*: This office is responsible for recording the quantity of material on the jobs, the takeoff, pricing labor and material, analyzing subcontractor bids, and bidding on projects.
- *Operations*: This office plans, controls, and supervises all project-related activities.

Other divisions of responsibility a company may create involve architectural and engineering design functions. These divisions usually become separate departments.

3.2.0 Authority, Responsibility, and Accountability

As an organization grows, the manager must ask others to perform many duties so that the manager can concentrate on management tasks. Managers typically assign (delegate) activities to their subordinates. When delegating activities, the crew leader assigns others the responsibility to perform the designated tasks.

Responsibility means obligation, so once the responsibility is delegated, the person to whom it is assigned is obligated to perform the duties.

Along with responsibility comes authority. *Authority* is the power to act or make decisions in carrying out an assignment. The type and amount of authority a supervisor or worker has depends on the company for which he or she works. Authority and responsibility must be balanced so employees can carry out their tasks. In addition, delegation of sufficient authority is needed to make an employee accountable to the crew leader for the results.

Accountability is the act of holding an employee responsible for completing the assigned activities. Even though authority and responsibility may be delegated to crew members, the final responsibility always rests with the crew leader.

3.3.0 Job Descriptions

Many companies furnish each employee with a written job description that explains the job in detail. Job descriptions set a standard for the employee. They make judging performance easier, clarify the tasks each person should handle, and simplify the training of new employees.

Each new employee should understand all the duties and responsibilities of the job after reviewing the job description. Thus, the time it takes for the employee to make the transition from being a new and uninformed employee to a more experienced member of a crew is shortened.

A job description need not be long, but it should be detailed enough to ensure there is no misunderstanding of the duties and responsibilities of the position. The job description should contain all the information necessary to evaluate the employee's performance.

A job description should contain, at minimum, the following:

- Job title
- General description of the position
- Minimum qualifications for the job
- Specific duties and responsibilities
- The supervisor to whom the position reports
- Other requirements, such as qualifications, certifications, and licenses

A sample job description is shown in *Figure 3*.

3.4.0 Policies and Procedures

Most companies have formal policies and procedures established to help the crew leader carry out his or her duties. A *policy* is a general state-

Position:
Crew Leader

General Summary:
First line of supervision on a construction crew installing concrete formwork.

Reports To:
Job Superintendent

Physical and Mental Responsibilities:
- Ability to stand for long periods
- Ability to solve basic math and geometry problems

Duties and Responsibilities:
- Oversee crew
- Provide instruction and training in construction tasks as needed
- Make sure proper materials and tools are on the site to accomplish tasks
- Keep project on schedule
- Enforce safety policies and procedures

Knowledge, Skills, and Experience Required:
- Extensive travel throughout the Eastern United States, home base in Atlanta
- Ability to operate a backhoe and trencher
- Valid commercial driver's license with no DUI violations
- Ability to work under deadlines with the knowledge and ability to foresee problem areas and develop a plan of action to solve the situation

46101-11_F03.EPS

Figure 3 Example of a job description.

ment establishing guidelines for a specific activity. Examples include policies on vacations, breaks, workplace safety, and checking out tools.

Procedures are the ways that policies are carried out. For example, a procedure written to implement a policy on workplace safety would include guidelines for reporting accidents and general safety procedures that all employees are expected to follow.

A crew leader must be familiar with the company policies and procedures, especially with regard to safety practices. When OSHA inspectors visit a job site, they often question employees and crew leaders about the company policies related to safety. If they are investigating an accident, they will want to verify that the responsible crew leader knew the applicable company policy and followed it.

Did you know?

The Americans with Disabilities Act (ADA) is a law that prohibits employers from discriminating against disabled persons in their hiring practices. The law is specifically titled: An Act to establish a clear and comprehensive prohibition of discrimination on the basis of disability. While the ADA does not require written job descriptions, existing job descriptions will be used as evidence in determining whether discrimination occurred. This means that job descriptions must specifically define the job duties and the abilities needed to perform them. In making these determinations, consideration is given to whether a reasonable accommodation could have qualified the disabled person to fill a job.

Review Questions

1. The construction industry should provide training for craftworkers and supervisors _____.
   a. to ensure that there are enough future workers
   b. to avoid discrimination lawsuits
   c. in order to update the skills of older workers who are retiring at a later age than they previously did
   d. even though younger workers are now less likely to seek jobs in other areas than they were 10 years ago

2. Companies traditionally offer craftworker training _____.
   a. that a supervisor leads in a classroom setting
   b. that a craftworker leads in a classroom setting
   c. in a hands-on setting, where craftworkers learn from a co-worker or supervisor
   d. on a self-study basis to allow craftworkers to proceed at their own pace

3. One way to provide effective training is to _____.
   a. avoid giving negative feedback until trainees are more experienced in doing the task
   b. tailor the training to the career goals and needs of trainees
   c. choose one training method and use it for all trainees
   d. encourage trainees to listen, saving their questions for the end of the session

4. One way to prevent sexual harassment in the workplace is to _____.
   a. require employee training in which the potentially offensive subject of stereotypes is carefully avoided
   b. develop a consistent policy with appropriate consequences for engaging in sexual harassment
   c. communicate to workers that the victim of sexual harassment is the one who is being directly harassed, not those affected in a more indirect way
   d. educate workers to recognize sexual harassment for what it is—unwelcome conduct by the opposite sex

5. Employers can minimize all types of workplace discrimination by hiring based on a consistent list of job-related requirements.
   a. True
   b. False
6. Members tend to function best within an organization when they _____.
   a. are allowed to select their own style of clothing for each project
   b. understand their role
   c. do not disagree with the statements of other workers or supervisors
   d. are able to work without supervision
7. A formal organization is defined as a group of individuals who work independently, but share the same goal
   a. True
   b. False
8. A formal organization uses an organizational chart to _____.
   a. depict all companies with which it conducts business
   b. show all customers with which it conducts business
   c. track projects between departments
   d. show the relationships among the existing positions in the company
9. Which of the following is a function typically performed by the operations department of a company?
   a. Purchase materials
   b. Plan projects
   c. Prepare payrolls
   d. Recruiting and screening new hires
10. The company department that manages employee benefits and personnel recruiting is _____.
    a. Engineering
    b. Human Resources
    c. Purchasing
    d. Contract Administration
11. The power to make decisions and act on them in carrying out an assignment is _____.
    a. delegating
    b. responsibility
    c. decisiveness
    d. authority
12. Accountability is defined as _____.
    a. the power to act or make decisions in carrying out assignments
    b. giving an employee a particular task to perform
    c. the act of an employee responsible for the completion and results of a particular duty
    d. having the power to promote someone
13. A good job description should include _____.
    a. a complete organization chart
    b. any information needed to judge job performance
    c. the company dress code
    d. the company's sexual harassment policy
14. The purpose of a policy is to _____.
    a. establish company guidelines regarding a particular activity
    b. specify what tools and equipment are required for a job
    c. list all information necessary to judge an employee's performance
    d. inform employees about the future plans of the company
15. One example of a procedure would be the rules for taking time off.
    a. True
    b. False

SECTION TWO
LEADERSHIP SKILLS

Objectives

Upon completion of this section, you should be able to:

1. Describe the role of a crew leader.
2. List the characteristics of effective leaders.
3. Be able to discuss the importance of ethics in a supervisor's role.
4. Identify the three styles of leadership.
5. Describe the forms of communication.
6. Describe the four parts of verbal communication.
7. Describe the importance of active listening.
8. Explain how to overcome the barriers to communication.
9. List ways that leaders can motivate their employees.
10. Explain the importance of delegating and implementing policies and procedures.
11. Distinguish between problem solving and decision making.

1.0.0 INTRODUCTION TO LEADERSHIP

For the purpose of this program, it is important to define some of the positions that will be discussed. The term *craftworker* refers to a person who performs the work of his or her trade(s). The crew leader is a person who supervises one or more craftworkers on a crew. A superintendent is essentially an on-site supervisor who is responsible for one or more crew leaders or front-line supervisors. Finally, a project manager or general superintendent may be responsible for managing one or more projects. This training will concentrate primarily on the role of the crew leader.

Craftworkers and crew leaders differ in that the crew leader manages the activities that the skilled craftworkers on the crews actually perform. In order to manage a crew of craftworkers, a crew leader must have first-hand knowledge and experience in the activities being performed. In addition, he or she must be able to act directly in organizing and directing the activities of the various crew members.

This section explains the importance of developing effective leadership skills as a new crew leader. Effective ways to communicate with all levels of employees and co-workers, build teams, motivate crew members, make decisions, and resolve problems are covered in depth.

2.0.0 THE SHIFT IN WORK ACTIVITIES

The crew leader is generally selected and promoted from a work crew. The selection will often be based on that person's ability to accomplish tasks, to get along with others, to meet schedules, and to stay within the budget. The crew leader must lead the team to work safely and provide a quality product.

Making the transition from a craftworker to a crew leader can be difficult, especially when the new crew leader is in charge of supervising a group of peers. Crew leaders are no longer responsible for their work alone; rather, they are accountable for the work of an entire crew of people with varying skill levels and abilities, a multitude of personalities and work styles, and different cultural and educational backgrounds. Crew leaders must learn to put their personal relationships aside and work for the common goals of the team.

New crew leaders are often placed in charge of workers who were formerly their friends and peers on a crew. This situation can create some conflicts. For example, some of the crew may try to take advantage of the friendship by seeking special favors. They may also want to be privy to information that should be held closely. These problems can be overcome by working with the crew to set mutual performance goals and by freely communicating with them within permitted limits. Use their knowledge and strengths along with your own so that they feel like they are key players on the team.

As an employee moves from a craftworker position to the role of a crew leader, he or she will find that more hours will be spent supervising the work of others than actually performing the technical skill for which he or she has been trained. *Figure 4* represents the percentage of time craftworkers, crew leaders, superintendents, and project managers spend on technical and supervisory work as their management responsibilities increase.

The success of the new crew leader is directly related to the ability to make the transition from crew member into a leadership role.

3.0.0 BECOMING A LEADER

A crew leader must have leadership skills to be successful. Therefore, one of the primary goals of a person who wants to become a crew leader should be to develop strong leadership skills and learn to use them effectively.

There are many ways to define a leader. One straightforward definition is a person who influences other people in the achievement of a goal.

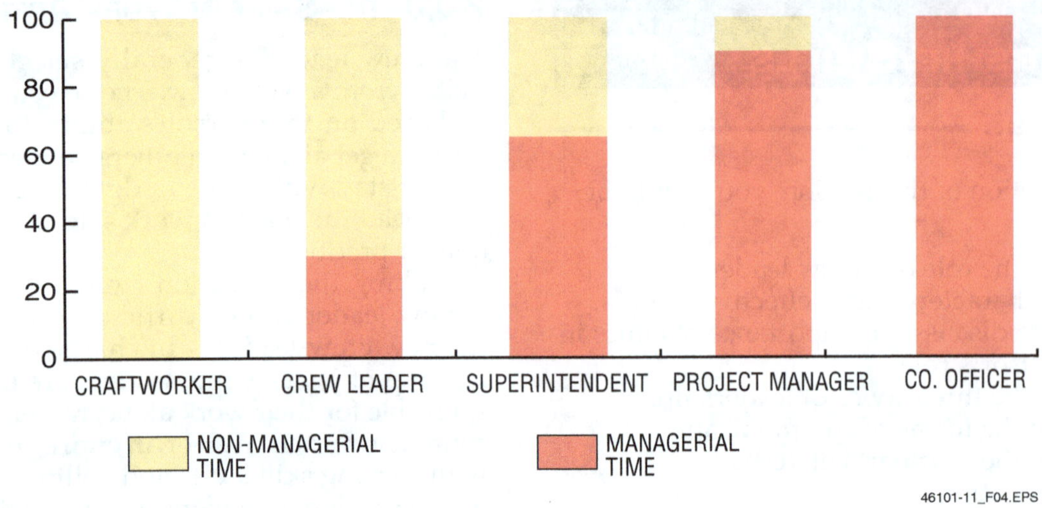

Figure 4 Percentage of time spent on technical and supervisory work.

Some people may have inherited leadership qualities or may have developed traits that motivate others to follow and perform. Research shows that people who possess such talents are likely to succeed as leaders.

3.1.0 Characteristics of Leaders

Leadership traits are similar to the skills that a crew leader needs in order to be effective. Although the characteristics of leadership are many, there are some definite commonalities among effective leaders.

First and foremost, effective leaders lead by example. In other words, they work and live by the standards that they establish for their crew members or followers, making sure they set a positive example.

Effective leaders also tend to have a high level of drive and determination, as well as a stick-to-it attitude. When faced with obstacles, effective leaders don't get discouraged; instead, they identify the potential problems, make plans to overcome them, and work toward achieving the intended goal. In the event of failure, effective leaders learn from their mistakes and apply that knowledge to future situations. They also learn from their successes.

Effective leaders are typically effective communicators who clearly express the goals of a project to their crew members. Accomplishing this may require that the leader overcome issues such as language barriers, gender bias, or differences in personalities to ensure that each member of the crew understands the established goals of the project.

Effective leaders have the ability to motivate their crew members to work to their full potential and become effective members of the team. Crew leaders try to develop crew member skills and encourage them to improve and learn as a means to contribute more to the team effort. Effective leaders strive for excellence from themselves and their team, so they work hard to provide the skills and leadership necessary to do so.

In addition, effective leaders must possess organizational skills. They know what needs to be accomplished, and they use their resources to make it happen. Because they cannot do it alone, leaders enlist the help of their team members to share in the workload. Effective leaders delegate work to their crew members, and they implement company policies and procedures to ensure that the work is completed safely, effectively, and efficiently.

Finally, effective leaders have the authority and self-confidence that allows them to make decisions and solve problems. In order to accomplish their goals, leaders must be able to calculate risks, absorb and interpret information, assess courses of action, make decisions, and assume the responsibility for those decisions.

3.1.1 Leadership Traits

There are many other traits of effective leaders. Some other major characteristics of leaders include the following:

- Ability to plan and organize
- Loyalty to their company and crew
- Ability to motivate
- Fairness

- Enthusiasm
- Willingness to learn from others
- Ability to teach others
- Initiative
- Ability to advocate an idea
- Good communication skills

3.1.2 Expected Leadership Behavior

Followers have expectations of their leaders. They look to their leaders to:

- Lead by example
- Suggest and direct
- Plan and organize the work
- Communicate effectively
- Make decisions and assume responsibility
- Have the necessary technical knowledge
- Be a loyal member of the group
- Abide by company policies and procedures

3.2.0 Functions of a Leader

The functions of a leader will vary with the environment, the group being led, and the tasks to be performed. However, there are certain functions common to all situations that the leader will be called upon to perform. Some of the major functions are:

- Organize, plan, staff, direct, and control work
- Empower group members to make decisions and take responsibility for their work
- Maintain a cohesive group by resolving tensions and differences among its members and between the group and those outside the group
- Ensure that all group members understand and abide by company policies and procedures
- Accept responsibility for the successes and failures of the group's performance
- Represent the group
- Be sensitive to the differences of a diverse workforce

3.3.0 Leadership Styles

There are three main styles of leadership. At one extreme is the autocratic or commander style of leadership, where the crew leader makes all of the decisions independently, without seeking the opinions of crew members. At the other extreme is the hands-off or facilitator style, where the crew leader empowers the employees to make decisions. In the middle is the democratic or collaborative style, where the crew leader seeks crew member opinions and makes the appropriate decisions based on their input.

The following are some characteristics of each of the three leadership styles:

Commander types:

- Expect crew members to work without questioning procedures
- Seldom seek advice from crew members
- Insist on solving problems alone
- Seldom permit crew members to assist each other
- Praise and criticize on a personal basis
- Have no sincere interest in creatively improving methods of operation or production

Partner types:

- Discuss problems with their crew members
- Listen to suggestions from crew members
- Explain and instruct
- Give crew members a feeling of accomplishment by commending them when they do a job well
- Are friendly and available to discuss personal and job-related problems

Facilitator types:

- Believe no supervision is best
- Rarely give orders
- Worry about whether they are liked by their crew members

Effective leadership takes many forms. The correct style for a particular situation or operation depends on the nature of the crew as well as the work it has to accomplish. For example, if the crew does not have enough experience for the job ahead, then a commander style may be appropriate. The autocratic style of leadership is also effective when jobs involve repetitive operations that require little decision-making.

However, if a worker's attitude is an issue, a partner style may be appropriate. In this case, providing the missing motivational factors may increase performance and result in the improvement of the worker's attitude. The democratic style of leadership is also used when the work is of a creative nature, because brainstorming and exchanging ideas with such crew members can be beneficial.

The facilitator style is effective with an experienced crew on a well-defined project.

The company must give a crew leader sufficient authority to do the job. This authority must be commensurate with responsibility, and it must be made known to crew members when they are hired so that they understand who is in charge.

A crew leader must have an expert knowledge of the activities to be supervised in order to be effective. This is important because the crew members need to know that they have someone to turn to when they have a question or a problem, when they need some guidance, or when modifications or changes are warranted by the job.

Respect is probably the most useful element of authority. Respect usually derives from being fair to employees, by listening to their complaints and suggestions, and by using incentives and rewards appropriately to motivate crew members. In addition, crew leaders who have a positive attitude and a favorable personality tend to gain the respect of their crew members as well as their peers. Along with respect comes a positive attitude from the crew members.

3.4.0 Ethics in Leadership

The crew leader should practice the highest standards of ethical conduct. Every day the crew leader has to make decisions that may have ethical implications. When an unethical decision is made, it not only hurts the crew leader, but also other workers, peers, and the company for which he or she works.

There are three basic types of ethics:

1. Business or legal
2. Professional or balanced
3. Situational

Business, or legal, ethics concerns adhering to all laws and regulations related to the issue.

Professional, or balanced, ethics relates to carrying out all activities in such a manner as to be honest and fair to everyone concerned.

Situational ethics pertains to specific activities or events that may initially appear to be a gray area. For example, you may ask yourself, "How will I feel about myself if my actions are published in the newspaper or if I have to justify my actions to my family, friends, and colleagues?"

The crew leader will often be put into a situation where he or she will need to assess the ethical consequences of an impending decision. For instance, should a crew leader continue to keep one of his or her crew working who has broken into a cold sweat due to overheated working conditions just because the superintendent says the activity is behind schedule? Or should a crew leader, who is the only one aware that the reinforcing steel placed by his or her crew was done incorrectly, correct the situation before the concrete is placed in the form? If a crew leader is ever asked to carry through on an unethical decision, it is up to him or her to inform the superintendent of the unethical nature of the issue, and if still requested to follow through, refuse to act.

4.0.0 Communication

Successful crew leaders learn to communicate effectively with people at all levels of the organization. In doing so, they develop an understanding of human behavior and acquire communication skills that enable them to understand and influence others.

There are many definitions of communication. Communication is the act of accurately and effectively conveying or transmitting facts, feelings, and opinions to another person. Simply stated, communication is the method of exchanging information and ideas.

Just as there are many definitions of communication, it also comes in many forms, including verbal, nonverbal, and written. Each of these forms of communication are discussed in this section.

4.1.0 Verbal Communication

Verbal communication refers to the spoken words exchanged between two or more people. Verbal communication consists of four distinct parts:

1. Sender
2. Message
3. Receiver
4. Feedback

Figure 5 depicts the relationship of these four parts within the communication process. In verbal communication, the focus is on feedback, which is used to verify that the sender's message was received as intended.

Did you know?

Research shows that the typical supervisor spends about 80 percent of his or her day communicating through writing, speaking, listening, or using body language. Of that time, studies suggest that approximately 20 percent of communication is written, and 80 percent involves speaking or listening.

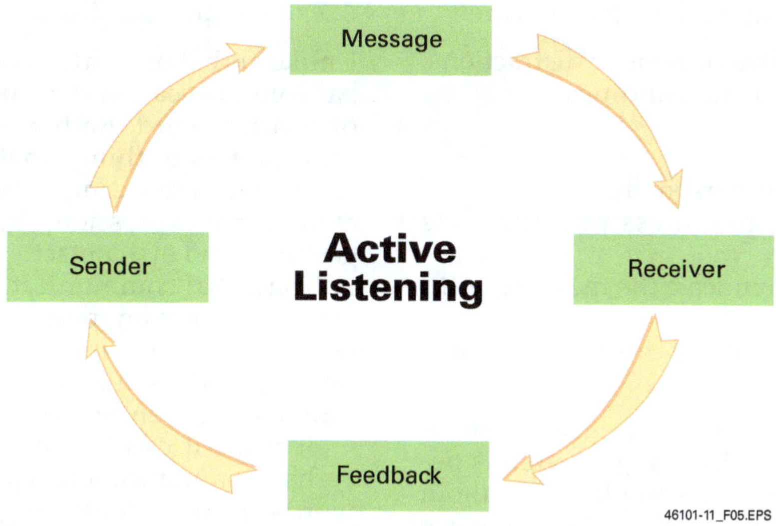

Figure 5 Communication process.

4.1.1 The Sender

The sender is the person who creates the message to be communicated. In verbal communication, the sender actually says the message aloud to the person(s) for whom it is intended.

The sender must be sure to speak in a clear and concise manner that can be easily understood by others. This is not an easy task; it takes practice. Some basic speaking tips are:

- Avoid talking with anything in your mouth (food, gum, etc.).
- Avoid swearing and acronyms.
- Don't speak too quickly or too slowly. In extreme cases, people tend to focus on the rate of speech rather than what is being communicated.
- Pronounce words carefully to prevent misunderstandings.
- Speak with enthusiasm. Avoid speaking in a harsh voice or in a monotone.

4.1.2 The Message

The message is what the sender is attempting to communicate to the audience. A message can be a set of directions, an opinion, or a feeling. Whatever its function, a message is an idea or fact that the sender wants the audience to know.

Before speaking, determine what must be communicated, then take the time to organize what to say, ensuring that the message is logical and complete. Taking the time to clarify your thoughts prevents rambling, not getting the message across effectively, or confusing the audience. It also permits the sender to get to the point quickly.

In delivering the message, the sender should assess the audience. It is important not to talk down to them. Remember that everyone, whether in a senior or junior position, deserves respect and courtesy. Therefore, the sender should use words and phrases that the audience can understand and avoid technical language or slang. In addition, the sender should use short sentences, which gives the audience time to understand and digest one point or fact at a time.

4.1.3 The Receiver

The receiver is the person to whom the message is communicated. For the communication process to be successful, it is important that the receiver understands the message as the sender intended. Therefore, the receiver must listen to what is being said.

There are many barriers to effective listening, particularly on a busy construction job site. Some of these obstacles include the following:

- Noise, visitors, cell phones, or other distractions
- Preoccupation, being under pressure, or daydreaming
- Reacting emotionally to what is being communicated
- Thinking about how to respond instead of listening
- Giving an answer before the message is complete
- Personal biases to the sender's communication style
- Finishing the sender's sentence

Some tips for overcoming these barriers are:

- Take steps to minimize or remove distractions; learn to tune out your surroundings
- Listen for key points
- Take notes
- Try not to take things personally
- Allow yourself time to process your thoughts before responding
- Let the sender communicate the message without interruption
- Be aware of your personal biases, and try to stay open-minded

There are many ways for a receiver to show that he or she is actively listening to what is being said. This can even be accomplished without saying a word. Examples include maintaining eye contact, nodding your head, and taking notes. It may also be accomplished through feedback.

4.1.4 Feedback

Feedback refers to the communication that occurs after the message has been sent by the sender and received by the receiver. It involves the receiver responding to the message.

Feedback is a very important part of the communication process because it allows the receiver to communicate how he or she interpreted the message. It also allows the sender to ensure that the message was understood as intended. In other words, feedback is a checkpoint to make sure the receiver and sender are on the same page.

The receiver can use the opportunity of providing feedback to paraphrase back what was heard. When paraphrasing what you heard, it is best to use your own words. That way, you can show the sender that you interpreted the message correctly and could explain it to others if needed.

In addition, the receiver can clarify the meaning of the message and request additional information when providing feedback. This is generally accomplished by asking questions.

One opportunity to provide feedback is in the performance of crew evaluations. Many companies have formal evaluation forms that are used on a yearly basis to evaluate workers for pay increases. These evaluations should not come as a once-a-year surprise. An effective crew leader provides constant performance feedback, which is ultimately reflected in the annual performance evaluation. It is also important to stress the importance of self-evaluation with your crew.

4.2.0 Nonverbal Communication

Unlike verbal or written communication, nonverbal communication does not involve the spoken or written word. Rather, non-verbal communication refers to things that you can actually see when communicating with others. Examples include facial expressions, body movements, hand gestures, and eye contact.

Nonverbal communication can provide an external signal of an individual's inner emotions. It occurs simultaneously with verbal communication; often, the sender of the nonverbal communication is not even aware of it.

Because it can be physically observed, nonverbal communication is just as important as the words used in conveying the message. Often, people are influenced more by nonverbal signals than by spoken words. Therefore, it is important to be conscious of nonverbal cues because you don't want the receiver to interpret your message incorrectly based on your posture or an expression on your face. After all, these things may have nothing to do with the communication exchange; instead, they may be carrying over from something else going on in your day.

4.3.0 Written or Visual Communication

Some communication will have to be written or visual. Written or visual communication refers to communication that is documented on paper or transmitted electronically using words or visuals.

Many messages on a job have to be communicated in text form. Examples include weekly reports, requests for changes, purchase orders, and correspondence on a specific subject. These items are written because they must be recorded for business and historical purposes. In addition, some communication on the job will have to be visual. Items that are difficult to explain verbally or by the written word can best be explained through diagrams or graphics. Examples include the plans or drawings used on a job.

When writing or creating a visual message, it is best to assess the reader or the audience before beginning. The reader must be able to read the message and understand the content; otherwise, the communication process will be unsuccessful. Therefore, the writer should consider the actual meaning of words or diagrams and how others might interpret them. In addition, the writer should make sure that all handwriting is legible if the message is being handwritten.

Here are some basic tips for writing:

- Avoid emotion-packed words or phrases.
- Be positive whenever possible.
- Avoid using technical language or jargon.
- Stick to the facts.
- Provide an adequate level of detail.
- Present the information in a logical manner.
- Avoid making judgments unless asked to do so.
- Proofread your work; check for spelling and grammatical errors.
- Make sure that the document is legible.
- Avoid using acronyms.
- Make sure the purpose of the message is clearly stated.
- Be prepared to provide a verbal or visual explanation, if needed.

Here are some basic tips for creating visuals:

- Provide an adequate level of detail.
- Ensure that the diagram is large enough to be seen.
- Avoid creating complex visuals; simplicity is better.
- Present the information in a logical order.
- Be prepared to provide a written or verbal explanation of the visual, if needed.

4.4.0 Communication Issues

It is important to note that each person communicates a little differently; that is what makes us unique as individuals. As the diversity of the workforce changes, communication will become even more challenging because the audience may include individuals from different ethnic groups, cultural backgrounds, educational levels, and economic status groups. Therefore, it is necessary to assess the audience in order to determine how to communicate effectively with each individual.

The key to effective communication is to acknowledge that people are different and to be able to adjust the communication style to meet the needs of the audience or the person on the receiving end of your message. This involves relaying the message in the simplest way possible, avoiding the use of words that people may find confusing. Be aware of how you use technical language, slang, jargon, and words that have multiple meanings. Present the information in a clear, concise manner. Avoid rambling and always speak clearly, using good grammar.

In addition, be prepared to communicate the message in multiple ways or adjust your level of detail or terminology to ensure that everyone understands the meaning as intended. For instance, a visual person who cannot comprehend directions in a verbal or written form may need a map. It may be necessary to overcome language barriers on the job site by using graphics or visual aids to relay the message.

Figure 6 shows how to tailor the message to the audience.

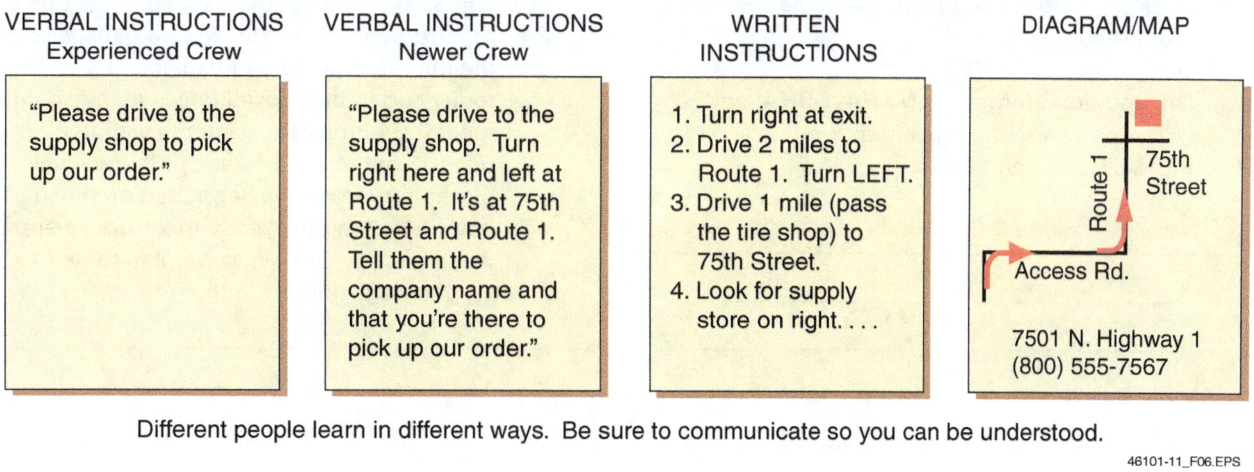

Figure 6 Tailor your message.

Participant Exercise A, Part I

Read the following verbal conversations, and identify any problems:

Conversation I:

Judy: Hey, Roger...
Roger: What's up?
Judy: Has the site been prepared for the job trailer yet?
Roger: Job trailer?
Judy: The job trailer—it's coming in today. What time will the job site be prepared?
Roger: The trailer will be here about 1:00 PM.
Judy: The job site! What time will the job site be prepared?

____

Conversation II:

John: Hey, Mike, I need your help.
Mike: What is it?
John: You and Joey go over and help Al's crew finish laying out the site.
Mike: Why me? I can't work with Joey. He can't understand a word I say.
John: Al's crew needs some help, and you and Joey are the most qualified to do the job.
Mike: I told you, I can't work with Joey.

____

Conversation III:

Ed: Hey, Jill.
Jill: Sir?
Ed: Have you received the latest DOL, EEO requirement to be sure the OFCP administrator finds our records up to date when he reviews them in August?
Jill: DOL, EEO, and OFCP?
Ed: Oh, and don't forget the MSHA, OSHA, and EPA reports are due this afternoon.
Jill: MSHA, OSHA, and EPA?

____

Conversation IV:

Susan: Hey, Bob, would you do me a favor?
Bob: Okay, Sue. What is it?
Susan: I was reading the concrete inspection report and found the concrete in Bays 4A, 3B, 6C, and 5D didn't meet the 3,000 psi strength requirements. Also, the concrete inspector on the job told me the two batches that came in today had to be refused because they didn't meet the slump requirements as noted on page 16 of the spec. I need to know if any placement problems happened on those bays, how long the ready mix trucks were waiting today, and what we plan to do to stop these problems in the future.

____

Participant Exercise A, Part II

Read the following written memos, and identify any problems:

Memo I:

Let's start with the transformer vault $285.00 due. For what you ask? Answer: practically nothing I admit, but here is the story. Paul the superintendent decided it was not the way good ole Comm Ed wanted it, we took out the ladder and part of the grading (as Paul instructed us to do) we brought it back here to change it. When Comm Ed the architect or Doe found out that everything would still work the way it was, Paul instructed us to reinstall the work. That is the whole story there is please add the $285.00 to my next payout.

____

____

____

____

____

____

____

____

Memo II:

Let's take rooms C 307-C-312 and C-313 we made the light track supports and took them to the job to erect them when we tried to put them in we found direct work in the way, my men spent all day trying to find out what to do so ask your Superintendent (Frank) he will verify seven hours pay for these men as he went back and forth while my men waited. Now the Architect has changed the system of hanging and has the gall to say that he has made my work easier, I can't see how. Anyway, we want an extra two (2) men for seven (7) hours for April 21 at $55.00 per hour or $385.00 on April 28th Doe Reference 197 finally resolved this problem. We will have no additional charges on Doe Reference 197, please note.

____

____

____

____

____

____

5.0.0 Motivation

The ability to motivate others is a key skill that leaders must develop. Motivation is the ability to influence. It also describes the amount of effort that a person is willing to put forth to accomplish something. For example, a crew member who skips breaks and lunch in an effort to complete a job on time is thought to be highly motivated, but a crew member who does the bare minimum or just enough to keep his or her job is considered unmotivated.

Employee motivation has dimension because it can be measured. Examples of how motivation can be measured include determining the level of absenteeism, the percentage of employee turnover, and the number of complaints, as well as the quality and quantity of work produced.

5.1.0 Employee Motivators

Different things motivate different people in different ways. Consequently, there is no one-size-fits-all approach to motivating crew members. It is important to recognize that what motivates one crew member may not motivate another. In addition, what works to motivate a crew member once may not motivate that same person again in the future.

Frequently, the needs that motivate individuals are the same as those that create job satisfaction. They include the following:

- Recognition and praise
- Accomplishment
- Opportunity for advancement
- Job importance
- Change
- Personal growth
- Rewards

A crew leader's ability to satisfy these needs increases the likelihood of high morale within a crew. Morale refers to an individual's attitude toward the tasks he or she is expected to perform. High morale, in turn, means that employees will be motivated to work hard, and they will have a positive attitude about coming to work and doing their jobs.

5.1.1 Recognition and Praise

Recognition and praise refer to the need to have good work appreciated, applauded, and acknowledged by others. This can be accomplished by simply thanking employees for helping out on a project, or it can entail more formal praise, such as an award for Employee of the Month.

Some tips for giving recognition and praise include the following:

- Be available on the job site so that you have the opportunity to witness good work.
- Know good work and praise it when you see it.
- Look for good work and look for ways to praise it.
- Give recognition and praise only when truly deserved; otherwise, it will lose its meaning.
- Acknowledge satisfactory performance, and encourage improvement by showing confidence in the ability of the crew members to do above-average work.

5.1.2 Accomplishment

Accomplishment refers to a worker's need to set challenging goals and achieve them. There is nothing quite like the feeling of achieving a goal, particularly a goal one never expected to accomplish in the first place.

Crew leaders can help their crew members attain a sense of accomplishment by encouraging them to develop performance plans, such as goals for the year that will be used in performance evaluations. In addition, crew leaders can provide the support and tools (such as training and coaching) necessary to help their crew members achieve these goals.

5.1.3 Opportunity for Advancement

Opportunity for advancement refers to an employee's need to gain additional responsibility and develop new skills and abilities. It is important that employees know that they are not limited to their current jobs. Let them know that they have a chance to grow with the company and to be promoted as recognition for excelling in their work.

Effective leaders encourage their crew members to work to their full potentials. In addition, they share information and skills with their employees in an effort to help them to advance within the organization.

5.1.4 Job Importance

Job importance refers to an employee's need to feel that his or her skills and abilities are valued and make a difference. Employees who do not feel valued tend to have performance and attendance issues. Crew leaders should attempt to make every crew member feel like an important part of the team, as if the job wouldn't be possible without their help.

5.1.5 Change

Change refers to an employee's need to have variety in work assignments. Change is what keeps things interesting or challenging. It prevents the boredom that results from doing the same task day after day with no variety.

5.1.6 Personal Growth

Personal growth refers to an employee's need to learn new skills, enhance abilities, and grow as a person. It can be very rewarding to master a new competency on the job. Similar to change, personal growth prevents the boredom associated with doing the same thing day after day without developing any new skills.

Crew leaders should encourage the personal growth of their employees as well as themselves. Learning should be a two-way street on the job site; crew leaders should teach their crew members and learn from them as well. In addition, crew members should be encouraged to learn from each other.

5.1.7 Rewards

Rewards are compensation for hard work. Rewards can include a crew member's base salary or go beyond that to include bonuses or other incentives. They can be monetary in nature (salary raises, holiday bonuses, etc.), or they can be nonmonetary, such as free merchandise (shirts, coffee mugs, jackets, etc.) or other prizes. Attendance at training courses can be another form of reward.

5.2.0 Motivating Employees

To increase motivation in the workplace, crew leaders must individualize how they motivate different crew members. It is important that crew leaders get to know their crew members and determine what motivates them as individuals. Once again, as diversity increases in the workforce, this becomes even more challenging; therefore, effective communication skills are essential.

Did you know?

Education doesn't stop the day a person receives a diploma or certificate. It is a lifelong activity. Continuing education has long been recognized as a pathway to advancement, but in reality, it is essential to simply remaining in place. Regardless of what you do, new materials, methods, and processes are constantly emerging. Those who don't make the effort to keep up will fall behind.

Here is a list of some tips for motivating employees:

- Keep jobs challenging and interesting. Boredom is a guaranteed de-motivator.
- Communicate your expectations. People need clear goals in order to feel a sense of accomplishment when the goals are achieved.
- Involve the employees. Feeling that their opinions are valued leads to pride in ownership and active participation.
- Provide sufficient training. Give employees the skills and abilities they need to be motivated to perform.
- Mentor the employees. Coaching and supporting employees boosts their self-esteem, their self-confidence, and ultimately their motivation.
- Lead by example. Become the kind of leader employees admire and respect, and they will be motivated to work for you.
- Treat employees well. Be considerate, kind, caring, and respectful; treat employees the way that you want to be treated.
- Avoid using scare tactics. Threatening employees with negative consequences can backfire, resulting in employee turnover instead of motivation.
- Reward your crew for doing their best by giving them easier tasks from time to time. It is tempting to give your best employees the hardest or dirtiest jobs because you know they will do the jobs correctly.
- Reward employees for a job well done.

Participant Exercise B

You are the crew leader of a masonry crew. Sam Williams is the person whom the company holds responsible for ensuring that equipment is operable and distributed to the jobs in a timely manner.

Occasionally, disagreements with Sam have resulted in tools and equipment arriving late. Sam, who has been with the company 15 years, resents having been placed in the job and feels that he outranks all the crew leaders.

Sam figured it was about time he talked with someone about the abuse certain tools and other items of equipment were receiving on some of the jobs. Saws were coming back with guards broken and blades chewed up, bits were being sheared in half, motor housings were bent or cracked, and a large number of tools were being returned covered with mud. Sam was out on your job when he observed a mason carrying a portable saw by the cord. As he watched, he saw the mason bump the swinging saw into a steel column. When the man arrived at his workstation, he dropped the saw into the mud.

You are the worker's crew leader. Sam approached as you were coming out of the work trailer. He described the incident. He insisted, as crew leader, you are responsible for both the work of the crew and how its members use company property. Sam concluded, "You'd better take care of this issue as soon as possible! The company is sick and tired of having your people mess up all the tools!"

You are aware that some members of your crew have been mistreating the company equipment.

1. How would you respond to Sam's accusations?

____

____

____

____

2. What action would you take regarding the misuse of the tools?

____

____

____

____

3. How can you motivate the crew to take better care of their tools? Explain.

____

____

____

____

6.0.0 Team Building

Organizations are making the shift from the traditional boss-worker mentality to one that promotes teamwork. The manager becomes the team leader, and the workers become team members. They all work together to achieve the common goals of the team.

There are a number of benefits associated with teamwork. They include the ability to complete complex projects more quickly and effectively, higher employee satisfaction, and a reduction in turnover.

6.1.0 Successful Teams

Successful teams are made up of individuals who are willing to share their time and talents in an effort to reach a common goal—the goal of the team. Members of successful teams possess an *Us* or *We* attitude rather than an *I* or *You* attitude; they consider what's best for the team and put their egos aside.

Some characteristics of successful teams include the following:

- Everyone participates and every team member counts.
- There is a sense of mutual trust and interdependence.
- Team members are empowered.
- They communicate.
- They are creative and willing to take risks.
- The team leader develops strong people skills and is committed to the team.

6.2.0 Building Successful Teams

To be successful in the team leadership role, the crew leader should contribute to a positive attitude within the team.

There are several ways in which the team leader can accomplish this. First, he or she can work with the team members to create a vision or purpose of what the team is to achieve. It is important that every team member is committed to the purpose of the team, and the team leader is instrumental in making this happen.

Team leaders within the construction industry are typically assigned a crew. However, it can be beneficial for the team leader to be involved in selecting the team members. Selection should be based on a willingness of people to work on the team and the resources that they are able to bring to the team.

When forming a new team, team leaders should do the following:

- Explain the purpose of the team. Team members need to know what they will be doing, how long they will be doing it (if they are temporary or permanent), and why they are needed.
- Help the team establish goals or targets. Teams need a purpose, and they need to know what it is they are responsible for accomplishing.
- Define team member roles and expectations. Team members need to know how they fit into the team and what is expected of them as members of the team.
- Plan to transfer responsibility to the team as appropriate. Teams should be responsible for the tasks to be accomplished.

7.0.0 Getting the Job Done

Crew leaders must implement policies and procedures to make sure that the work is done correctly. Construction jobs have crews of people with various experiences and skill levels available to perform the work. The crew leader's job is to draw from this expertise to get the job done well and in a timely manner.

7.1.0 Delegating

Once the various activities that make up the job have been determined, the crew leader must identify the person or persons who will be responsible for completing each activity. This requires that the crew leader be aware of the skills and abilities of the people on the crew. Then, the crew leader must put this knowledge to work in matching the crew's skills and abilities to specific tasks that must be accomplished to complete the job.

After matching crew members to specific activities, the crew leader must then delegate the assignments to the responsible person(s). Delegation is generally communicated verbally by the crew leader talking directly to the person who has been assigned the activity. However, there may be times when work is assigned indirectly through written instructions or verbally through someone other than the crew leader.

When delegating work, remember to:

- Delegate work to a crew member who can do the job properly. If it becomes evident that he or she does not perform to the standard desired, either teach the crew member to do the work correctly or turn it over to someone else who can.

- Make sure the crew member understands what to do and the level of responsibility. Make sure desired results are clear, specify the boundaries and deadlines for accomplishing the results, and note the available resources.
- Identify the standards and methods of measurement for progress and accomplishment, along with the consequences of not achieving the desired results. Discuss the task with the crew member and check for understanding by asking questions. Allow the crew member to contribute feedback or make suggestions about how the task should be performed in a safe and quality manner.
- Give the crew member the time and freedom to get started without feeling the pressure of too much supervision. When making the work assignment, be sure to tell the crew member how much time there is to complete it, and confirm that this time is consistent with the job schedule.
- Examine and evaluate the result once a task is complete. Then, give the crew member some feedback as to how well it has been done. Get the crew member's comments. The information obtained from this is valuable and will enable the crew leader to know what kind of work to assign that crew member in the future. It will also give the crew leader a means of measuring his or her own effectiveness in delegating work.

7.2.0 Implementing Policies and Procedures

Every company establishes policies and procedures that employees are expected to follow and the crew leaders are expected to implement. Company policies and procedures are essentially guides for how the organization does business. They can also reflect organizational philosophies such as putting safety first or making the customer the top priority. Examples of policies and procedures include safety guidelines, credit standards, and billing processes.

Here are some tips for implementing policies and procedures:

- Learn the purpose of each policy. That way, you can follow it and apply it properly and fairly.
- If you're not sure how to apply a company policy or procedure, check the company manual or ask your supervisor.
- Apply company policies and procedures. Remember that they combine what's best for the customer and the company. In addition, they provide direction on how to handle specific situations and answer questions.
- If you are uncertain how to apply a policy, check with your supervisor.

Crew leaders may need to issue orders to their crew members. Basically, an order initiates, changes, or stops an activity. Orders may be general or specific, written or oral, and formal or informal. The decision of how an order will be issued is up to the crew leader, but it is governed by the policies and procedures established by the company.

When issuing orders:

- Make them as specific as possible.
- Avoid being general or vague unless it is impossible to foresee all of the circumstances that could occur in carrying out the order.
- Recognize that it is not necessary to write orders for simple tasks unless the company requires that all orders be written.
- Write orders for more complex tasks that will take considerable time to complete or orders that are permanent.
- Consider what is being said, the audience to whom it applies, and the situation under which it will be implemented to determine the appropriate level of formality for the order.

8.0.0 Problem Solving and Decision Making

Problem solving and decision making are a large part of every crew leader's daily work. There will always be problems to be resolved and decisions to be made, especially in fast-paced, deadline-oriented industries.

8.1.0 Decision Making vs. Problem Solving

Sometimes, the difference between decision making and problem solving is not clear. Decision making refers to the process of choosing an alternative course of action in a manner appropriate for the situation. Problem solving involves determining the difference between the way things are and the way things should be, and finding out how to bring the two together. The two activities are interrelated because in order to make a decision, you may also have to use problem-solving techniques.

8.2.0 Types of Decisions

Some decisions are routine or simple. Such decisions can be made based on past experiences. An example would be deciding how to get to and from work. If you've worked at the same place for a long time, you are already aware of the options for traveling to and from work (take the bus, drive a car, carpool with a co-worker, take a taxi, etc.). Based on past experiences with the options identified, you can make a decision about how best to get to and from work.

On the other hand, some decisions are more difficult. These decisions require more careful thought about how to carry out an activity by using a formal problem-solving technique. An example is planning a trip to a new vacation spot. If you are not sure how to get there, where to stay, what to see, etc., one option is to research the area to determine the possible routes, hotel accommodations, and attractions. Then, you will have to make a decision about which route to take, what hotel to choose, and what sites to visit, without the benefit of direct past experience.

8.3.0 Problem Solving

The ability to solve problems is an important skill in any workplace. It's especially important for craftworkers, whose workday is often not predictable or routine. In this section, you will learn a five-step process for solving problems, which you can apply to both workplace and personal issues. Review the following steps and then see how they can be applied to a job-related problem. Keep in mind that a problem will not be solved until everyone involved admits that there is a problem.

Step 1 ***Define the problem.*** This isn't as easy as it sounds. Thinking through the problem often uncovers additional problems.

Step 2 ***Think about different ways to solve the problem.*** There is often more than one solution to a problem, so you must think through each possible solution and pick the best one. The best solution might be taking parts of two different solutions and combining them to create a new solution.

Step 3 ***Pick the solution that seems best and figure out an action plan.*** It is best to receive input both from those most affected by the problem and from those who will be most affected by any potential solution.

Step 4 ***Test the solution to determine whether it actually works.*** Many solutions sound great in theory but in practice don't turn out to be effective. On the other hand, you might discover from trying to apply a solution that it is acceptable with a little modification. If a solution does not work, think about how you could improve it, and then implement your new plan.

Step 5 ***Evaluate the process.*** Review the steps you took to discover and implement the solution. Could you have done anything better? If the solution turns out to be satisfactory, you can add the solution to your knowledge base.

Next, you will see how to apply the problem-solving process to a workplace problem. Read the following situation and apply the five-step problem-solving process to come up with a solution to the issues posed by the situation.

Situation:

You are part of a team of workers assigned to a new shopping mall project. The project will take about 18 months to complete. The only available parking is half a mile from the job site. The crew has to carry heavy toolboxes and safety equipment from their cars and trucks to the work area at the start of the day, and then carry them back at the end of their shifts.

Step 1 ***Define the problem.*** Workers are wasting time and energy hauling all their equipment to and from the work site.

Step 2 ***Think about different ways to solve the problem.*** Several solutions have been proposed:

- Install lockers for tools and equipment closer to the work site.
- Have workers drive up to the work site to drop off their tools and equipment before parking.
- Bring in another construction trailer where workers can store their tools and equipment for the duration of the project.
- Provide a round-trip shuttle service to ferry workers and their tools.

NOTE

Each solution will have pros and cons, so it is important to receive input from the workers affected by the problem. For example, workers will probably object to any plan (like the drop-off plan) that leaves their tools vulnerable to theft.

Step 3 ***Pick the solution that seems best and figure out an action plan.*** The workers decide that the shuttle service makes the most sense. It should solve the time and energy problem, and workers can keep their tools with them. To put the plan into effect, the project supervisor arranges for a large van and driver to provide the shuttle service.

Step 4 ***Test the solution to determine whether it actually works.*** The solution works, but there is a problem. All the workers are scheduled to start and leave at the same time, so there is not enough room in the van for all the workers and their equipment. To solve this problem, the supervisor schedules trips spaced 15 minutes apart. The supervisor also adjusts worker schedules to correspond with the trips. That way, all the workers will not try to get on the shuttle at the same time.

Step 5 ***Evaluate the process.*** This process gave both management and workers a chance to express an opinion and discuss the various solutions. Everyone feels pleased with the process and the solution.

8.4.0 Special Leadership Problems

Because they are responsible for leading others, it is inevitable that crew leaders will encounter problems and be forced to make decisions about how to respond to the problem. Some problems will be relatively simple to resolve, like covering for a sick crew member who has taken a day off from work. Other problems will be complex and much more difficult to handle.

Some complex problems are relatively common. A few of the major employee problems include:

- Inability to work with others
- Absenteeism and turnover
- Failure to comply with company policies and procedures

8.4.1 Inability to Work with Others

Crew leaders will sometimes encounter situations where an employee has a difficult time working with others on the crew. This could be a result of personality differences, an inability to communicate, or some other cause. Whatever the reason, the crew leader must address the issue and get the crew working as a team.

The best way to determine the reason for why individuals don't get along or work well together is to talk to the parties involved. The crew leader should speak openly with the employee, as well as the other individual(s) to uncover the source of the problem and discuss its resolution.

Once the reason for the conflict is found, the crew leader can determine how to respond. There may be a way to resolve the problem and get the workers communicating and working as a team again. On the other hand, there may be nothing that can be done that will lead to a harmonious solution. In this case, the crew leader would either have to transfer the employee to another crew or have the problem crew member terminated. This latter option should be used as a last measure and should be discussed with one's superiors or Human Resources Department.

8.4.2 Absenteeism and Turnover

Absenteeism and turnover are big problems. Without workers available to do the work, jobs are delayed, and money is lost.

Absenteeism refers to workers missing their scheduled work time on a job. Absenteeism has many causes, some of which are inevitable. For instance, people get sick, they have to take time off for family emergencies, and they have to attend family events such as funerals. However, there are some causes of absenteeism that can be prevented by the crew leader.

The most effective way to control absenteeism is to make the company's policy clear to all employees. Companies that do this find that chronic absenteeism is reduced. New employees should have the policy explained to them. This explanation should include the number of absences allowed and the reasons for which sick or personal days can be taken. In addition, all workers should know how to inform their crew leaders when they miss work and understand the consequences of exceeding the number of sick or personal days allowed.

Once the policy on absenteeism is explained to employees, crew leaders must be sure to implement it consistently and fairly. If the policy is administered equally, employees will likely follow it. However, if the policy is not administered equally and some employees are given exceptions, then it will not be effective. Consequently, the rate of absenteeism is likely to increase.

Despite having a policy on absenteeism, there will always be employees who are chronically late or miss work. In cases where an employee abuses the absenteeism policy, the crew leader should discuss the situation directly with the employee. The crew leader should confirm that the employee understands the company's policy and insist that the employee comply with it. If the employee's behavior continues, disciplinary action may be in order.

Turnover refers to the loss of an employee that is initiated by that employee. In other words, the employee quits and leaves the company to work elsewhere or is fired for cause.

Like absenteeism, there are some causes of turnover that cannot be prevented and others that can. For instance, it is unlikely that a crew leader could keep an employee who finds a job elsewhere earning twice as much money. However, crew leaders can prevent some employee turnover situations. They can work to ensure safe working conditions for their crew, treat their workers fairly and consistently, and help promote good working conditions. The key is communication. Crew leaders need to know the problems if they are going to be able to successfully resolve them.

Some of the major causes of turnover include the following:

- Unfair/inconsistent treatment by the immediate supervisor
- Unsafe project sites
- Lack of job security

For the most part, the actions described for absenteeism are also effective for reducing turnover. Past studies have shown that maintaining harmonious relationships on the job site goes a long way in reducing both turnover and absenteeism. This requires effective leadership on the part of the crew leader.

8.4.3 Failure to Comply With Company Policies and Procedures

Policies are the rules that define the relationship between the company, its employees, its clients, and its subcontractors. Procedures are the instructions for carrying out the policies. Some companies have dress codes that are reflected in their policies. The dress code may be partly to ensure safety, and partly to define the image a company wants to project to the outside world.

Companies develop procedures to ensure that everyone who performs a task does it safely and efficiently. Many procedures deal with safety. A lockout/tagout procedure is an example. In this procedure, the company defines who may perform a lockout, how it is done, and who has the authority to remove or override it. Workers who fail to follow the procedure endanger themselves, as well as their co-workers.

Among a typical company's policies is the policy on disciplinary action. This policy defines steps to be taken in the event that an employee violates the company's policies or procedures. The steps range from counseling by a supervisor for the first offense, to a written warning, to dismissal for repeat offenses. This will vary from one company to another. For example, some companies will fire an employee for any violation of safety procedures.

The crew leader has the first-line responsibility for enforcing company policies and procedures. The crew leader should take the time with a new crew member to discuss the policies and procedures and show the crew member how to access them. If a crew member shows a tendency to neglect a policy or procedure, it is up to the crew leader to counsel that individual. If the crew member continues to violate a policy or procedure, the crew leader has no choice but to refer that individual to the appropriate authority within the company for disciplinary action.

Participant Exercise C

Case I:

On the way over to the job trailer, you look up and see a piece of falling scrap heading for one of the laborers. Before you can say anything, the scrap material hits the ground about five feet in front of the worker. You notice the scrap is a piece of conduit. You quickly pick it up, assuring the worker you will take care of this matter.

Looking up, you see your crew on the third floor in the area from which the material fell. You decide to have a talk with them. Once on the deck, you ask the crew if any of them dropped the scrap. The men look over at Bob, one of the electricians in your crew. Bob replies, "I guess it was mine. It slipped out of my hand."

It is a known fact that the Occupational Safety and Health Administration (OSHA) regulations state that an enclosed chute of wood shall be used for material waste transportation from heights of 20 feet or more. It is also known that Bob and the laborer who was almost hit have been seen arguing lately.

1. Assuming Bob's action was deliberate, what action would you take?

___

___

___

2. Assuming the conduit accidentally slipped from Bob's hand, how can you motivate him to be more careful?

___

___

___

3. What follow-up actions, if any, should be taken relative to the laborer who was almost hit?

___

___

___

4. Should you discuss the apparent OSHA violation with the crew? Why or why not?

___

___

___

5. What acts of leadership would be effective in this case? To what leadership traits are they related?

___

___

___

Case II:

Mike has just been appointed crew leader of a tile-setting crew. Before his promotion into management, he had been a tile setter for five years. His work had been consistently of superior quality.

Except for a little good-natured kidding, Mike's co-workers had wished him well in his new job. During the first two weeks, most of them had been cooperative while Mike was adjusting to his supervisory role.

At the end of the second week, a disturbing incident took place. Having just completed some of his duties, Mike stopped by the job-site wash station. There he saw Steve and Ron, two of his old friends who were also in his crew, washing.

"Hey, Ron, Steve, you should not be cleaning up this soon. It's at least another thirty minutes until quitting time," said Mike. "Get back to your work station, and I'll forget I saw you here."

"Come off it, Mike," said Steve. "You used to slip up here early on Fridays. Just because you have a little rank now, don't think you can get tough with us." To this Mike replied, "Things are different now. Both of you get back to work, or I'll make trouble." Steve and Ron said nothing more, and they both returned to their work stations.

From that time on, Mike began to have trouble as a crew leader. Steve and Ron gave him the silent treatment. Mike's crew seemed to forget how to do the most basic activities. The amount of rework for the crew seemed to be increasing. By the end of the month, Mike's crew was behind schedule.

1. How do you think Mike should have handled the confrontation with Ron and Steve?

___

___

2. What do you suggest Mike could do about the silent treatment he got from Steve and Ron?

___

___

3. If you were Mike, what would you do to get your crew back on schedule?

___

___

4. What acts of leadership could be used to get the crew's willing cooperation?

___

___

5. To which leadership traits do they correspond?

___

___

Review Questions

1. A crew leader differs from a craftworker in that a _____.
   a. crew leader need not have direct experience in those job duties that a craftworker typically performs
   b. crew leader can expect to oversee one or more craftworkers in addition to performing some of the typical duties of the craftworker
   c. crew leader is exclusively in charge of overseeing, since performing technical work is not part of this role
   d. crew leader's responsibilities do not include being present on the job site

2. Among the many traits effective leaders should have is _____.
   a. the ability to communicate the goals of a project
   b. the drive necessary to carry the workload by themselves in order to achieve a goal
   c. a perfectionist nature, which ensures that they will not make useless mistakes
   d. the ability to make decisions without needing to listen to the opinions of others

3. Of the three styles of leadership, the _____ style would be effective in dealing with a craftworker's negative attitude.
   a. facilitator
   b. commander
   c. partner
   d. dictator

4. One way to overcome barriers to effective communication is to _____.
   a. avoid taking notes on the content of the message, since this can be distracting
   b. avoid reacting emotionally to the message
   c. anticipate the content of the message and interrupt if necessary in order to show interest
   d. think about how to respond to the message while listening

5. Feedback is important in verbal communication because it requires the _____.
   a. sender to repeat the message
   b. receiver to restate the message
   c. sender to avoid technical jargon
   d. sender to concentrate on the message

6. A good way to motivate employees is to use a one-size-fits-all approach, since employees are members of a team with a common goal.
   a. True
   b. False

7. A crew leader can effectively delegate responsibilities by _____.
   a. refraining from evaluating the employee's performance once the task is completed, since it is a new task for the employee
   b. doing the job for the employee to make sure the task is done correctly
   c. allowing the employee to give feedback and suggestions about the task
   d. communicating information to the employee, generally in written form

8. Problem solving differs from decision making in that _____.
   a. problem solving involves identifying discrepancies between the way a situation is and the way it should be
   b. decision making involves separating facts from non-facts
   c. decision making involves eliminating differences
   d. problem solving involves determining an alternative course of action for a given situation

SECTION THREE
SAFETY

Objectives

Upon completion of this section, you will be able to:

1. Explain the importance of safety.
2. Give examples of direct and indirect costs of workplace accidents.
3. Identify safety hazards of the construction industry.
4. Explain the purpose of OSHA.
5. Discuss OSHA inspection procedures.
6. Identify the key points of a safety program.
7. List steps to train employees on how to perform new tasks safely.
8. Identify a crew leader's safety responsibilities.
9. Explain the importance of having employees trained in first aid and cardiopulmonary resuscitation (CPR).
10. Describe the indications of substance abuse.
11. List the essential parts of an accident investigation.
12. Describe ways to maintain employee interest in safety. Distinguish between company policies and procedures.

1.0.0 SAFETY OVERVIEW

Businesses lose millions of dollars every year because of on-the-job accidents. Work-related injuries, sickness, and deaths have caused untold suffering for workers and their families. Project delays and budget overruns from injuries and fatalities result in huge losses for employers, and work-site accidents erode the overall morale of the crew.

Craftworkers are exposed to hazards as part of the job. Examples of these hazards include falls from heights, working on scaffolds, using cranes in the presence of power lines, operating heavy machinery, and working on electrically-charged or pressurized equipment. Despite these hazards, experts believe that applying preventive safety measures could drastically reduce the number of accidents.

As a crew leader, one of your most important tasks is to enforce the company's safety program and make sure that all workers are performing their tasks safely. To be successful, the crew leader should:

- Be aware of the costs of accidents.
- Understand all federal, state, and local governmental safety regulations.
- Be involved in training workers in safe work methods.
- Conduct training sessions.
- Get involved in safety inspections, accident investigations, and fire protection and prevention.

Crew leaders are in the best position to ensure that all jobs are performed safely by their crew members. Providing employees with a safe working environment by preventing accidents and enforcing safety standards will go a long way towards maintaining the job schedule and enabling a job's completion on time and within budget.

Did you know?

When OSHA inspects a job site, they focus on the types of safety hazards that are most likely to cause fatal injuries. These hazards fall into the following classifications:

- Falls from elevations
- Struck-by hazards
- Caught in/between hazards
- Electrical shock hazards

1.1.0 Accident Statistics

Each day, workers in construction and industrial occupations face the risk of falls, machinery accidents, electrocutions, and other potentially fatal occupational hazards.

The National Institute of Occupational Safety and Health (NIOSH) statistics show that about 1,000 construction workers are killed on the job each year, more fatalities than in any other industry. Falls are the leading cause of deaths in the construction industry, accounting for over 40 percent of the fatalities. Nearly half of the fatal falls occurred from roofs, scaffolds, or ladders. Roofers, structural metal workers, and painters experienced the greatest number of fall fatalities.

In addition to the number of fatalities that occur each year, there are a staggering number of work-related injuries. In 2007, for example, more than 135,000 job-related injuries occurred in the construction industry. NIOSH reports that approximately 15 percent of all worker's compensation costs are spent on injured construction workers. The causes of injuries on construction sites include falls, coming into contact with electric current, fires, and mishandling of machinery or equipment. According to NIOSH, back injuries are the leading safety problem in workplaces.

2.0.0 Costs of Accidents

Occupational accidents are estimated to cost more than $100 billion every year. These costs affect the employee, the company, and the construction industry as a whole.

Organizations encounter both direct and indirect costs associated with workplace accidents. Examples of direct costs include workers' compensation claims and sick pay; indirect costs include increased absenteeism, loss of productivity, loss of job opportunities due to poor safety records, and negative employee morale attributed to workplace injuries. There are many other related costs involved with workplace accidents. A company can be insured against some of them, but not others. To compete and survive, companies must control these as well as all other employment-related costs.

2.1.0 Insured Costs

Insured costs are those costs either paid directly or reimbursed by insurance carriers. Insured costs related to injuries or deaths include the following:

- Compensation for lost earnings (known as worker's comp)
- Medical and hospital costs
- Monetary awards for permanent disabilities
- Rehabilitation costs
- Funeral charges
- Pensions for dependents

Insurance premiums or charges related to property damages include:

- Fire
- Loss and damage
- Use and occupancy
- Public liability
- Replacement cost of equipment, material, and structures

2.2.0 Uninsured Costs

The costs related to accidents can be compared to an iceberg, as shown in *Figure 7*. The tip of the iceberg represents direct costs, which are the visible costs. The more numerous indirect costs are not readily measureable, but they can represent a greater burden than the direct costs.

Uninsured costs related to injuries or deaths include the following:

- First aid expenses
- Transportation costs
- Costs of investigations
- Costs of processing reports
- Down time on the job site
- Costs to train replacement workers

Uninsured costs related to wage losses include:

- Idle time of workers whose work is interrupted
- Time spent cleaning the accident area
- Time spent repairing damaged equipment
- Time lost by workers receiving first aid
- Costs of training injured workers in a new career

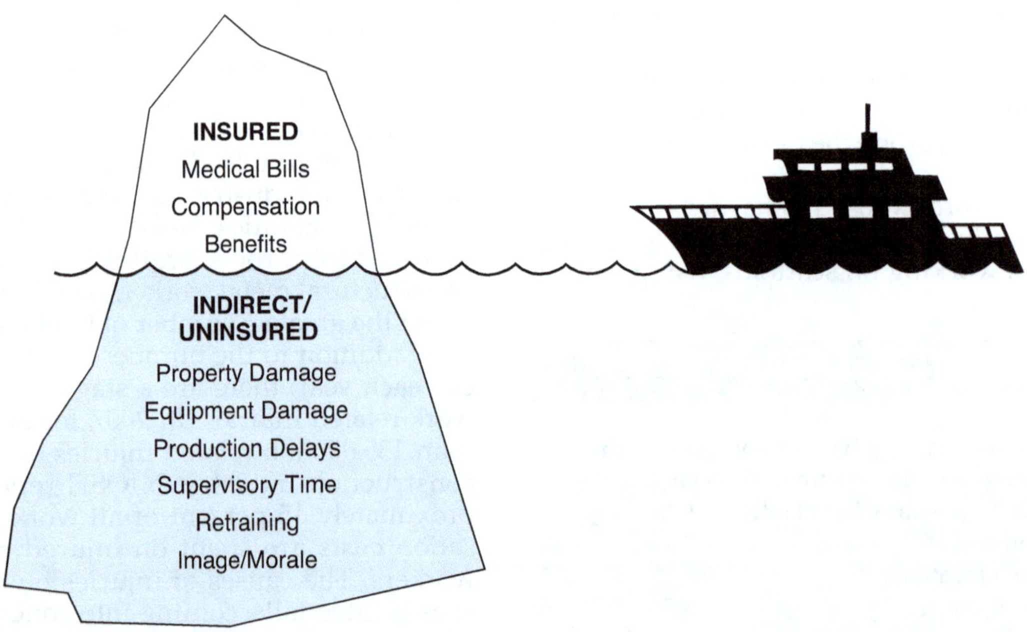

Figure 7 Costs associated with accidents.

> **Did you know?**
>
> If you are injured as the result of an accident, and it can be shown that you were negligent, you cannot sue your employer for damages. In addition, workers' compensation insurance companies may decline to pay claims for employees injured or killed due to their own negligence.

Uninsured costs related to production losses include:

- Product spoiled by accident
- Loss of skill and experience
- Lowered production or worker replacement
- Idle machine time

Associated costs may include the following:

- Difference between actual losses and amount recovered
- Costs of rental equipment used to replace damaged equipment
- Costs of new workers used to replace injured workers
- Wages or other benefits paid to disabled workers
- Overhead costs while production is stopped
- Impact on schedule
- Loss of bonus or payment of forfeiture for delays

Uninsured costs related to off-the-job activities include:

- Time spent on injured workers' welfare
- Loss of skill and experience of injured workers
- Costs of training replacement workers

Uninsured costs related to intangibles include:

- Lowered employee morale
- Increased labor conflict
- Unfavorable public relations
- Loss of bid opportunities because of poor safety records
- Loss of client goodwill

3.0.0 Safety Regulations

To reduce safety and health risks and the number of injuries and fatalities on the job, the federal government has enacted laws and regulations, including the *Occupational Safety and Health Act of 1970*. The purpose of OSHA is "to assure so far as possible every working man and woman in the Nation safe and healthful working conditions and to preserve our human resources."

To promote a safe and healthy work environment, OSHA issues standards and rules for working conditions, facilities, equipment, tools, and work processes. It does extensive research into occupational accidents, illnesses, injuries, and deaths in an effort to reduce the number of occurrences and adverse effects. In addition, OSHA regulatory agencies conduct workplace inspections to ensure that companies follow the standards and rules.

3.1.0 Workplace Inspections

To enforce OSHA regulations, the government has granted regulatory agencies the right to enter public and private properties to conduct workplace safety investigations. The agencies also have the right to take legal action if companies are not in compliance with the Act. These regulatory agencies employ OSHA Compliance Safety and Health Officers (CSHOs), who are chosen for their knowledge in the occupational safety and health field. The CSHOs are thoroughly trained in OSHA standards and in recognizing safety and health hazards.

States with their own occupational safety and health programs conduct inspections. To do so, they enlist the services of qualified state CSHOs.

Companies are inspected for a multitude of reasons. They may be randomly selected, or they may be chosen because of employee complaints, due to an imminent danger, or as a result of major accidents or fatalities.

OSHA can assess significant financial penalties for safety violations. In some cases, a superintendent or crew leader can be held criminally liable for repeat violations.

3.2.0 Penalties for Violations

OSHA has established monetary fines for the violation of their regulations. The penalties as of 2010 are shown in *Table 1*.

In addition to the fines, there are possible criminal charges for willful violations resulting in death or serious injury. There can also be personal liability for failure to comply with OSHA regulations. The attitude of the employer and their safety history can have a significant effect on the outcome of a case.

Table 1 OSHA Penalties for Violations

Violation	Penalty
Willful Violations	Maximum $70,000
Repeated Violations	Minimum $70,000
Serious, Other-than-Serious, Other Specific Violations	Minimum $7,000
OSHA Notice Violation	$1,000
Failure to Post *OSHA 300A Summary of Work-Related Injuries and Illnesses*	$1,000
Failure to Properly Maintain *OSHA 300 Log of Work-Related Injuries and Illnesses*	$1,000
Failure to Promptly and Properly Report Fatality/Catastrophe	$5,000
Failure to Permit Access to Records Under *OSHA 1904* Regulations	$1,000
Failure to Follow Advance Notification Requirements Under *OSHA 1903.6* Regulations	$2,000
Failure to Abate – for Each Calendar Day Beyond Abatement Date	$7,000
Retaliation Against Individual for Filing OSHA Complaint	$10,000

Did you know?

Nearly half the states in the U.S. have state-run OSHA programs. These programs are set up under federal OSHA guidelines and must establish job health and safety standards that are at least as effective as the federal standards. The states have the option of adopting more stringent standards or setting standards for hazards not addressed in the federal program. Of the 22 states with state-run OSHA programs, eight of them limit their coverage to public employees.

4.0.0 Employer Safety Responsibilities

Each employer must set up a safety and health program to manage workplace safety and health and to reduce work-related injuries, illnesses, and fatalities. The program must be appropriate for the conditions of the workplace. It should consider the number of workers employed and the hazards to which they are exposed while at work.

To be successful, the safety and health program must have management leadership and employee participation. In addition, training and informational meetings play an important part in effective programs. Being consistent with safety policies is the key. Regardless of the employer's responsibility, however, the individual worker is ultimately responsible for his or her own safety.

4.1.0 Safety Program

The crew leader plays a key role in the successful implementation of the safety program. The crew leader's attitudes toward the program set the standard for how crew members view safety. Therefore, the crew leader should follow all program guidelines and require crew members to do the same.

Safety programs should consist of the following:

- Safety policies and procedures
- Hazard identification and assessment
- Safety information and training
- Safety record system
- Accident investigation procedures
- Appropriate discipline for not following safety procedures
- Posting of safety notices

4.1.1 Safety Policies and Procedures

Employers are responsible for following OSHA and state safety standards. Usually, they incorporate OSHA and state regulations into a safety policies and procedures manual. Such a manual is presented to employees when they are hired.

Basic safety requirements should be presented to new employees during their orientation to the company. If the company has a safety manual, the new employee should be required to read it and sign a statement indicating that it is understood. If the employee cannot read, the employer should have someone read it to the employee and answer any questions that arise. The employee should then sign a form stating that he or she understands the information.

It is not enough to tell employees about safety policies and procedures on the day they are hired and never mention them again. Rather, crew leaders should constantly emphasize and reinforce the importance of following all safety policies and procedures. In addition, employees should play an active role in determining job safety hazards and find ways that the hazards can be prevented and controlled.

4.1.2 Hazard Identification and Assessment

Safety policies and procedures should be specific to the company. They should clearly present the hazards of the job. Therefore, crew leaders should also identify and assess hazards to which employees are exposed. They must also assess compliance with OSHA and state standards.

To identify and assess hazards, OSHA recommends that employers conduct inspections of the workplace, monitor safety and health information logs, and evaluate new equipment, materials, and processes for potential hazards before they are used.

Crew leaders and employees play important roles in identifying hazards. It is the crew leader's responsibility to determine what working conditions are unsafe and to inform employees of hazards and their locations. In addition, they should encourage their crew members to tell them about hazardous conditions. To accomplish this, crew leaders must be present and available on the job site.

The crew leader also needs to help the employee be aware of and avoid the built-in hazards to which craftworkers are exposed. Examples include working at elevations, working in confined spaces such as tunnels and underground vaults, on caissons, in excavations with earthen walls, and other naturally dangerous projects.

In addition, the crew leader can take safety measures, such as installing protective railings to prevent workers from falling from buildings, as well as scaffolds, platforms, and shoring.

4.1.3 Safety Information and Training

The employer must provide periodic information and training to new and long-term employees. This must be done as often as necessary so that all employees are adequately trained. Special training and informational sessions must be provided when safety and health information changes or workplace conditions create new hazards. It is important to note that safety-related information must be presented in a manner that the employee will understand.

Whenever a crew leader assigns an experienced employee a new task, the crew leader must ensure that the employee is capable of doing the work in a safe manner. The crew leader can accomplish this by providing safety information or training for groups or individuals.

The crew leader should do the following:

- Define the task.
- Explain how to do the task safely.
- Explain what tools and equipment to use and how to use them safely.
- Identify the necessary personal protective equipment.
- Explain the nature of the hazards in the work and how to recognize them.
- Stress the importance of personal safety and the safety of others.
- Hold regular safety training sessions with the crew's input.
- Review material safety data sheets (MSDSs) that may be applicable.

4.1.4 Safety Record Systems

OSHA regulations (*CFR 29, Part 1904*) require that employers keep records of hazards identified and document the severity of the hazard. The information should include the likelihood of employee exposure to the hazard, the seriousness of the harm associated with the hazard, and the number of exposed employees.

In addition, the employer must document the actions taken or plans for action to control the hazards. While it is best to take corrective action immediately, it is sometimes necessary to develop a plan for the purpose of setting priorities and deadlines and tracking progress in controlling hazards.

Employers who are subject to the recordkeeping requirements of the *Occupational Safety and Health Act of 1970* must maintain a log of all recordable occupational injuries and illnesses. This is known as the *OSHA 300/300A* form.

An MSDS is designed to provide both workers and emergency personnel with the proper procedures for handling or working with a substance that may be dangerous. An MSDS will include information such as physical data (melting point, boiling point, flash point, etc.), toxicity, health effects, first aid, reactivity, storage, disposal, protective equipment, and spill/leak procedures. These sheets are of particular use if a spill or other accident occurs.

Did you know?

Any company with 11 or more employees must post an *OSHA 300A* form, *Log of Work-Related Injuries and Illnesses,* between February 1 and April 30 of each year. Employees have the right to review this form. Check your company's policies with regard to these reports.

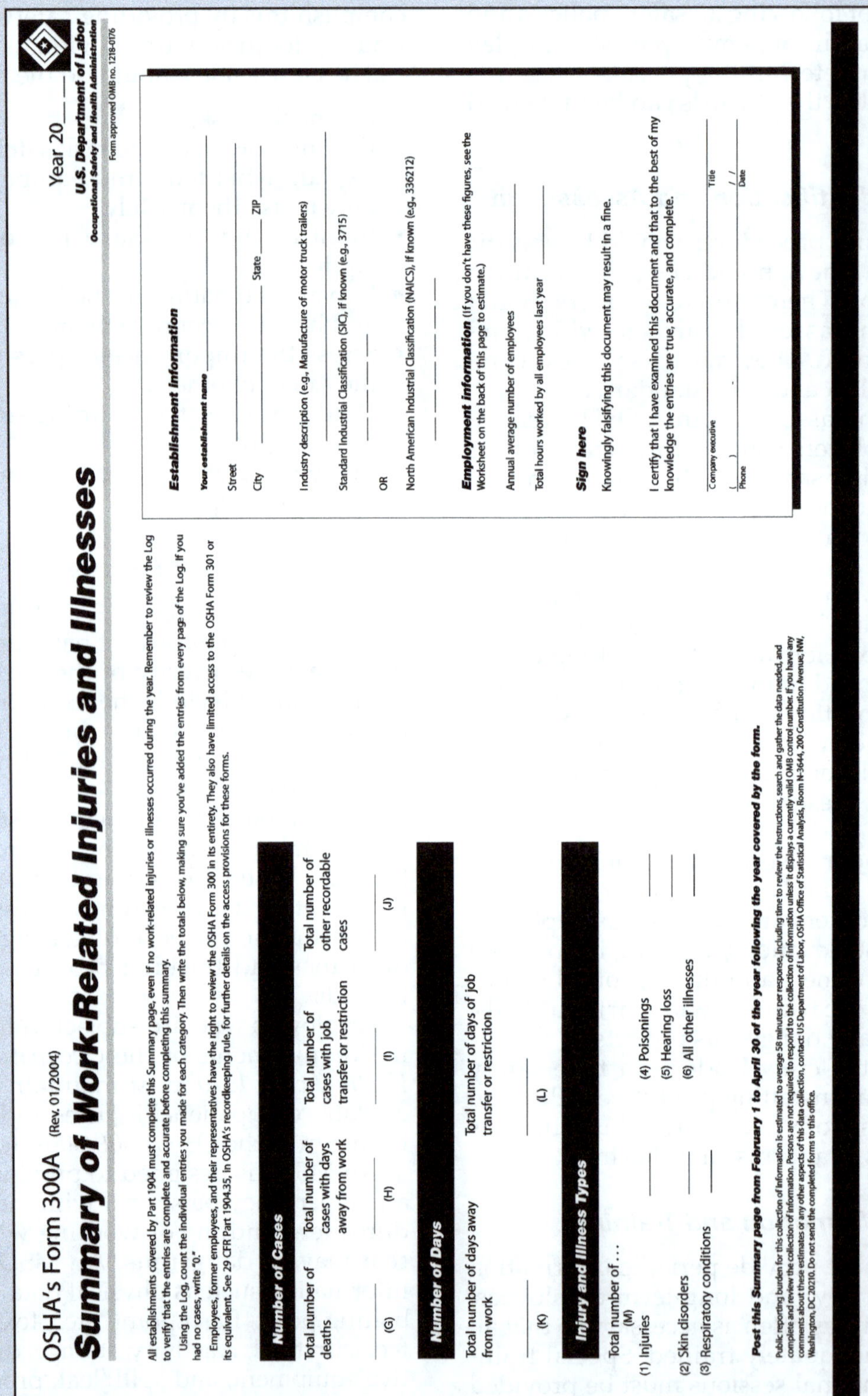
OSHA's Form 300A (Rev. 01/2004)

Summary of Work-Related Injuries and Illnesses

Year 20__ __

U.S. Department of Labor
Occupational Safety and Health Administration

Form approved OMB no. 1218-0176

All establishments covered by Part 1904 must complete this Summary page, even if no work-related injuries or illnesses occurred during the year. Remember to review the Log to verify that the entries are complete and accurate before completing this summary.

Using the Log, count the individual entries you made for each category. Then write the totals below, making sure you've added the entries from every page of the Log. If you had no cases, write "0."

Employees, former employees, and their representatives have the right to review the OSHA Form 300 in its entirety. They also have limited access to the OSHA Form 301 or its equivalent. See 29 CFR Part 1904.35, in OSHA's recordkeeping rule, for further details on the access provisions for these forms.

Number of Cases

Total number of deaths	Total number of cases with days away from work	Total number of cases with job transfer or restriction	Total number of other recordable cases
____ (G)	____ (H)	____ (I)	____ (J)

Number of Days

Total number of days away from work	Total number of days of job transfer or restriction
____ (K)	____ (L)

Injury and Illness Types

Total number of . . .
(M)

(1) Injuries	____	(4) Poisonings	____
		(5) Hearing loss	____
(2) Skin disorders	____	(6) All other illnesses	____
(3) Respiratory conditions	____		

Post this Summary page from February 1 to April 30 of the year following the year covered by the form.

Public reporting burden for this collection of information is estimated to average 58 minutes per response, including time to review the instructions, search and gather the data needed, and complete and review the collection of information. Persons are not required to respond to the collection of information unless it displays a currently valid OMB control number. If you have any comments about these estimates or any other aspects of this data collection, contact: US Department of Labor, OSHA Office of Statistical Analysis, Room N-3644, 200 Constitution Avenue, NW, Washington, DC 20210. Do not send the completed forms to this office.

Establishment information

Your establishment name ____________

Street ____________

City ____________ State ____ ZIP ____

Industry description (e.g., Manufacture of motor truck trailers) ____________

Standard Industrial Classification (SIC), if known (e.g., 3715) __ __ __ __

OR

North American Industrial Classification (NAICS), if known (e.g., 336212) __ __ __ __ __ __

Employment information (If you don't have these figures, see the Worksheet on the back of this page to estimate.)

Annual average number of employees ____________

Total hours worked by all employees last year ____________

Sign here

Knowingly falsifying this document may result in a fine.

I certify that I have examined this document and that to the best of my knowledge the entries are true, accurate, and complete.

____________ Company executive ____________ Title

( ) - ____________ Phone / / ____________ Date

46101-11_SA01.EPS

Logs must be maintained and retained for five years following the end of the calendar year to which they relate. Logs must be available (normally at the establishment) for inspection and copying by representatives of the Department of Labor, the Department of Health and Human Services, or states accorded jurisdiction under the Act. Employees, former employees, and their representatives may also have access to these logs.

4.1.5 Accident Investigation

In the event of an accident, the employer is required to investigate the cause of the accident and determine how to avoid it in the future. According to OSHA, the employer must investigate each work-related death, serious injury or illness, or incident having the potential to cause death or serious physical harm. The employer should document any findings from the investigation, as well as the action plan to prevent future occurrences. This should be done immediately, with photos or video if possible. It is important that the investigation uncover the root cause of the accident so that it can be avoided in the future. In many cases, the root cause can be traced to a flaw in the system that failed to recognize the unsafe condition or the potential for an unsafe act (*Figure 8*).

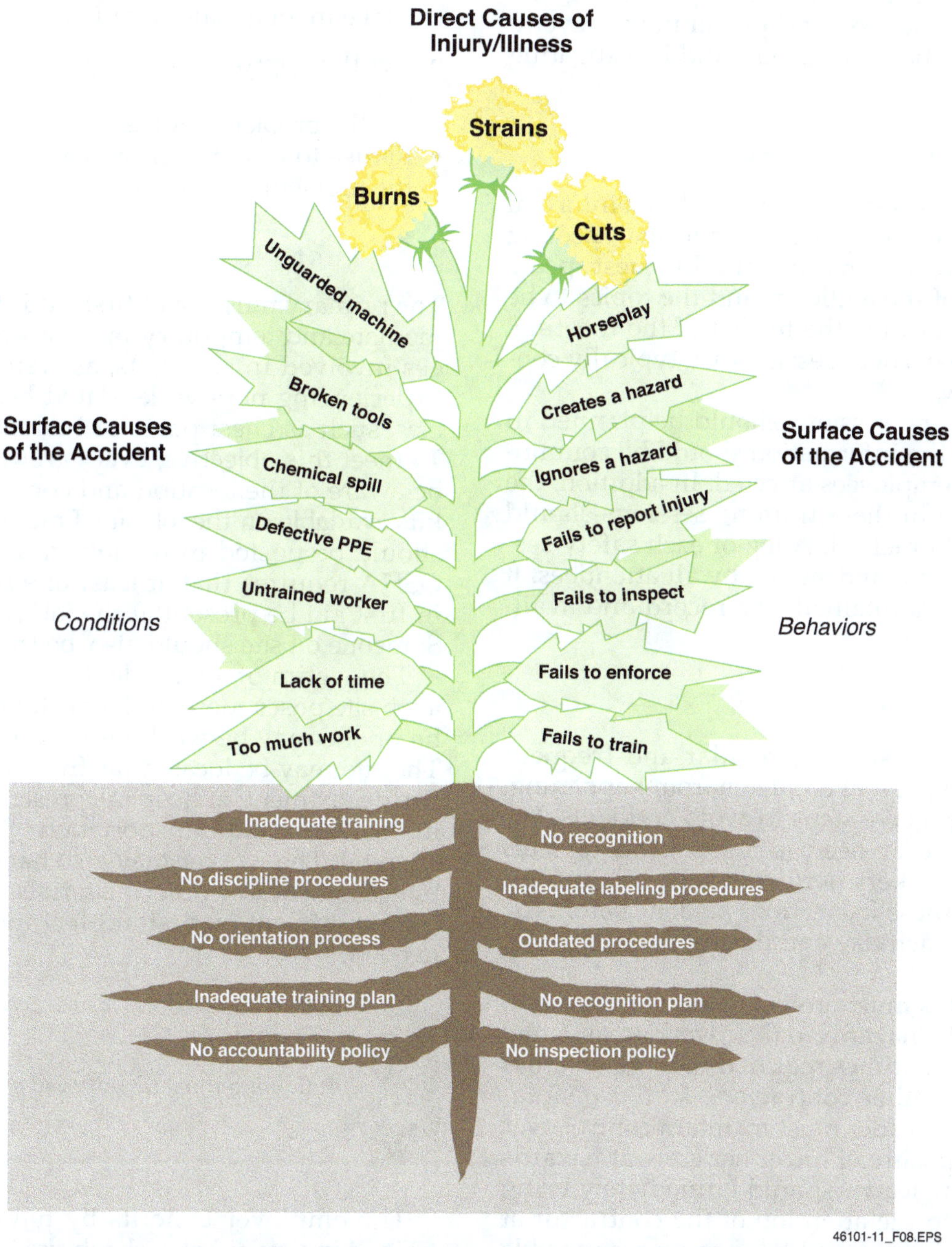

Figure 8 Root causes of accidents.

Did you know?

Any accident that involves a fatality or three or more hospital admissions must be reported to OSHA within eight hours.

5.0.0 CREW LEADER INVOLVEMENT IN SAFETY

To be effective leaders, crew leaders must be actively involved in the safety program. Crew leader involvement includes conducting frequent safety training sessions and inspections; promoting first aid and fire protection and prevention; preventing substance abuse on the job; and investigating accidents.

5.1.0 Safety Training Sessions

A safety training session may be a brief, informal gathering of a few employees or a formal meeting with instructional films and talks by guest speakers. The size of the audience and the topics to be addressed determine the format of the meeting. Small, informal safety sessions are typically conducted weekly.

Safety training sessions should be planned in advance, and the information should be communicated to all employees affected. In addition, the topics covered in these training sessions should be timely and practical. A log of each safety session must be kept and signed by all attendees. It must then be maintained as a record and available for inspection.

5.2.0 Inspections

Crew leaders must make regular and frequent inspections to prevent accidents from happening. They must also take steps to avoid accidents. For that purpose, they need to inspect the job sites where their workers perform tasks. It is recommended that these inspections be done before the start of work each day and during the day at different times.

Crew leaders must protect workers from existing or potential hazards in their work areas. Crew leaders are sometimes required to work in areas controlled by other contractors. In these situations, the crew leader must maintain control over the safety exposure of his or her crew. If hazards exist, the crew leader should immediately bring the hazards to the attention of the contractor at fault, their superior, and the person responsible for the job site.

Crew leader inspections are only valuable if action is taken to correct potential hazards. Therefore, crew leaders must be alert for unsafe acts on their work sites. When an employee performs an unsafe action, the crew leader must explain to the employee why the act was unsafe, ask that the employee not do it again, and request cooperation in promoting a safe working environment. The crew leader must document what happened and what the employee was asked to do to correct the situation. It is then very important that crew leaders follow up to make certain the employee is complying with the safety procedures. Never allow a safety violation to go uncorrected. There are three courses of action that you, as a crew leader, can take in an unsafe situation:

- Get the appropriate party to correct the problem.
- Fix the problem yourself.
- Refuse to have the crew work in the area until the problem is corrected.

5.3.0 First Aid

The primary purpose of first aid is to provide immediate and temporary medical care to employees involved in accidents, as well as employees experiencing non-work-related health emergencies, such as chest pains or breathing difficulty. To meet this objective, every crew leader should be aware of the location and contents of first aid kits available on the job site. Emergency numbers should be posted in the job trailer. In addition, OSHA requires that at least one person trained in first aid be present at the job site at all times. Someone on site should also be trained in CPR.

The victim of an accident or sudden illness at a job site poses more problems than normal since he or she may be working in a remote location. The site may be located far from a rescue squad, fire department, or hospital, presenting a problem in the rescue and transportation of the victim to a hospital. The worker may also have been injured by falling rock or other materials, so special rescue equipment or first-aid techniques are often needed.

NOTE

CPR training must be renewed every two years.

The employer benefits by having personnel trained in first aid at each job site in the following ways:

- The immediate and proper treatment of minor injuries may prevent them from developing into more serious conditions. As a result, medical expenses, lost work time, and sick pay may be eliminated or reduced.
- It may be possible to determine if professional medical attention is needed.
- Valuable time can be saved when a trained individual prepares the patient for treatment when professional medical care arrives. As a result, lives can be saved.

The American Red Cross, Medic First Aid, and the United States Bureau of Mines provide basic and advanced first aid courses at nominal costs. These courses include both first aid and CPR. The local area offices of these organizations can provide further details regarding the training available.

5.4.0 Fire Protection and Prevention

Fires and explosions kill and injure many workers each year, so it is important that crew leaders understand and practice fire-prevention techniques as required by company policy.

The need for protection and prevention is increasing as new building materials are introduced. Some building materials are highly flammable. They produce great amounts of smoke and gases, which cause difficulties for fire fighters, and can quickly overcome anyone present. Other materials melt when ignited and may spread over floors, preventing fire-fighting personnel from entering areas where this occurs.

OSHA has specific standards for fire safety. They require that employers provide proper exits, fire-fighting equipment, and employee training on fire prevention and safety. For more information, consult OSHA guidelines.

5.5.0 Substance Abuse

Unfortunately, drug and alcohol abuse is a continuing problem in the workplace. Drug abuse means inappropriately using drugs, whether they are legal or illegal. Some people use illegal street drugs, such as cocaine or marijuana. Others use legal prescription drugs incorrectly by taking too many pills, using other people's medications, or self-medicating. Others consume alcohol to the point of intoxication.

It is essential that crew leaders enforce company policies and procedures regarding substance abuse. Crew leaders must work with management to deal with suspected drug and alcohol abuse and should not handle these situations themselves. These cases are normally handled by the Human Resources Department or designated manager. There are legal consequences of drug and alcohol abuse and the associated safety implications. If you suspect that an employee is suffering from drug or alcohol abuse, immediately contact your supervisor and/or Human Resources Department for assistance. That way, the business and the employee's safety are protected.

It is the crew leader's responsibility to make sure that safety is maintained at all times. This may include removing workers from a work site where they may be endangering themselves or others.

For example, suppose several crew members go out and smoke marijuana or have a few beers during lunch. Then, they return to work to erect scaffolding for a concrete pour in the afternoon. If you can smell marijuana on the crew member's clothing or alcohol on their breath, you must step in and take action. Otherwise, they might cause an accident that could delay the project or cause serious injury or death to themselves or others.

It is often difficult to detect drug and alcohol abuse because the effects can be subtle. The best way is to look for identifiable effects, such as those mentioned above or sudden changes in behavior that are not typical of the employee. Some examples of such behaviors include the following:

- Unscheduled absences; failure to report to work on time
- Significant changes in the quality of work
- Unusual activity or lethargy
- Sudden and irrational temper flare-ups
- Significant changes in personal appearance, cleanliness, or health

There are other more specific signs that should arouse suspicion, especially if more than one is exhibited. Among them are:

- Slurring of speech or an inability to communicate effectively
- Shiftiness or sneaky behavior, such as an employee disappearing to wooded areas, storage areas, or other private locations
- Wearing sunglasses indoors or on overcast days to hide dilated or constricted pupils
- Wearing long-sleeved garments, particularly on hot days, to cover marks from needles used to inject drugs
- Attempting to borrow money from co-workers
- The loss of an employee's tools and company equipment

Did you know?

An employee who is involved in an accident while under the influence of drugs or alcohol may be denied workers compensation insurance benefits.

5.6.0 Job-Related Accident Investigations

There are two times when a crew leader may be involved with an accident investigation. The first time is when an accident, injury, or report of work-connected illness takes place. If present on site, the crew leader should proceed immediately to the accident location to ensure that proper first aid is being provided. He or she will also want to make sure that other safety and operational measures are taken to prevent another incident.

If mandated by company policy, the crew leader will need to make a formal investigation and submit a report after an incident. An investigation looks for the causes of the accident by examining the situation under which it occurred and talking to the people involved. Investigations are perhaps the most useful tool in the prevention of future accidents.

There are four major parts to an accident investigation. The crew leader will be concerned with each one. They are:

- Describing the accident
- Determining the cause of the accident
- Determining the parties involved and the part played by each
- Determining how to prevent re-occurrences

Case Study

For years, a prominent safety engineer was confused as to why sheet metal workers fractured their toes frequently. The crew leader had not performed thorough accident investigations, and the injured workers were embarrassed to admit how the accidents really occurred. It was later discovered they used the metal reinforced cap on their safety shoes as a "third hand" to hold the sheet metal vertically in place when they fastened it. The sheet metal was inclined to slip and fall behind the safety cap onto the toes, causing fractures. Several injuries could have been prevented by performing a proper investigation after the first accident.

6.0.0 Promoting Safety

The best way for crew leaders to encourage safety is through example. Crew leaders should be aware that their behavior sets standards for their crew members. If a crew leader cuts corners on safety, then the crew members may think that it is okay to do so as well.

The key to effectively promote safety is good communication. It is important to plan and coordinate activities and to follow through with safety programs. The most successful safety promotions occur when employees actively participate in planning and carrying out activities.

Some activities used by organizations to help motivate employees on safety and help promote safety awareness include:

- Safety training sessions
- Contests
- Recognition and awards
- Publicity

6.1.0 Safety Training Sessions

Safety training sessions can help keep workers focused on safety and give them the opportunity to discuss safety concerns with the crew. This topic was addressed in a previous section.

6.2.0 Safety Contests

Contests are a great way to promote safety in the workplace. Examples of safety-related contests include the following:

- Sponsoring housekeeping contests for the cleanest job site or work area
- Challenging employees to come up with a safety slogan for the company or department
- Having a poster contest that involves employees or their children creating safety-related posters
- Recording the number of accident-free workdays or person-hours
- Giving safety awards (hats, T-shirts, other promotional items or prizes)

One of the positive aspects of contests is their ability to encourage employee participation. It is important, however, to ensure that the contest has a valid purpose. For example, the posters or slogans created in a poster contest can be displayed throughout the organization as safety reminders.

6.3.0 Incentives and Awards

Incentives and awards serve several purposes. Among them are acknowledging and encouraging good performance, building goodwill, reminding employees of safety issues, and publicizing the importance of practicing safety standards. There are countless ways to recognize and award safety. Examples include the following:

- Supplying food at the job site when a certain goal is achieved
- Providing a reserved parking space to acknowledge someone for a special achievement
- Giving gift items such as T-shirts or gift certificates to reward employees
- Giving plaques to a department or an individual (*Figure 9*)
- Sending a letter of appreciation
- Publicly honoring a department or an individual for a job well done

Creativity can be used to determine how to recognize and award good safety on the work site. The only precautionary measure is that the award should be meaningful and not perceived as a bribe. It should be representative of the accomplishment.

6.4.0 Publicity

Publicizing safety is the best way to get the message out to employees. An important aspect of publicity is to keep the message accurate and current. Safety posters that are hung for years on end tend to lose effectiveness. It is important to keep ideas fresh.

Examples of promotional activities include posters or banners, advertisements or information on bulletin boards, payroll mailing stuffers, and employee newsletters. In addition, merchandise can be purchased that promotes safety, including buttons, hats, T-shirts, and mugs.

46101-11_F09.EPS

Figure 9 Examples of safety plaques.

Participant Exercise D

Described here are three scenarios that reflect unsafe practices by craft workers. For each of these scenarios write down how you would deal with the situation, first as the crew leader of the craft worker, and then as the leader of another crew.

1. You observe a worker wearing his hard hat backwards and his safety glasses hanging around his neck. He is using a concrete saw.

2. As you are supervising your crew on the roof deck of a building under construction, you notice that a section of guard rail has been removed. Another contractor was responsible for installing the guard rail.

3. Your crew is part of plant shutdown at a power station. You observe that a worker is welding without a welding screen in an area where there are other workers.

Review Questions

1. One of a crew leader's responsibilities is to enforce company safety policies.

   a. True
   b. False

2. Which of the following is an indirect cost of an accident?

   a. Medical bills
   b. Production delays
   c. Compensation
   d. Employee benefits

3. A crew leader can be held criminally liable for repeat safety violations.

   a. True
   b. False

4. OSHA inspection of a business or job site _____.

   a. can be done only by invitation
   b. is done only after an accident
   c. can be conducted at random
   d. can be conducted only if a safety violation occurs

5. The *OSHA 300* form deals with _____.

   a. penalties for safety violations
   b. workplace illnesses and injuries
   c. hazardous material spills
   d. safety training sessions

6. A crew leader's responsibilities include all of the following, *except* _____.

   a. conducting safety training sessions
   b. developing a company safety program
   c. performing safety inspections
   d. participating in accident investigations

7. In order to ensure workplace safety, the crew leader should _____.

   a. hold formal safety training sessions
   b. have crew members conduct on-site safety inspections
   c. notify contractors and their supervisor of hazards in a situation where a job is being performed in an unsafe area controlled by other contractors
   d. hold crew members responsible for making a formal report and investigation following an accident

8. Prohibitions on the abuse of drugs deals only with illegal drugs such as cocaine and marijuana.

   a. True
   b. False

Section Four
Project Control

Objectives

Upon completion of this section, you will be able to:

1. Describe the three phases of a construction project.
2. Define the three types of project delivery systems.
3. Define planning and describe what it involves.
4. Explain why it is important to plan.
5. Describe the two major stages of planning.
6. Explain the importance of documenting job site work.
7. Describe the estimating process.
8. Explain how schedules are developed and used.
9. Identify the two most common schedules.
10. Explain how the critical path method (CPM) of scheduling is used.
11. Describe the different costs associated with building a job.
12. Explain the crew leader's role in controlling costs.
13. Illustrate how to control the main resources of a job: materials, tools, equipment, and labor.
14. Explain the differences between production and productivity and the importance of each.

Performance Tasks

1. Develop and present a look-ahead schedule.
2. Develop an estimate for a given work activity

1.0.0 Project Control Overview

The contractor, project manager, superintendent, and crew leader each have management responsibilities for their assigned jobs. For example, the contractor's responsibility begins with obtaining the contract, and it does not end until the client takes ownership of the project. The project manager is generally the person with overall responsibility for coordinating the project. Finally, the superintendent and crew leader are responsible for coordinating the work of one or more workers, one or more crews of workers within the company and, on occasion, one or more crews of subcontractors. The crew leader directs a crew in the performance of work tasks.

This section describes methods of effective and efficient project control. It examines estimating, planning and scheduling, and resource and cost control. All the workers who participate in the job are responsible at some level for controlling cost and schedule performance and for ensuring that the project is completed according to plans and specifications.

> **NOTE**
> The material in this section is based largely on building-construction projects. However, the project control principles described here apply generally to all types of projects.

Construction projects are made up of three phases: the development phase, the planning phase, and the construction phase.

1.1.0 Development Phase

A new building project begins when an owner has decided to build a new facility or add to an existing facility. The development process is the first stage of planning for a new building project. This process involves land research and feasibility studies to ensure that the project has merit. Architects or engineers develop the conceptual drawings that define the project graphically. They then provide the owner with sketches of room layouts and elevations and make suggestions about what construction materials should be used.

During the development phase, an estimate for the proposed project is developed in order to establish a preliminary budget. Once that budget has been established, the financing of the project is discussed with lending institutions. The architects/engineers and/or the owner begins preliminary reviews with government agencies. These reviews include zoning, building restrictions, landscape requirements, and environmental impact studies.

Also during the development phase, the owner must analyze the project's cost and potential retun on investment (ROI) to ensure that its costs will not exceed its market value and that the project provides a good return on investment. If the project passes this test, the architect/engineer will proceed to the planning phase.

1.2.0 Planning Phase

When the architect/engineer begins to develop the project drawings and specifications, other design professionals such as structural, mechanical, and electrical engineers are brought in. They perform the calculations, make a detailed technical analysis, and check details of the project for accuracy.

The design professionals create drawings and specifications. These drawings and specifications are used to communicate the necessary information to the contractors, subcontractors, suppliers, and workers that contribute to a project.

During the planning phase, the owners hold many meetings to refine estimates, adjust plans to conform to regulations, and secure a construction loan. If the project is a condominium, an office building, or a shopping center, then a marketing program must be developed. In such cases, the selling of the project is often started before actual construction begins.

Next, a complete set of drawings, specifications, and bid documents is produced. Then the owner will select the method to obtain contractors. The owner may choose to negotiate with several contractors or select one through competitive bidding. Note that safety must also be considered as part of the planning process. A safety crew leader may walk through the site as part of the pre-bid process.

Contracts can take many forms. The three basic types from which all other types are derived are firm fixed price, cost reimbursable, and guaranteed maximum price.

- *Firm fixed price* – In this type of contract, the buyer generally provides detailed drawings and specifications, which the contractor uses to calculate the cost of materials and labor. To these costs, the contractor adds a percentage representing company overhead expenses such as office rent, insurance, and accounting/payroll costs. On top of all this, the contractor adds a profit factor. When submitting the bid, the contractor will state very specifically the conditions and assumptions on which the bid is based. These conditions and assumptions form the basis from which changes can be priced. Because the price is established in advance, any changes in the job requirements once the job is started will impact the contractor's profit margin. This is where the crew leader can play an important role by identifying problems that increase the amount of labor or material that was planned. By passing this information up the chain of command, the crew leader allows the company to determine if the change is outside the scope of the bid. If so, they can submit a change order request to cover the added cost.
- *Cost reimbursable* – In this type of contract, the buyer reimburses the contractor for labor, materials, and other costs encountered in the performance of the contract. Typically, the contractor and buyer agree in advance on hourly or daily labor rates for different categories of worker. These rates include an amount representing the contractor's overhead expense. The buyer also reimburses the contractor for the cost of materials and equipment used on the job. The buyer and contractor also negotiate a profit margin. On this type of contract, the profit margin is likely to be lower than that of a fixed-price contract because the contractor's cost risk is significantly reduced. The profit margin is often subject to incentive or penalty clauses that make the amount of profit awarded subject to performance by the contractor. Performance is usually tied to project schedule milestones.
- *Guaranteed maximum price (GMP)* – This form of contract, also called a not-to-exceed contract, is most often used on projects that have been negotiated with the owner. Involvement in the process usually includes preconstruction, and the entire team develops the parameters that define the basis for the work. In some instances, the owner will require a competitively-bid GMP. In such cases, the scope of work has not been fully defined, but bids are taken for general conditions (direct costs) and fee based on an assumed volume of work. The advantages of the GMP contract vehicle are:
  - Reduced design time
  - Allows for phased construction
  - Uses a team approach to a project
  - Reduction in changes related to incomplete drawings

1.3.0 Construction Phase

The designated contractor enlists the help of mechanical, electrical, elevator, and other specialty subcontractors to complete the construction phase. The contractor may perform one or more parts of the construction, and rely on subcontractors for the remainder of the work. However, the general contractor is responsible for managing all the trades necessary to complete the project.

As construction nears completion, the architect/engineer, owner, and government agencies start their final inspections and acceptance of the project. If the project has been managed by the general contractor, the subcontractors have performed their work, and the architect/ engineers have regularly inspected the project to ensure that local codes have been followed, then the inspection procedure can be completed in a timely manner. This results in a satisfied client and a profitable project for all.

On the other hand, if the inspection reveals faulty workmanship, poor design or use of materials, or violation of codes, then the inspection and acceptance will become a lengthy battle and may result in a dissatisfied client and an unprofitable project.

Figure 10 shows the flow of a typical project.

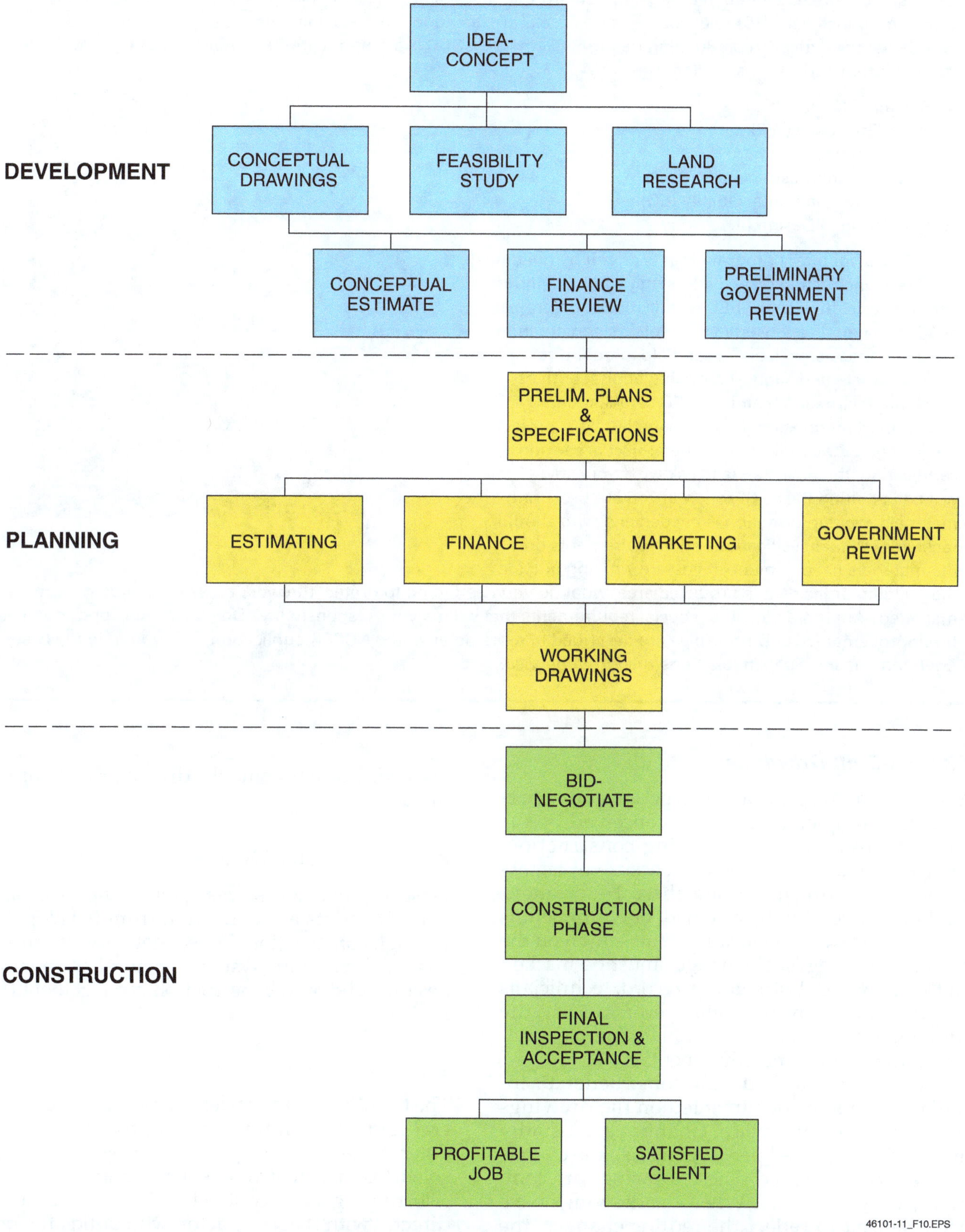

Figure 10 Project flow.

Did you know?

LEED stands for Leadership in Energy and Environmental Design. It is an initiative started by the U.S. Green Building Council (USGBC) to encourage and accelerate the adoption of sustainable construction standards worldwide through a Green Building Rating System™. USGBC is a non-government, not-for-profit group. Their rating system addresses six categories:

1. Sustainable Sites (SS)
2. Water Efficiency (WE)
3. Energy and Atmosphere (EA)
4. Materials and Resources (MR)
5. Indoor Environmental Quality (EQ)
6. Innovation in Design (ID)

46101-11_SA02.EPS

LEED is a voluntary program that is driven by building owners. Construction crew leaders may not have input into the decision to seek LEED certification for a project, or what materials are used in the project's construction. However, these crew leaders can help to minimize material waste and support recycling efforts, both of which are factors in obtaining LEED certification.

An important question to ask is whether your project is seeking LEED certification. If the project is seeking certification, the next step is to ask what your role will be in getting the certification. If you are procuring materials, what information is needed and who should receive it? What specifications and requirements do the materials need to meet? If you are working outside the building or inside in a protected area, what do you need to do to protect the work area? How should waste be managed? Are there any other special requirements that will be your responsibility? Do you see any opportunities for improvement? LEED principles are described in more detail in the NCCER publications, *Your Role in the Green Environment* and *Sustainable Construction Supervisor.*

1.3.1 As-Built Drawings

A set of drawings for a construction project reflects the completed project as conceived by the architect and engineers. During construction, changes usually are necessary because of factors unforeseen during the design phase. For example, if cabling or conduit is re-routed, or equipment is installed in a different location than shown on the original drawing, such changes must be marked on the drawings. Without this record, technicians called to perform maintenance or modify the equipment at a later date will have trouble locating all the cabling and equipment.

Any changes made during construction or installation must be documented on the drawings as the changes are made. Changes are usually made using a colored pen or pencil, so the change can be readily spotted. These drawings are commonly called redlines. When the drawings have been revised to reflect the redline changes, the final drawings are called as-builts, and are so marked. These become the drawings of record for the project.

2.0.0 Project Delivery Systems

Project delivery systems refer to the process by which projects are delivered, from development through construction. Project delivery systems focus on three main systems: general contracting, design-build, and construction management (*Figure 11*).

2.1.0 General Contracting

The traditional project delivery system uses a general contractor. In this type of project, the owner determines the design of the project, and then solicits proposals from general contractors. After selecting a general contractor, the owner contracts directly with that contractor, who builds the project as the prime, or controlling, contractor.

	GENERAL CONTRACTING	DESIGN-BUILD	CONSTRUCTION MANAGEMENT
OWNER	Designs project (or hires architect)	Hires general contractor	Hires construction management company
GENERAL CONTRACTOR	Builds project (with owner's design)	Involved in project design, builds project	Builds, may design (hired by construction management company)
CONSTRUCTION MANAGEMENT COMPANY			Hires and manages general contractor and architect

46101-11_F11.EPS

Figure 11 Project delivery systems.

2.2.0 Design-Build

The design-build project delivery system is different from the general contracting delivery system. In the design-build system, both the design and construction of a project are managed by a single entity. GMP contracts are commonly used in these situations.

2.3.0 Construction Management

The construction management project delivery system uses a construction manager to facilitate the design and construction of a project. Construction managers are very involved in project control; their main concerns are controlling time, cost, and the quality of the project.

3.0.0 Cost Estimating and Budgeting

Before a project is built, an estimate must be prepared. An estimate is the process of calculating the cost of a project. There are two types of costs to consider, including direct and indirect costs. Direct costs, also known as general conditions, are those that can clearly be assigned to a budget. Indirect costs are overhead costs that are shared by all projects. These costs are generally applied as an overhead percentage to labor and material costs.

Direct costs include the following:

- Materials
- Labor
- Tools
- Equipment

Indirect costs refer to overhead items such as:

- Office rent
- Utilities
- Telecommunications
- Accounting
- Office supplies, signs

The bid price includes the estimated cost of the project as well as the profit. Profit refers to the amount of money that the contractor will make after all of the direct and indirect costs have been paid. If the direct and indirect costs exceed those estimated to perform the job, the difference between the actual and estimated costs must come out of the company's profit. This reduces what the contractor makes on the job.

Profit is the fuel that powers a business. It allows the business to invest in new equipment and facilities, provide training, and to maintain a reserve fund for times when business is slow. In the case of large companies, profitability attracts investors who provide the capital necessary for the business to grow. For these reasons, contractors cannot afford to lose money on a consistent basis. Those who cannot operate profitably are forced out of business. Crew leaders can help their companies remain profitable by managing budget, schedule, quality, and safety adhering to the drawings, specifications, and project schedule.

3.1.0 The Estimating Process

The cost estimate must consider a number of factors. Many companies employ professional cost estimators to do this work. They also maintain performance data for previous projects. This data can be used as a guide in estimating new projects. A complete estimate is developed as follows:

Step 1 Using the drawings and specifications, an estimator records the quantity of the materials needed to construct the job. This is called a quantity takeoff. The information is placed on a takeoff sheet like the one shown in *Figure 12*.

Step 2 Productivity rates are used to estimate the amount of labor required to complete the project. Most companies keep records of these rates for the type and size of the jobs that they perform. The company's estimating department keeps these records.

Step 3 The amount of work to be done is divided by the productivity rate to determine labor hours. For example, if the productivity rate for concrete finishing is 40 square feet per hour, and there are 10,000 square feet of concrete to be finished, then 250 hours of concrete finishing labor is required. This number would be multiplied by the hourly rate for concrete finishing to determine the cost of that labor category. If this work is subcontracted, then the subcontractor's cost estimate, raised by an overhead factor, would be used in place of direct labor cost.

Step 4 The total material quantities are taken from the quantity takeoff sheet and placed on a summary or pricing sheet, an example of which is shown in *Figure 13*. Material prices are obtained from local suppliers, and the total cost of materials is calculated.

Step 5 The cost of equipment needed for the project is determined. This number could reflect rental cost or a factor used by the company when their own equipment is to be used.

Step 6 The total cost of all resources—materials, equipment, tools, and labor—is then totaled on the summary sheet. The unit cost—the total cost divided by the total number of units of material to be put into place—can also be calculated.

Step 7 The cost of taxes, bonds, insurance, subcontractor work, and other indirect costs are added to the direct costs of the materials, equipment, tools, and labor.

Step 8 Direct and indirect costs are summed to obtain the total project cost. The contractor's expected profit is then added to that total.

NOTE

There are software programs available to simplify the cost estimating process. Many of them are tailored to specific industries such as construction and manufacturing, and to specific trades within the industries. For example, there are programs available for electrical and HVAC contractors. Estimating programs are typically set up to include a takeoff form and a form for estimating labor by category. Most of these programs include a data base that contains current prices for labor and materials, so they automatically price the job and produce a bid. Once the job is awarded, the programs can generate purchase orders for materials.

3.1.1 Estimating Material Quantities

The crew leader may be required to estimate quantities of materials.

A set of construction drawings and specifications is needed in order to estimate the amount of a certain type of material required to perform a job. The appropriate section of the technical specifications and page(s) of drawings should be carefully reviewed to determine the types and quantities of materials required. The quantities are then placed on the worksheet. For example, the specification section on finished carpentry should be reviewed along with the appropriate pages of drawings before taking off the linear feet of door and window trim.

If an estimate is required because not enough materials were ordered to complete the job, the estimator must also determine how much more work is necessary. Once this is known, the crew leader can then determine the materials needed. The construction drawings will also be used in this process.

WORKSHEET

Takeoff By:

Checked By:

DATE ________

SHEET ____ of ____

PROJECT ________

ARCHITECT ________

PAGE #

REF.	DESCRIPTION	DIMENSIONS				EXTENSION	QUANTITY	UNIT	TOTAL		REMARKS
		NO.	LENGTH	WIDTH	HEIGHT				QUANTITY	UNIT	

46101-11_F12.EPS

Figure 12 Quantity takeoff sheet.

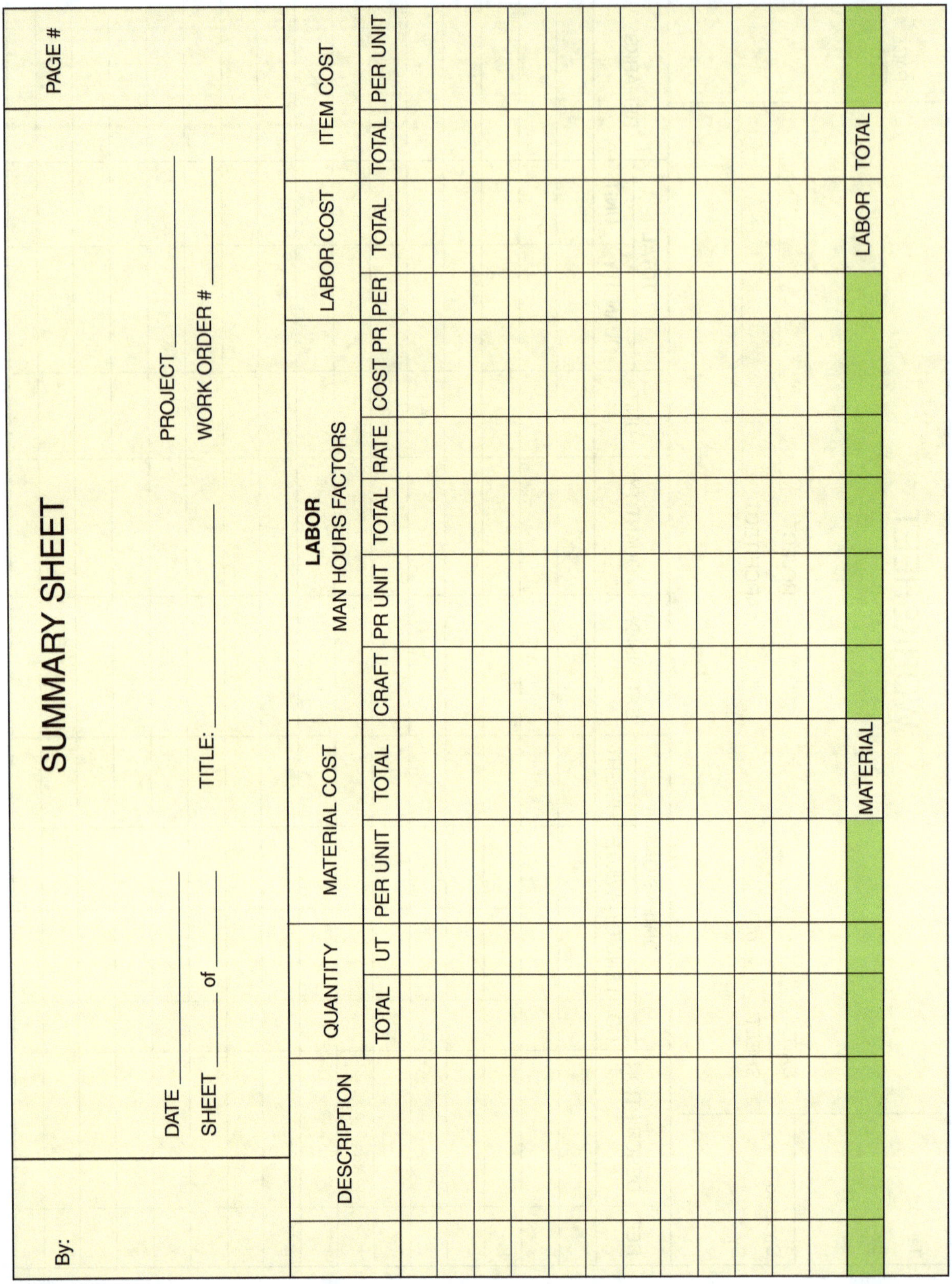

SUMMARY SHEET

By:

PAGE #

DATE ________ PROJECT ________

SHEET ______ of ______ TITLE: ________ WORK ORDER # ________

DESCRIPTION	QUANTITY		MATERIAL COST		LABOR MAN HOURS FACTORS					LABOR COST		ITEM COST	
	TOTAL	UT	PER UNIT	TOTAL	CRAFT	PR UNIT	TOTAL	RATE	COST PR	PER	TOTAL	TOTAL	PER UNIT
				MATERIAL							LABOR	TOTAL	

46101-11_F13.EPS

Figure 13 Summary sheet.

Participant Exercise E

Assume you are the leader of a crew building footing formwork for the construction shown in *Figure 14*. You have used all of the materials provided for the job, yet you have not completed it. You study the drawings and see that the formwork consists of two side forms, each 12" high. The total length of footing for the entire project is 115'-0". You have completed 88'-0" to date; therefore, you have 27'-0" remaining (115'-0" – 88'-0" = 27'-0"). Your job is to prepare an estimate of materials that you will need to complete the job. In this case, only the side forms will be estimated (the miscellaneous materials will not be considered here).

- Footing length to complete: 27'-0"
- Footing height: 1'-0"

Refer to the worksheet in *Figure 15* for a final tabulation of the side forms needed to complete the job.

1. Using the same footing as described in the example above, calculate the quantity (square feet) of formwork needed to finish 203 linear feet of the footing. Place this information directly on the worksheet.
2. You are the crew leader of a carpentry crew whose task is to side a warehouse with plywood sheathing. The wall height is 16 feet, and there is a total of 480 linear feet of wall to side. You have done 360 linear feet of wall and have run out of materials. Calculate how many more feet of plywood you will need to complete the job. If you are using 4' × 8' plywood panels, how many will you need to order, assuming no waste? Write your estimate on the worksheet.

   Show your calculations to the instructor.

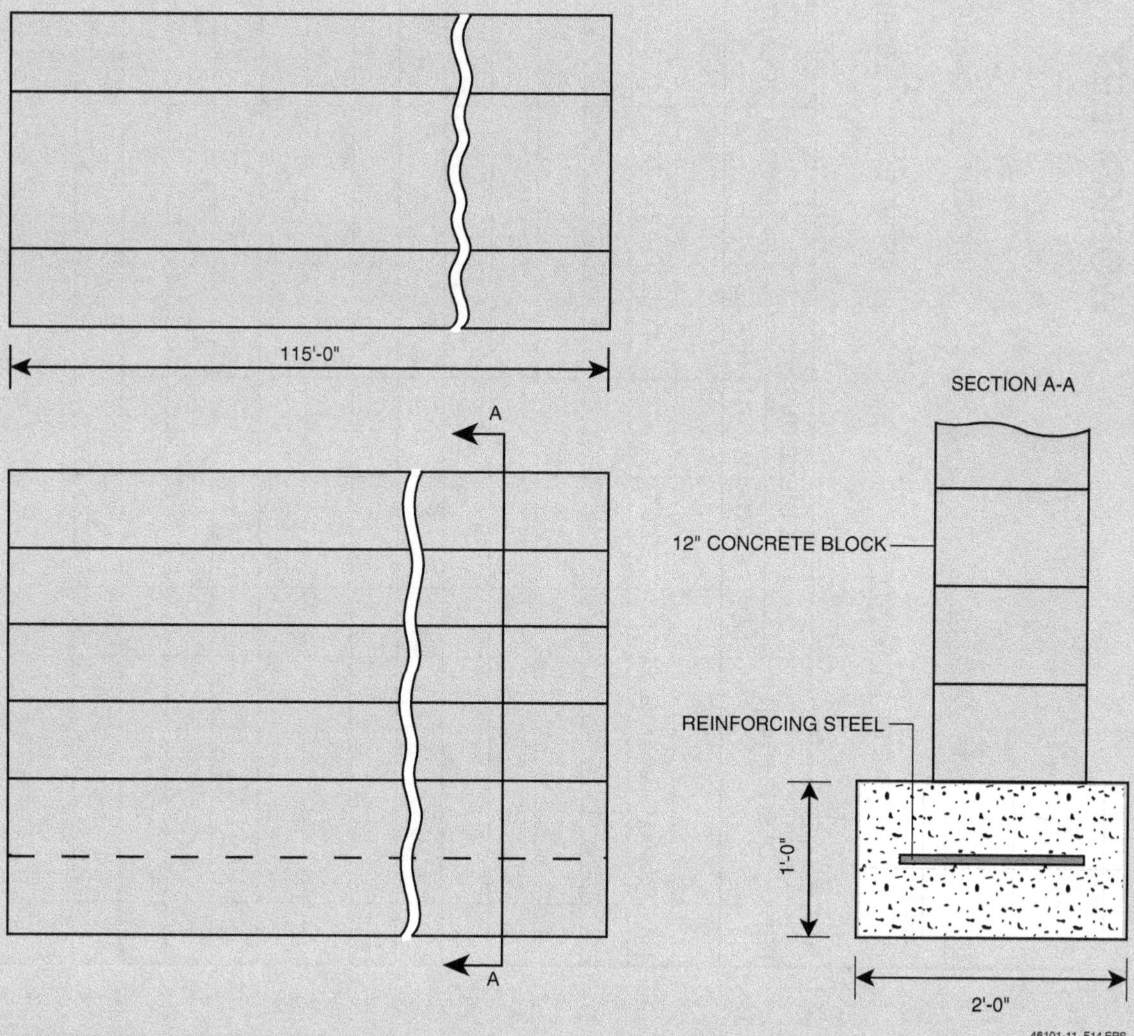

Figure 14 Footing formwork detail.

Takeoff By: RWH

Checked By:

WORKSHEET

PAGE #1

DATE 2/1/15

PROJECT Sam's Diner

SHEET 01 of 01

ARCHITECT 654b

REF.	DESCRIPTION	DIMENSIONS				EXTENSION	QUANTITY	UNIT	TOTAL		REMARKS
		NO	LENGTH	WIDTH	HEIGHT				QUANTITY	UNIT	
	Footing Side Forms	2	27'0"		1'0"	2x27x1	54	SF	54	SF	

46101-11_F15.EPS

Figure 15 Worksheet with entries.

4.0.0 Planning

Planning can be defined as determining the method used to carry out the different tasks to complete a project. It involves deciding what needs to be done and coming up with an organized sequence of events or plan for doing the work.

Planning involves the following:

- Determining the best method for performing the job
- Identifying the responsibilities of each person on the work crew
- Determining the duration and sequence of each activity
- Identifying what tools and equipment are needed to complete a job
- Ensuring that the required materials are at the work site when needed
- Making sure that heavy construction equipment is available when required
- Working with other contractors in such a way as to avoid interruptions and delays

4.1.0 Why Plan?

With a plan, a crew leader can direct work efforts efficiently and can use resources such as personnel, materials, tools, equipment, work area, and work methods to their full potential.

Some reasons for planning include the following:

- Controlling the job in a safe manner so that it is built on time and within cost
- Lowering job costs through improved productivity
- Preparing for bad weather or unexpected occurrences
- Promoting and maintaining favorable employee morale
- Determining the best and safest methods for performing the job

Participant Exercise F

1. In your own words, define planning, and describe how a job can be done better if it is planned. Give an example.

____

____

____

____

2. Consider a job that you recently worked on to answer the following:
   a. List the material(s) used.
   b. List each member of the crew with whom you worked and what each person did.
   c. List the kinds of equipment used.

____

____

____

____

____

3. List some suggestions for how the job could have been done better, and describe how you would plan for each of the suggestions.

____

____

____

____

4.2.0 Stages of Planning

There are various times when planning is done for a construction job. The two most important occur in the pre-construction phase and during the construction work.

4.2.1 Pre-Construction Planning

The pre-construction stage of planning occurs before the start of construction. Except in a fairly small company or for a relatively small job, the crew leader usually does not get directly involved in the pre-construction planning process, but it is important to understand what it involves.

There are two phases of pre-construction planning. The first is when the proposal, bid, or negotiated price for the job is being developed. This is when the estimator, the project manager, and the field superintendent develop a preliminary plan for how the work will be done. This is accomplished by applying experience and knowledge from previous projects. It involves determining what methods, personnel, tools, and equipment will be used and what level of productivity they can achieve.

The second phase occurs after the contract is awarded. This phase requires a thorough knowledge of all project drawings and specifications. During this stage, the actual work methods and resources needed to perform the work are selected. Here, crew leaders might get involved, but their planning must adhere to work methods, production rates, and resources that fit within the estimate prepared before the contract was awarded. If the project requires a method of construction different from what is normal, the crew leader will usually be informed of what method to use.

4.2.2 Construction Planning

During construction, the crew leader is directly involved in planning on a daily basis. This planning consists of selecting methods for completing tasks before beginning work. Effective planning exposes likely difficulties, and enables the crew leader to minimize the unproductive use of personnel and equipment. Effective planning also provides a gauge to measure job progress. Effective crew leaders develop what is known as look-ahead (short-term) schedules. These schedules consider actual circumstances as well as projections two to three weeks into the future. Developing a look-ahead schedule helps ensure that all resources are available on the project when needed.

Did you know?

A good crew leader will always have a backup plan in case circumstances prevent the original plan from working. There are many circumstances that can cause a plan to go awry, including adverse weather, equipment failure, absent crew members, and schedule slippage by other crafts.

One of the characteristics of an effective crew leader is the ability to reduce each job to its simpler parts and organize a plan for handling each task.

Project planners establish time and cost limits for the project; the crew leader's planning must fit within those constraints. Therefore, it is important to consider the following factors that may affect the outcome:

- Site and local conditions, such as soil types, accessibility, or available staging areas
- Climate conditions that should be anticipated during the project
- Timing of all phases of work
- Types of materials to be installed and their availability
- Equipment and tools required and their availability
- Personnel requirements and availability
- Relationships with the other contractors and their representatives on the job

On a simple job, these items can be handled almost automatically. However, larger or more complex jobs require the planner to give these factors more formal consideration and study.

5.0.0 The Planning Process

The planning process consists of the following five steps:

Step 1 Establish a goal.

Step 2 Identify the work activities that must be completed in order to achieve the goal.

Step 3 Identify the tasks that must be done to accomplish those activities.

Step 4 Communicate responsibilities.

Step 5 Follow up to see that the goal is achieved.

5.1.0 Establish a Goal

The term *goal* has different meanings for different people. In general, a goal is a specific outcome that one works toward. For example, the project superintendent of a home construction project could establish the goal to have a house dried-in by a certain date. (Dried-in means ready for the application of roofing and siding.) In order to meet that goal, the leader of the framing crew and the superintendent would need to agree to a goal to have the framing completed by a given date. The crew leader would then establish sub-goals (objectives) for the crew to complete each element of the framing (floors, walls, roof) by a set time. The superintendent would need to set similar goals with the crews that install sheathing, building wrap, windows, and exterior doors. However, if the framing crew does not meet its goal, the other crews will be delayed.

5.2.0 Identify the Work to be Done

The second step in planning is to identify the work to be done to achieve the goal. In other words, it is a series of activities that must be done in a certain sequence. The topic of breaking down a job into activities is covered later in this section. At this point, the crew leader should know that, for each activity, one or more objectives must be set.

An objective is a statement of what is desired at a specific time. An objective must:

- Mean the same thing to everyone involved
- Be measurable, so that everyone knows when it has been reached
- Be achievable
- Have everyone's full support

Examples of practical objectives include the following:

- By 4:30 P.M. today, the crew will have completed installation of the floor joists.
- By closing time Friday, the roof framing will be complete.

Notice that both examples meet the first three requirements of an objective. In addition, it is assumed that everyone involved in completing the task is committed to achieving the objective. The advantage in developing objectives for each work activity is that it allows the crew leader to evaluate whether or not the plan and schedules are being followed. In addition, objectives serve as sub-goals that are usually under the crew leader's control.

Some construction work activities, such as installing 12"-deep footing forms, are done so often that they require little planning. However, other jobs, such as placing a new type of mechanical equipment, require substantial planning. This type of job requires that the crew leader set specific objectives.

Whenever faced with a new or complex activity, take the time to establish objectives that will serve as guides for accomplishing the job. These guides can be used in the current situation, as well as in similar situations in the future.

5.3.0 Identify Tasks to be Performed

To plan effectively, the crew leader must be able to break a work activity assignment down into smaller tasks. Large jobs include a greater number of tasks than small ones, but all jobs can be broken down into manageable components.

When breaking down an assignment into tasks, make each task identifiable and definable. A task is identifiable when the types and amounts of resources it requires are known. A task is definable if it has a specific duration. For purposes of efficiency, the job breakdown should not be too detailed or complex, unless the job has never been done before or must be performed with strictest efficiency.

For example, a suitable breakdown for the work activity to install 12" × 12" vinyl floor tile in a cafeteria might be the following:

Step 1 Prepare the floor.

Step 2 Lay out the tile.

Step 3 Spread the adhesive.

Step 4 Lay the tile.

Step 5 Clean the tile.

Step 6 Wax the floor.

The crew leader could create even more detail by breaking down any one of the tasks, such as lay the tile, into subtasks. In this case, however, that much detail is unnecessary and wastes the crew leader's time and the project's money. However, breaking tasks down further might be necessary in a case where the job is very complex or the analysis of the job needs to be very detailed.

Every work activity can be divided into three general parts:

- Preparing
- Performing
- Cleaning up

One of the most frequent mistakes made in the planning process is forgetting to prepare and to clean up. The crew leader must be certain that preparation and cleanup are not overlooked.

After identifying the various tasks that make up the job and developing an objective for each task, the crew leader must determine what resources the job requires. Resources include labor, equipment, materials, and tools. In most jobs, these resources are identified in the job estimate. The crew leader must make sure that these resources are available on the site when needed.

5.4.0 Communicating Responsibilities

A crew leader is unable to complete all of the activities within a job independently. Other people must be relied upon to get everything done. Therefore, most jobs have a crew of people with various experiences and skill levels to assist in the work. The crew leader's job is to draw from this expertise to get the job done well and in a safe and timely manner.

Once the various activities that make up the job have been determined, the crew leader must identify the person or persons responsible for completing each activity. This requires that the crew leader be aware of the skills and abilities of the people on the crew. Then, the crew leader must put this knowledge to work in matching the crew's skills and abilities to specific tasks that must be performed to complete the job.

After matching crew members to specific activities, the crew leader must then communicate the assignments to the crew. Communication of responsibilities is generally handled verbally; the crew leader often talks directly to the person to which the activity has been assigned. There may be times when work is assigned indirectly through written instructions or verbally through someone other than the crew leader. Either way, the crew members should know what it is they are responsible for accomplishing on the job.

5.5.0 Follow-Up Activities

Once the activities have been delegated to the appropriate crew members, the crew leader must follow up to make sure that they are completed effectively and efficiently. Follow-up work involves being present on the job site to make sure all the resources are available to complete the work; ensuring that the crew members are working on their assigned activities; answering any questions; and helping to resolve any problems that occur while the work is being done. In short, follow-up activity means that the crew leader is aware of what's going on at the job site and is doing whatever is necessary to make sure that the work is completed on schedule.

Figure 16 reviews the planning steps.

The crew leader should carry a small note pad or electronic device to be used for planning and note taking. That way, thoughts about the project can be recorded as they occur, and pertinent details will not be forgotten. The crew leader may also choose to use a planning form such as the one illustrated in *Figure 17*.

As the job is being built, refer back to these plans and notes to see that the tasks are being done in sequence and according to plan. This is referred to as analyzing the job. Experience shows that jobs that are not built according to work plans usually end up costing more and taking more time; therefore, it is important that crew leaders refer back to the plans periodically.

The crew leader is involved with many activities on a day-to-day basis. As a result, it is easy to forget important events if they are not documented. To help keep track of events such as job changes, interruptions, and visits, the crew leader should keep a job diary.

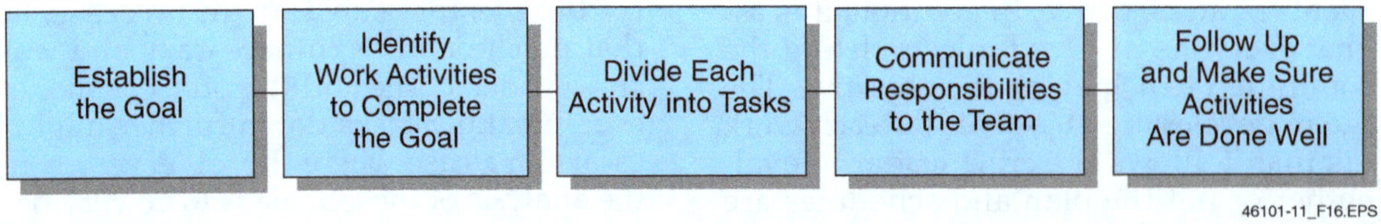

Figure 16 Steps to effective planning.

DAILY WORK PLAN

"PLAN YOUR WORK AND WORK YOUR PLAN = EFFICIENCY"

Plan of ______________________ Date __________

PRIORITY	DESCRIPTION	✓ When Completed X Carried Forward

46101-11_F17.EPS

Figure 17 Planning form.

A job diary is a notebook in which the crew leader records activities or events that take place on the job site that may be important later. When recording in a job diary, make sure that the information is accurate, factual, complete, consistent, organized, and up to date. Follow company policy in determining which events should be recorded. However, if there is a doubt about what to include, it is better to have too much information than too little.

Figure 18 shows a sample page from a job diary.

6.0.0 Planning Resources

Once a job has been broken down into its tasks or activities, the next step is to assign the various resources needed to perform them.

6.1.0 Safety Planning

Using the company safety manual as a guide, the crew leader must assess the safety issues associated with the job and take necessary measures to minimize any risk to the crew. This may involve working with the company or site safety officer and may require a formal hazard analysis.

6.2.0 Materials Planning

The materials required for the job are identified during pre-construction planning and are listed on the job estimate. The materials are usually ordered from suppliers who have previously provided quality materials on schedule and within estimated cost.

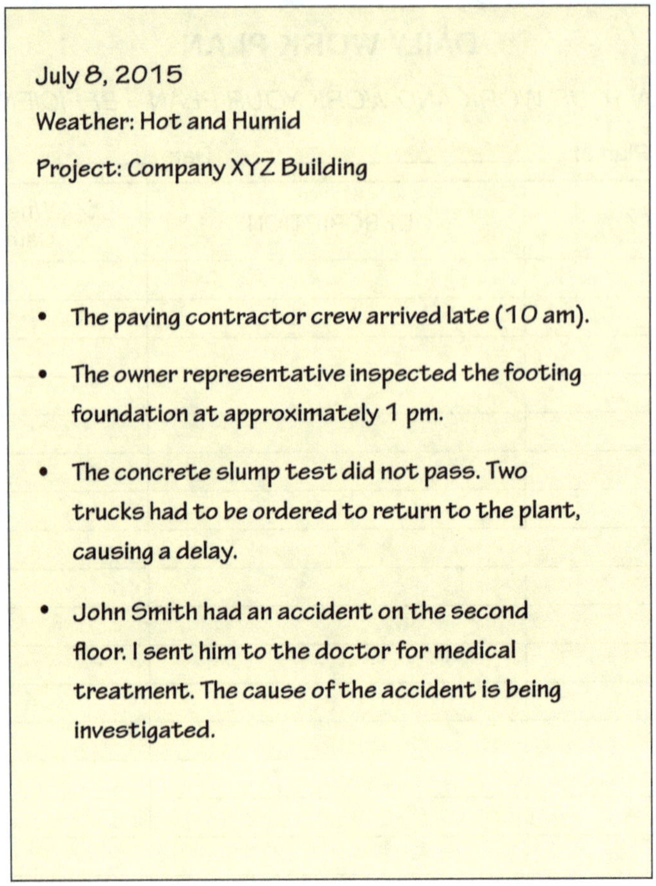

July 8, 2015

Weather: Hot and Humid

Project: Company XYZ Building

- The paving contractor crew arrived late (10 am).
- The owner representative inspected the footing foundation at approximately 1 pm.
- The concrete slump test did not pass. Two trucks had to be ordered to return to the plant, causing a delay.
- John Smith had an accident on the second floor. I sent him to the doctor for medical treatment. The cause of the accident is being investigated.

46101-11_F18.EPS

Figure 18 Sample page from a job diary.

The crew leader is usually not involved in the planning and selection of materials, since this is done in the pre-construction phase. The crew leader does, however, have a major role to play in the receipt, storage, and control of the materials after they reach the job site.

The crew leader is also involved in planning materials for tasks such as job-built formwork and scaffolding. In addition, the crew leader may run out of a specific material, such as fasteners, and need to order more. In such cases, a higher authority should be consulted, since most companies have specific purchasing policies and procedures.

6.3.0 Site Planning

There are many planning elements involved in site work. The following are some of the key elements:

- Emergency procedures
- Access roads
- Parking
- Stormwater runoff
- Sedimentation control
- Material and equipment storage
- Material staging
- Site security

6.4.0 Equipment Planning

Much of the planning for use of construction equipment is done during the pre-construction phase. This planning includes the types of equipment needed, the use of the equipment, and the length of time it will be on the site. The crew leader must work with the home office to make certain that the equipment reaches the job site on time. The crew leader must also ensure that sure equipment operators are properly trained.

Coordinating the use of the equipment is also very important. Some equipment is used in combination with other equipment. For example, dump trucks are generally required when loaders and excavators are used. The crew leader should also coordinate equipment with other contractors on the job. Sharing equipment can save time and money and avoid duplication of effort.

Did you know?

Habitat for Humanity, a charitable organization that builds homes for disadvantaged families, accepts surplus building materials for use in their projects. In some cities, they have stores called ReStores, which serve as retail outlets for such materials. They may also accept materials salvaged during demolition of a structure. LEED credits can be obtained through practices such as salvaging building materials, segregating scrap materials for recycling, and taking steps to minimize waste.

46101-11_SA03.EPS

The crew leader must designate time for equipment maintenance in order to prevent equipment failure. In the event of an equipment failure, the crew leader must know who to contact to resolve the problem. An alternate plan must be ready in case one piece of equipment breaks down, so that the other equipment does not sit idle. This planning should be done in conjunction with the home office or the crew leader's immediate superior.

6.5.0 Tool Planning

A crew leader is responsible for planning what tools will be used on the job. This task includes:

- Determining the tools required
- Informing the workers who will provide the tools (company or worker)
- Making sure the workers are qualified to use the tools safely and effectively
- Determining what controls to establish for tools

6.6.0 Labor Planning

All jobs require some sort of labor because the crew leader cannot complete all the work alone. When planning for labor, the crew leader must:

- Identify the skills needed to perform the work.
- Determine how many people having those specific skills are needed.
- Decide who will actually be on the crew.

In many companies, the project manager or job superintendent determines the size and makeup of the crew. Then, the crew leader is expected to accomplish the goals and objectives with the crew provided. Even though the crew leader may not be involved in staffing the crew, the crew leader is responsible for training the crew members to ensure that they have the skills needed to do the job.

In addition, the crew leader is responsible for keeping the crew adequately staffed at all times so that jobs are not delayed. This involves dealing with absenteeism and turnover, two common problems that affect industry today.

7.0.0 Scheduling

Planning and scheduling are closely related and are both very important to a successful job. Planning involves determining the activities that must be completed and how they should be accomplished. Scheduling involves establishing start and finish times or dates for each activity.

A schedule for a project typically shows:

- Operations listed in sequential order
- Units of construction
- Duration of activities
- Estimated date to start and complete each activity
- Quantity of materials to be installed

There are different types of schedules used today. They include the bar chart; the network schedule, which is sometimes called the critical path method (CPM) or precedence diagram; and the short-term, or look-ahead schedule.

7.1.0 The Scheduling Process

The following is a brief summary of the steps a crewleader must complete to develop a schedule.

Step 1 Make a list of all of the activities that will be performed to build the job, including individual work activities and special tasks, such as inspections or the delivery of materials.

At this point, the crew leader should just be concerned with generating a list, not with determining how the activities will be accomplished, who will perform them, how long they will take, or in what sequence they will be completed.

Step 2 Use the list of activities created in Step 1 to reorganize the work activities into a logical sequence.

When doing this, keep in mind that certain steps cannot happen until others have been completed. For example, footings must be excavated before concrete can be placed.

Step 3 Assign a duration or length of time that it will take to complete each activity and determine the start time for each. Each activity will then be placed into a schedule format. This step is important because it helps the crew leader ensure that the activities are being completed on schedule.

The crew leader must be able to read and interpret the job schedule. On some jobs, the beginning and expected end date for each activity, along with the expected crew or worker's production rate, is provided on the form. The crew leader can use this information to plan work more effectively, set realistic goals, and measure whether or not they were accomplished within the scheduled time.

Before starting a job, the crew leader must:

- Determine the materials, tools, equipment, and labor needed to complete the job.
- Determine when the various resources are needed.
- Follow up to ensure that the resources are available on the job site when needed.

Availability of needed resources should be verified three to four working days before the start of the job. It should be done even earlier for larger jobs. This advance preparation will help avoid situations that could potentially delay the job or cause it to fall behind schedule.

Did you know?

Many companies use project scheduling software. Among the scheduling applications are Microsoft Project, Zoho Projects, and FastTrack Schedule 9. Such programs are variations on spreadsheets that can respond to schedule changes by altering downstream events and producing a revised schedule.

7.2.0 Bar Chart Schedule

Bar chart schedules, also known as Gantt charts, can be used for both short-term and long-term jobs. However, they are especially helpful for jobs of short duration.

Bar charts provide management with the following:

- A visual concept of the overall time required to complete the job through the use of a logical method rather than a calculated guess
- A means to review each part of the job
- Coordination requirements between crafts
- Alternative methods of performing the work

A bar chart can be used as a control device to see whether the job is on schedule. If the job is not on schedule, immediate action can be taken in the office and the field to correct the problem and ensure that the activity is completed on schedule.

A bar chart is illustrated in *Figure 19*.

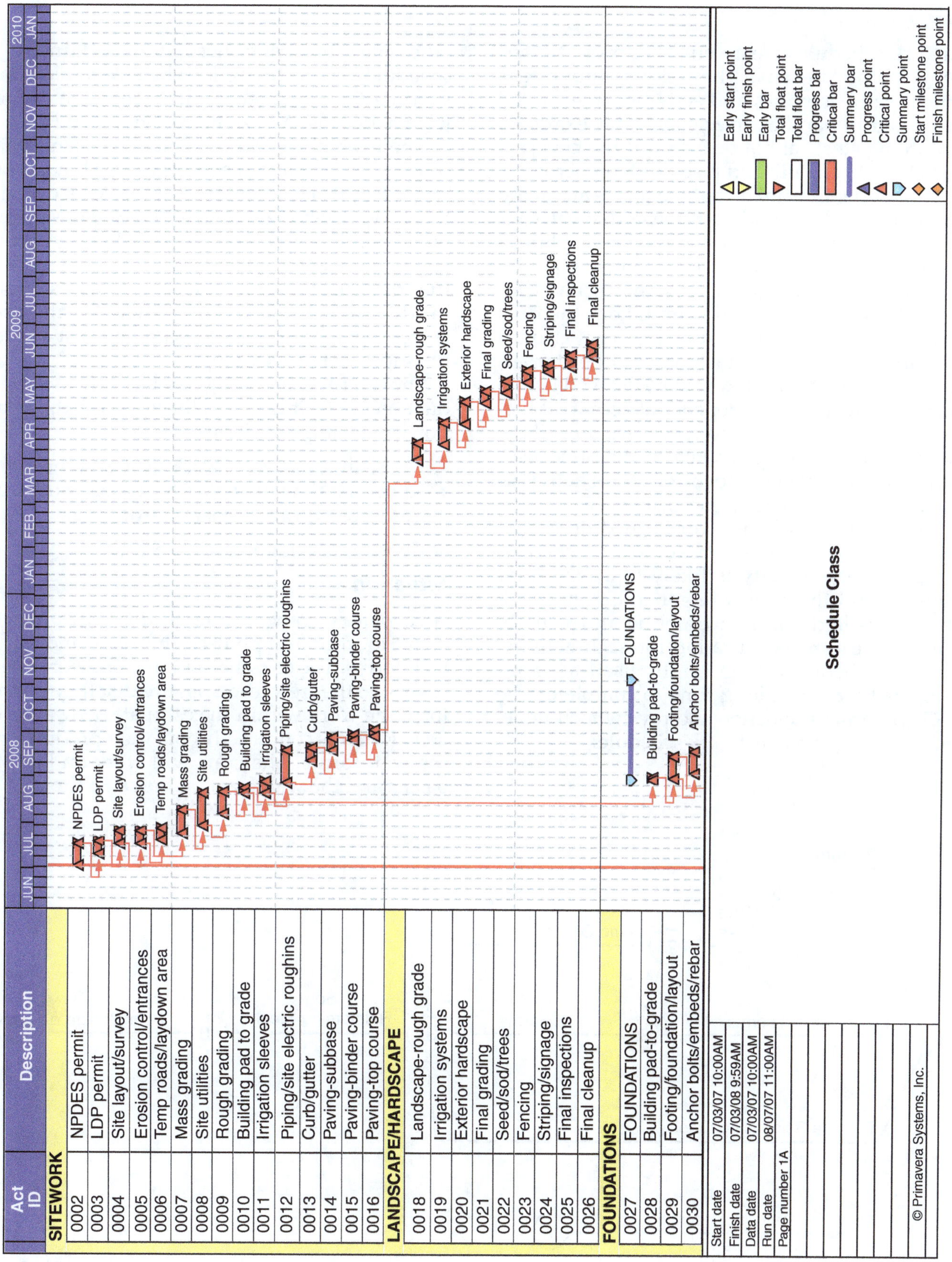

Figure 19 Example of a bar chart schedule.

7.3.0 Network Schedule

Network schedules are an effective project management tool because they show dependent (critical path) activities and activities that can be performed in parallel. In *Figure 20,* for example, reinforcing steel cannot be set until the concrete forms have been built and placed. Other activities are happening in parallel, but the forms are in the critical path. When building a house, drywall cannot be installed and finished until wiring, plumbing, and HVAC ductwork have been roughed-in. Because other activities, such as painting and trim work, depend on drywall completion, the drywall work is a critical-path function. That is, until it is complete, the other tasks cannot be started, and the project itself is likely to be delayed by the amount of delay in any dependent activity. Likewise, drywall work can't even be started until the rough-ins are complete. Therefore, the project superintendent is likely to focus on those activities when evaluating schedule performance.

The advantage of a network schedule is that it allows project leaders to see how a schedule change on one activity is likely to affect other activities and the project in general. A network schedule is laid out on a timeline and usually shows the estimated duration for each activity. Network schedules are generally used for complex jobs that take a long time to complete. The PERT (program evaluation and review technique) schedule is a form of network schedule.

7.4.0 Short-Term Scheduling

Since the crew leader needs to maintain the job schedule, he or she needs to be able to plan daily production. Short-term scheduling is a method used to do this. An example is shown in *Figure 21.*

The information to support short-term scheduling comes from the estimate or cost breakdown. The schedule helps to translate estimate data and the various job plans into a day-to-day schedule of events. The short-term schedule provides the crew leader with visibility over the project. If actual production begins to slip behind estimated production, the schedule will warn the crew leader that a problem lies ahead and that a schedule slippage is developing.

Short-term scheduling can be used to set production goals. It is generally agreed that production can be improved if workers:

- Know the amount of work to be accomplished
- Know the time they have to complete the work
- Can provide input when setting goals

Consider the following example:

Situation:

A carpentry crew on a retaining wall project is about to form and pour catch basins and put up wall forms. The crew has put in a number of catch basins, so the crew leader is sure that they can perform the work within the estimate. However, the crew leader is concerned about their production of the wall forms. The crew will work on both the basins and the wall forms at the same time.

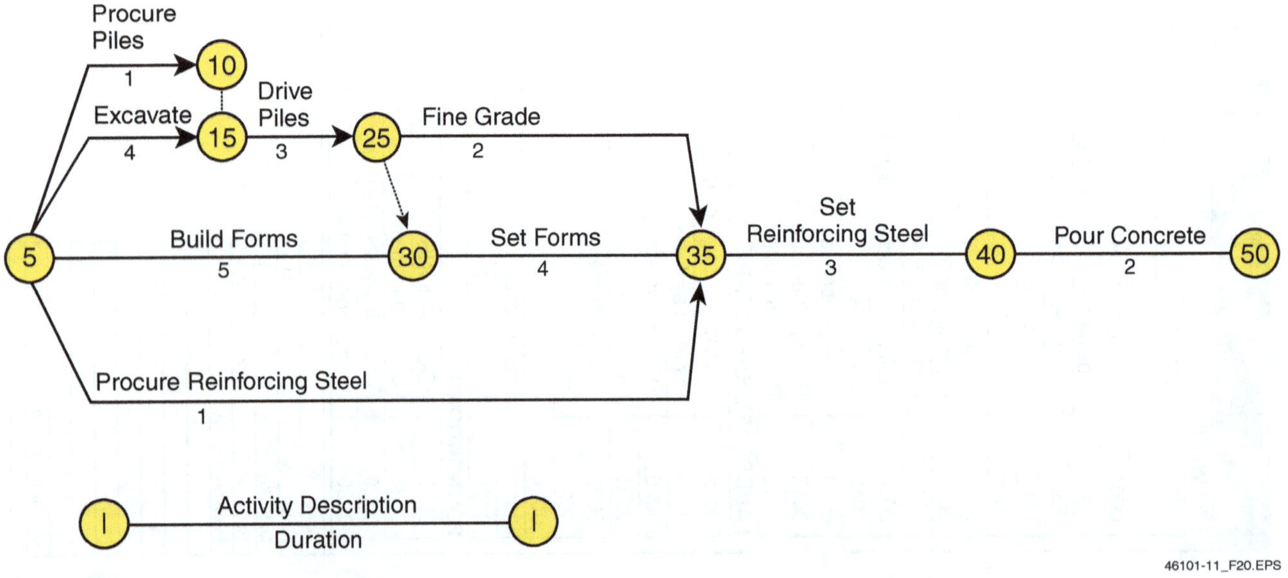

Figure 20 Example of a network schedule.

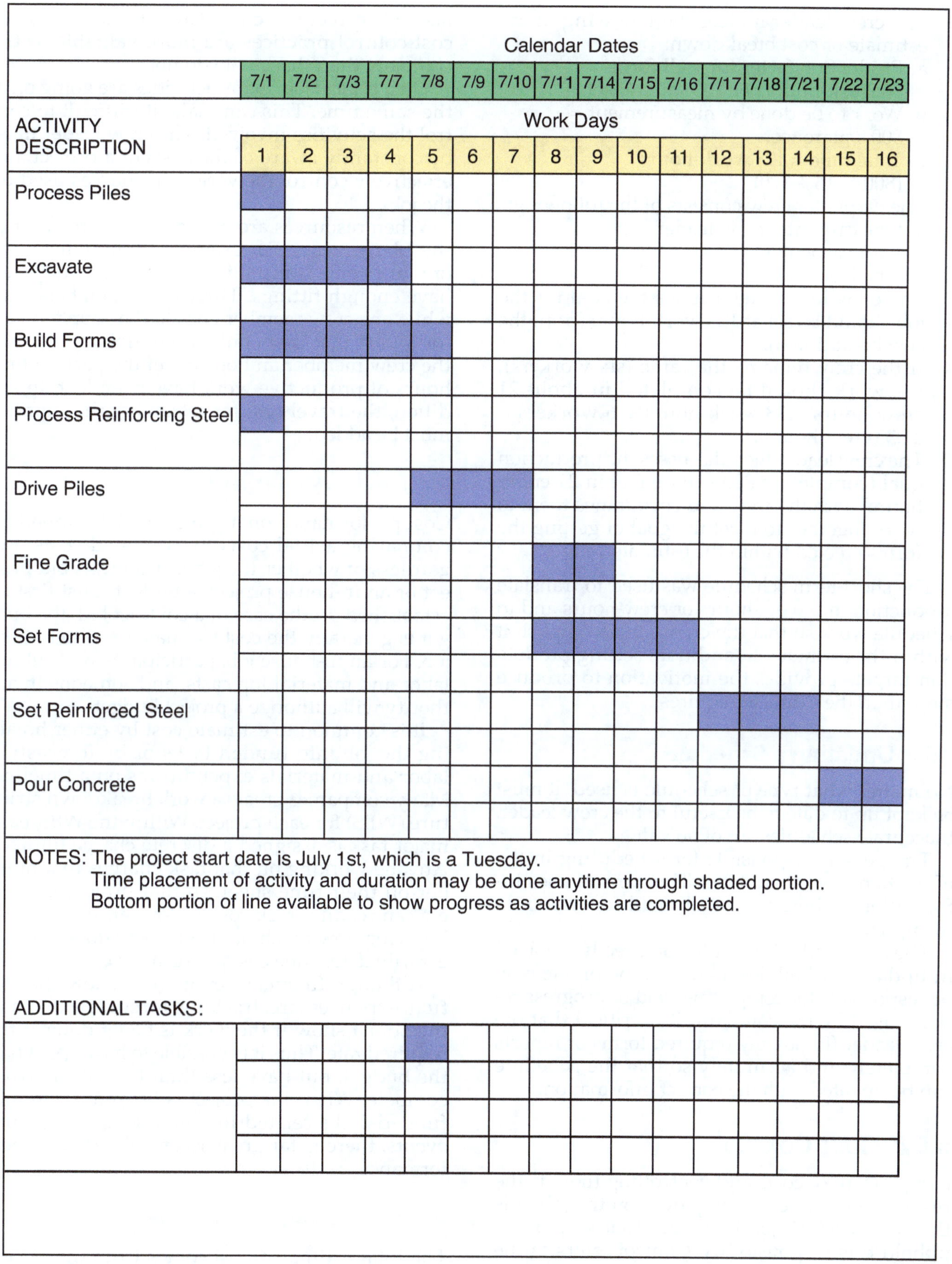

46101-11_F21.EPS

Figure 21 Short-term schedule.

1. The crew leader notices the following in the estimate or cost breakdown:
   a. Production factor for wall forms = 16 work-hours per 100 square feet
   b. Work to be done by measurement = 800 square feet
   c. Total time = 128 work-hours ($800 \times 16 \div 100$)
2. The carpenter crew consists of the following:
   a. One carpenter crew leader
   b. Four carpenters
   c. One laborer
3. The crew leader determines the goal for the job should be set at 128 work-hours (from the cost breakdown).
4. If the crew remains the same (six workers), the work should be completed in about 21 crew-hours (128 work hours ÷ 6 workers = 21.33 crew-hours).
5. The crew leader then discusses the production goal (completing 800 square feet in 21 crew-hours) with the crew and encourages them to work together to meet the goal of getting the forms erected within the estimated time.

The short-term schedule was used to translate production into work-hours or crew-hours and to schedule work so that the crew can accomplish it within the estimate. In addition, setting production targets provides the motivation to produce more than the estimate requires.

7.5.0 Updating a Schedule

No matter what type of schedule is used, it must be kept up to date to be useful to the crew leader. Inaccurate schedules are of no value.

The person responsible for scheduling in the office handles the updates. This person uses information gathered from job field reports to do the updates.

The crew leader is usually not directly involved in updating schedules. However, he or she may be responsible for completing field or progress reports used by the company. It is critical that the crew leader fill out any required forms or reports completely and accurately so that the schedule can be updated with the correct information.

8.0.0 Cost Control

Being aware of costs and controlling them is the responsibility of every employee on the job. It is the crew leader's job to ensure that employees uphold this responsibility. Control refers to the comparison of estimated performance against actual performance and following up with any needed corrective action. Crew leaders who use cost-control practices are more valuable to the company than those who do not.

On a typical job, many activities are going on at the same time. This can make it difficult to control the activities involved. The crew leader must be constantly aware of the costs of a project and effectively control the various resources used on the job.

When resources are not controlled, the cost of the job increases. For example, a plumbing crew of four people is installing soil pipe and does not have enough fittings. Three crew members wait while one crew member goes to the supply house for a part that costs only a few dollars. It takes the crew member an hour to get the part, so four hours of productive work have been lost. In addition, the travel cost for retrieving the supplies must be added.

8.1.0 Assessing Cost Performance

Cost performance on a project is determined by comparing actual costs to estimated costs. Regardless of whether the job is a contract bid project or an in-house project, a budget must first be established. In the case of a contract bid, the budget is generally the cost estimate used to bid the job. For an in-house job, participants will submit labor and material forecasts, and someone in authority will authorize a project budget.

It is common to estimate cost by either breaking the job into funded tasks or by forecasting labor and materials expenditures on a timeline. Many companies create a work breakdown structure (WBS) for each project. Within the WBS, each major task is assigned a discrete charge number. Anyone working on that task charges that number on their time sheet, so that project managers can readily track cost performance. However, knowing how much has been spent does not necessarily determine cost performance.

Although financial reports can show that actual expenses are tracking forecast expenses, they don't show if the work is being done at the required rate. Thus it is possible to have spent half the budget, but have less than half of the work compete. When the project is broken down into funded tasks related to schedule activities and events, there is far greater control over cost performance.

8.2.0 Field Reporting System

The total estimated cost comes from the job estimates, but the actual cost of doing the work is obtained from an effective field reporting system.

A field reporting system is made up of a series of forms, which are completed by the crew leader and others. Each company has its own forms and methods for obtaining information. The general information and the process of how they are used are described here.

First, the number of hours each person worked on each task must be known. This information comes from daily time cards. Once the accounting department knows how many hours each employee worked on an activity, it can calculate the total cost of the labor by multiplying the number of hours worked by the wage rate for each worker. The cost for the labor to do each task can be calculated as the job progresses. This cost will be compared with the estimated cost. This comparison will also be done at the end of the job.

When material is put in place, a designated person will measure the quantities from time to time and send this information to the home office and, possibly, the crew leader. This information, along with the actual cost of the material and the amount of hours it took the workers to install it, is compared to the estimated cost. If the cost is greater than the estimate, management and the crew leader will have to take action to reduce the cost.

A similar process is used to determine if the costs to operate equipment or the production rate are comparable to the estimated cost and production rate.

For this comparison process to be of use, the information obtained from field personnel must be correct. It is important that the crew leader be accurate in reporting. The crew leader is responsible for carrying out his or her role in the field reporting system. One of the best ways to do this is to maintain a daily diary, using a notebook or electronic device. In the event of a legal/contractual conflict with the client, such diaries are considered as evidence in court proceedings, and can be helpful in reaching a settlement.

Here is an example. You are running a crew of five concrete finishers for a subcontractor. When you and your crew show up to finish a slab, the GC says, "We're a day behind on setting the forms, so I need you and your crew to stand down until tomorrow." What do you do?

Of course, you would first call your office to let them know about the delay. Then, you would immediately record it in your job diary. A six-man crew for one day represents 48 labor hours. If your company charges $30 an hour, that's a potential loss of $1,440, which the company would want to recover from the GC. If there is a dispute, your entry in the job diary could result in a favorable decision for your employer.

8.3.0 Crew Leader's Role in Cost Control

The crew leader is often the company representative in the field, where the work takes place. Therefore, the crew leader has a great deal to do with determining job costs. When work is assigned to a crew, the crew leader should be given a budget and schedule for completing the job. It is then up to the crew leader to make sure the job is done on time and stays within budget. This is done by actively managing the use of labor, materials, tools, and equipment.

If the actual costs are at or below the estimated costs, the job is progressing as planned and scheduled, and the company will realize the expected profit. However, if the actual costs exceed the estimated costs, one or more problems may result in the company losing its expected profit, and maybe more. No company can remain in business if it continually loses money. One of the factors that can increase cost is client-related changes. The crew leaders must be able to assess the potential impact of such changes and, if necessary, confer with their employer to determine the course of action. If losses are occurring, the crew leader and superintendent will need to work together to get the costs back in line.

Noted below are some of the reasons why actual costs can exceed estimated costs and suggestions on what the crew leader can do to bring the costs back in line. Before starting any action, however, the crew leader should check with his or her superior to see that the action proposed is acceptable and within the company's policies and procedures.

- *Cause* – Late delivery of materials, tools, and/or equipment
  Corrective Action: Plan ahead to ensure that job resources will be available when needed
- *Cause* – Inclement weather
  Corrective Action: Work with the superintendent and have alternate plans ready
- *Cause* – Unmotivated workers
  Corrective Action: Counsel the workers
- *Cause* – Accidents
  Corrective Action: Enforce the existing safety program

There are many other methods to get the job done on time if it gets off schedule. Examples include working overtime, increasing the size of the crew, pre-fabricating assemblies, or working staggered shifts. However, these examples may increase the cost of the job, so they should not be done without the approval of the project manager.

9.0.0 RESOURCE CONTROL

The crew leader's job is to ensure that assigned tasks are completed safely according to the plans and specifications, on schedule, and within the scope of the estimate. To accomplish this, the crew leader must closely control how resources of materials, equipment, tools, and labor are used. Waste must be minimized whenever possible.

Control involves measuring performance and correcting deviations from plans and specifications to accomplish objectives. Control anticipates deviation from plans and specifications and takes measures to prevent it from occurring.

An effective control process can be broken down into the following steps:

Step 1 Establish standards and divide them into measurable units.

For example, a baseline can be created using experience gained on a typical job, where 2,000 LF of 1¼" copper water tube was installed in five days. Dividing 2,000 by 5 gives 400. Thus, the average installation rate in this case for 1¼" copper water tube was 400 LF/day.

Step 2 Measure performance against a standard.

On another job, 300 square feet of the same tube was placed during an average day. Thus, this average production of 300 LF/day did not meet the average rate of 400 LF/day.

Step 3 Adjust operations to ensure that the standard is met.

In Step 2 above, if the plan called for the job to be completed in five days, the crew leader would have to take action to ensure that this happens. If 300 LF/day is the actual average daily rate, it will have to be increased by 100 LF/day to meet the standard.

9.1.0 Materials Control

The crew leader's responsibility in materials control depends on the policies and procedures of the company. In general, the crew leader is responsible for ensuring on-time delivery, preventing waste, controlling delivery and storage, and preventing theft of materials.

9.1.1 Ensuring On-Time Delivery

It is essential that the materials required for each day's work be on the job site when needed. The crew leader should confirm in advance that all materials have been ordered and will be delivered on schedule. A week or so before the delivery date, follow-up is needed to make sure there will be no delayed deliveries or items on backorder.

To be effective in managing materials, the crew leader must be familiar with the plans and specifications to be used, as well as the activities to be performed. He or she can then determine how many and what types of materials are needed.

If other people are responsible for providing the materials for a job, the crew leader must follow up to make sure that the materials are available when they are needed. Otherwise, delays occur as crew members stand around waiting for the materials to be delivered.

9.1.2 Preventing Waste

Waste in construction can add up to loss of critical and costly materials and may result in job delays. The crew leader needs to ensure that every crew member knows how to use the materials efficiently. The crew should be monitored to make certain that no materials are wasted.

An example of waste is a carpenter who saws off a piece of lumber from a fresh piece, when the size needed could have been found in the scrap pile. Another example of waste involves installing a fixture or copper tube incorrectly. The time spent installing the item incorrectly is wasted because the task will need to be redone. In addition, the materials may need to be replaced if damaged during installation or removal.

Under LEED, waste control is very important. Credits are given for finding ways to reduce waste and for recycling waste products. Waste material should be separated for recycling if feasible (*Figure 22*).

Did you know?

Just-in-time (JIT) delivery is a strategy in which materials are delivered to the job site when they are needed. This means that the materials may be installed right off the truck. This method reduces the need for on-site storage and staging. It also reduces the risk of loss or damage as products are moved about the site. Other modern material management methods include the use of radio frequency identification tags (RFIDs) that make it easy to locate material in crowded staging areas.

9.1.3 Verifying Material Delivery

A crew leader may be responsible for the receipt of materials delivered to the work site. When this happens, the crew leader should require a copy of the shipping ticket and check each item on the shipping ticket against the actual materials to see that the correct amounts were received.

46101-11_F22.EPS

Figure 22 Waste material separated for recycling.

The crew leader should also check the condition of the materials to verify that nothing is defective before signing the shipping ticket. This can be difficult and time consuming because it means that cartons must be opened and their contents examined. However, it is necessary, because a signed shipping ticket indicates that all of the materials were received undamaged. If the crew leader signs for the materials without checking them, and then finds damage, no one will be able to prove that the materials came to the site in that condition.

Once the shipping ticket is checked and signed, the crew leader should give the original or a copy to the superintendent or project manager. The shipping ticket will then be filed for future reference because it serves as the only record the company has to check bills received from the supply house.

9.1.4 Controlling Delivery and Storage

Another very important element of materials control is where the materials will be stored on the job site. There are two factors in determining the appropriate storage location. The first is convenience. If possible, the materials should be stored near where they will be used. The time and effort saved by not having to carry the materials long distances will greatly reduce the installation costs.

Next, the materials must be stored in a secure area where they will not be damaged. It is important that the storage area suit the materials being stored. For instance, materials that are sensitive to temperature, such as chemicals or paints, should be stored in climate-controlled areas. Otherwise, waste may occur.

9.1.5 Preventing Theft and Vandalism

Theft and vandalism of construction materials increase costs because these materials are needed to complete the job. The replacement of materials and the time lost because the needed materials are missing can add significantly to the cost. In addition, the insurance that the contractor purchases will increase in cost as the theft and vandalism rate grows.

The best way to avoid theft and vandalism is a secure job site. At the end of each work day, store unused materials and tools in a secure location, such as a locked construction trailer. If the job site is fenced or the building can be locked, the materials can be stored within. Many sites have security cameras and/or intrusion alarms to help minimize theft and vandalism.

9.2.0 Equipment Control

The crew leader may not be responsible for long-term equipment control. However, the equipment required for a specific job is often the crew leader's responsibility. The first step is to identify when the required equipment must be transported from the shop or rental yard. The crew leader is responsible for informing the shop where it is being used and seeing that it is returned to the shop when the job is done.

It is common for equipment to lay idle at a job site because the job has not been properly planned and the equipment arrived early. For example, if wire-pulling equipment arrives at a job site before the conduit is in place, this equipment will be out of service while awaiting the conduit installation. In addition, it is possible that this unused equipment could be damaged, lost, or stolen.

The crew leader needs to control equipment use, ensure that the equipment is operated in accordance with its design, and that it is being used within time and cost guidelines. The crew leader must also ensure that equipment is maintained and repaired as indicated by the preventive maintenance schedule. Delaying maintenance and repairs can lead to costly equipment failures. The crew leader must also ensure that the equipment operators have the necessary credentials to operate the equipment, including applicable licenses.

The crew leader is responsible for the proper operation of all other equipment resources, including cars and trucks. Reckless or unsafe operation of vehicles will likely result in damaged equipment and a delayed or unproductive job. This, in turn, could affect the crew leader's job security.

The crew leader should also ensure that all equipment is secured at the close of each day's work in an effort to prevent theft. If the equipment is still being used for the job, the crew leader should ensure that it is locked in a safe place; otherwise, it should be returned to the shop.

9.3.0 Tool Control

Among companies, various policies govern who provides hand and power tools to employees. Some companies provide all the tools, while others furnish only the larger power tools. The crew leader should find out about and enforce any company policies related to tools.

Tool control is a twofold process. First, the crew leader must control the issue, use, and maintenance of all tools provided by the company. Next, the crew leader must control how the tools are being used to do the job. This applies to tools that are issued by the company as well as tools that belong to the workers.

Using the proper tools correctly saves time and energy. In addition, proper tool use reduces the chance of damage to the tool being used. Proper use also reduces injury to the worker using the tool, and to nearby workers.

Tools must be adequately maintained and properly stored. Making sure that tools are cleaned, dried, and lubricated prevents rust and ensures that the tools are in the proper working order.

In the event that tools are damaged, it is essential that they be repaired or replaced promptly. Otherwise, an accident or injury could result.

NOTE

Regardless of whether a tool is owned by a worker or the company, OSHA will hold the company responsible for it when it is used on a job site. The company can be held accountable if an employee is injured by a defective tool. Therefore, the crew leader needs to be aware of any defects in the tools the crew members are using.

Company-issued tools should be taken care of as if they are the property of the user. Workers should not abuse tools simply because they are not their own.

One of the major causes of time lost on a job is the time spent searching for a tool. To prevent this from occurring, a storage location for company-issued tools and equipment should be established. The crew leader should make sure that all company-issued tools and equipment are returned to this designated location after use. Similarly, workers should make sure that their personal toolboxes are organized so that they can readily find the appropriate tools and return their tools to their toolboxes when they are finished using them.

Studies have shown that the key to an effective tool control system lies in:

- Limiting the number of people allowed access to stored tools
- Limiting the number of people held responsible for tools
- Controlling the ways in which a tool can be returned to storage
- Making sure tools are available when needed

9.4.0 Labor Control

Labor typically represents more than half the cost of a project, and therefore has an enormous impact on profitability. For that reason, it is essential to manage a crew and their work environment in a way that maximizes their productivity. One of the ways to do that is to minimize the time spent on unproductive activities such as:

- Engaging in bull sessions
- Correcting drawing errors
- Retrieving tools, equipment, and materials
- Waiting for other workers to finish

If crew members are habitually goofing off, it is up to the crew leader to counsel those workers. The counseling should be documented in the crew leader's daily diary. Repeated violations will need to be referred to the attention of higher management as guided by company policy.

Errors will occur and will need to be corrected. Some errors, such as mistakes on drawings, may be outside of the crew leader's control. However, some drawing errors can be detected by carefully examining the drawings before work begins.

If crew members are making mistakes due to inexperience, the crew leader can help avoid these errors by providing on-the-spot training and by checking on inexperienced workers more often.

The availability and location of tools, equipment, and materials can have a profound effect on a crew's productivity. If the crew has to wait for these things, or travel a distance to get them, it reduces their productivity. The key to minimizing such problems is proactive management of these resources. As discussed earlier, practices such as checking in advance to be sure equipment and materials will be available when scheduled and placing materials close to the work site will help minimize unproductive time.

Delays caused by others can be avoided by carefully tracking the project schedule. By doing so, crew leaders can anticipate delays that will affect the work of their crews and either take action to prevent the delay or redirect the crew to another task.

Participant Exercise G

1. List the methods your company uses to minimize waste.

____

____

____

____

____

2. List the methods your company uses to control small tools on the job.

____

____

____

____

____

3. List five ways that you feel your company could control labor to maximize productivity.

____

____

____

____

____

10.0.0 Production and Productivity

Production is the amount of construction put in place. It is the quantity of materials installed on a job, such as 1,000 linear feet of waste pipe installed in a given day. On the other hand, productivity depends on the level of efficiency of the work. It is the amount of work done per hour or day by one worker or a crew.

Production levels are set during the estimating stage. The estimator determines the total amount of materials to be put in place from the plans and specifications. After the job is complete, the actual amount of materials installed can be assessed, and the actual production can be compared to the estimated production.

Productivity relates to the amount of materials put in place by the crew over a certain time period. The estimator uses company records during the estimating stage to determine how much time and labor it will take to place a certain quantity of materials. From this information, the estimator calculates the productivity necessary to complete the job on time.

For example, it might take a crew of two people ten days to paint 5,000 square feet. The crew's productivity per day is obtained by dividing 5,000 square feet by ten days. The result is 500 square feet per day. The crew leader can compare the daily production of any crew of two painters doing similar work with this average, as discussed previously.

Planning is essential to productivity. The crew must be available to perform the work, and have all of the required materials, tools, and equipment in place when the job begins.

The time on the job should be for business, not for taking care of personal problems. Anything not work-related should be handled after hours, away from the job site. Planning after-work activities, arranging social functions, or running personal errands should be handled after work or during breaks.

Organizing field work can save time. The key to effectively using time is to work smarter, not necessarily harder. For example, most construction projects require that the contractor submit a set of as-built plans at the completion of the work. These plans describe how the materials were actually installed. The best way to prepare these plans is to mark a set of working plans as the work is in progress. That way, pertinent details will not be forgotten and time will not be wasted trying to remember how the work was done.

The amount of material actually used should not exceed the estimated amount. If it does, either the estimator has made a mistake, undocumented changes have occurred, or rework has caused the need for additional materials. Whatever the case, the crew leader should use effective control techniques to ensure the efficient use of materials.

When bidding a job, most companies calculate the cost per labor hour. For example, a ten-day job might convert to 160 labor hours (two painters for ten days at eight hours per day). If the company charges a labor rate of $30/hour, the labor cost would be $4,800. The estimator then adds the cost of materials, equipment, and tools, along with overhead costs and a profit factor, to determine the price of the job.

After a job has been completed, information gathered through field reporting allows the home office to calculate actual productivity and compare it to the estimated figures. This helps to identify productivity issues and improves the accuracy of future estimates.

The following labor-related practices can help to ensure productivity:

- Ensure that all workers have the required resources when needed.
- Ensure that all personnel know where to go and what to do after each task is completed.
- Make reassignments as needed.
- Ensure that all workers have completed their work properly.

Review Questions

1. Which of these activities occurs during the development phase of a project?
   a. Architect/engineer sketches are prepared and a preliminary budget is developed.
   b. Government agencies give a final inspection of the design, adherence to codes, and materials used.
   c. Project drawings and specifications are prepared.
   d. Contracts for the project are awarded.

2. The type of contract in which the client pays the contractor for their actual labor and material expenses they incur is known as a _____.
   a. firm fixed-price contract
   b. time-spent contract
   c. cost-reimbursable contract
   d. performance-based contract

3. On-site changes in the original design that are made during construction are recorded in the _____.
   a. as-built plans
   b. takeoff sheet
   c. project schedule
   d. job specifications

4. On a design-build project, _____.
   a. the owner is responsible for providing the design
   b. the architect does the design and the general contractor builds the project
   c. the same contractor is responsible for both design and construction
   d. a construction manager is hired to oversee the project

5. One example of a direct cost when bidding a job is _____.
   a. office rent
   b. labor
   c. accounting
   d. utilities

6. The control method that a crew leader uses to plan a few weeks in advance is a ______.
   a. network schedule
   b. bar chart schedule
   c. daily diary
   d. look-ahead schedule

7. A job diary should typically indicate _____.
   a. items such as job interruptions and visits
   b. changes needed to project drawings
   c. the estimated time for each job task related to a particular project
   d. the crew leader's ideas for improving employee morale

8. Gantt charts can help crew leaders in the field by _____.
   a. offering a comparison of actual production to estimated production
   b. providing short-term and long-term schedule information
   c. stating the equipment and materials necessary to complete a task
   d. providing the information needed to develop an estimate or an estimate breakdown

9. What is the crew leader's responsibility with regard to cost control?
   a. Cost control is outside the scope of a crew leader's responsibility.
   b. The crew leader is responsible only for minimizing material waste.
   c. The crew leader must ensure that all team members are aware of project costs and how to control them.
   d. The crew leader typically prepares the company's cost estimate and is therefore responsible for cost performance.

10. Which of the following is a correct statement regarding project cost?
    a. Cost is handled by the accounting department and is not a concern of the crew leader.
    b. A company's profit on a project is affected by the difference between the estimated cost and the actual cost.
    c. Wasted material is factored into the estimate, and is therefore not a concern.
    d. The contractor's overhead costs are not included in the cost estimate.

11. The crew leader is responsible for ensuring that equipment used by his or her crew is properly maintained.
    a. True
    b. False
12. Who is responsible if a defect in an employee's tool results in an accident?
    a. The employee
    b. The company
    c. The crew leader
    d. The tool manufacturer
13. Productivity is defined as the amount of work accomplished.
    a. True
    b. False
14. If a crew of masons is needed to lay 1,000 concrete blocks, and the estimator determined that two masons could complete the job in one eight-hour day, what is the estimated productivity rate?
    a. 125 blocks per hour
    b. 62.5 blocks per hour
    c. 31.25 blocks per hour
    d. 16 blocks per hour

Additional Resources

This module presents thorough resources for task training. The following resource material is suggested for further study.

Aging Workforce News, **www.agingworkforcenews.com**.

American Society for Training and Development (ASTD), **www.astd.org**.

Architecture, Engineering, and Construction Industry (AEC), **www.aecinfo.com**.

CIT Group, **www.citgroup.com**.

Equal Employment Opportunity Commission (EEOC), **www.eeoc.gov**.

National Association of Women in Construction (NAWIC), **www.nawic.org**.

National Census of Fatal Occupational Injuries (NCFOI), **www.bls.gov**.

National Center for Construction Education and Research, **www.nccer.org**.

National Institute of Occupational Safety and Health (NIOSH), **www.cdc.gov/niosh**.

National Safety Council, **www.nsc.org**.

NCCER Publications:

Your Role in the Green Environment

Sustainable Construction Supervisor

Occupational Safety and Health Administration (OSHA), **www.osha.gov**.

Society for Human Resources Management (SHRM), **www.shrm.org**.

United States Census Bureau, **www.census.gov**.

United States Department of Labor, **www.dol.gov**.

USA Today, **www.usatoday.com**.

Figure Credits

© 2010 Photos.com, a division of Getty Images. All rights reserved., Module opener

United States Department of Labor, Occupational Safety and Health Administration, SA01

plaquemaster.com, Figure 9

© U.S. Green Building Council, SA02

Sushil Shenoy, SA03 and Figure 22

John Ambrosia, Figure 19

NCCER CURRICULA — USER UPDATE

NCCER makes every effort to keep its textbooks up-to-date and free of technical errors. We appreciate your help in this process. If you find an error, a typographical mistake, or an inaccuracy in NCCER's curricula, please fill out this form (or a photocopy), or complete the online form at **www.nccer.org/olf**. Be sure to include the exact module ID number, page number, a detailed description, and your recommended correction. Your input will be brought to the attention of the Authoring Team. Thank you for your assistance.

Instructors – If you have an idea for improving this textbook, or have found that additional materials were necessary to teach this module effectively, please let us know so that we may present your suggestions to the Authoring Team.

NCCER Product Development and Revision

13614 Progress Blvd., Alachua, FL 32615

Email: curriculum@nccer.org
Online: www.nccer.org/olf

❑ Trainee Guide ❑ Lesson Plans ❑ Exam ❑ PowerPoints Other ________________

Craft / Level: Copyright Date:

Module ID Number / Title:

Section Number(s):

Description:

Recommended Correction:

Your Name:

Address:

Email: Phone:

Glossary

Alloy: A metal that has had other elements added to it that substantially change its mechanical properties.

American Iron and Steel Institute (AISI): An industry organization responsible for preparing standards for steel and steel alloys that are based upon the Society of Automotive Engineers (SAE) code system.

American Society for Testing Materials (ASTM): An organization that developed a code system for identifying and labeling steels.

Anneal: To subject an alloy, metal, or glass to a process of heating and slow cooling to make it tougher and less brittle.

Austenitizing: Heating a steel up to or above the transformation temperature range to achieve partial or complete transformation to austenite grain structure.

Axially: In a straight line along an axis.

Axis: The center line of the weld or prepared base metal.

Camber: The rise of the crown of the main deck from side to side (athwartship).

Casting: The metal object produced by pouring molten metal into a mold.

Coefficient: A numerical measure of a physical or chemical property that is constant for a system under specified conditions, such as the coefficient of friction.

Degassing: Removing the gases formed during the welding process that can cause imperfections in the weld.

Ductile: The term used to describe a material that can be bent or shaped without breaking.

Ductility: The characteristic of metal that allows it to be stretched, drawn, or hammered without breaking.

Ferrous: Relating to iron or an alloy that is mostly iron; a metal containing iron.

Flange: The outer members of a structural beam, joist, or rail.

Floor: The upright structures used to support the tank top.

Gas metal arc welding (GMAW): In gas metal arc welding, sometimes called MIG, a consumable wire fed through the welding gun is used to make the weld; commonly used on steel and other metals.

Gas tungsten arc welding (GTAW): In gas tungsten arc welding, sometimes called TIG, a nonconsumable tungsten electrode is used to produce the weld; commonly used on stainless steel and light metals, such as aluminum and magnesium.

Globular transfer: A GMAW process in which the electrode wire does not touch the base metal, and the filler metal is transferred randomly in the arc as large, irregular globules.

Hardenability: A characteristic of a metal that enables it to become hard, usually through heat treatment.

Heat-affected zone (HAZ): The part of the base metal that has not been melted but has been altered structurally by the weld heat.

Hot shortness: The condition of metal when it proves to be very brittle and unbendable at red heat but can be bent without showing signs of brittleness when cold or at white heat. This condition is often a result of high sulfur and phosphorus content.

Interpass temperature: In a multiple-pass weld, the temperature of the base metal at the time the next weld pass is started; it is usually the same temperature that is used for preheating.

Malleable: Able to be hammered or pressed into another shape without breaking.

Mechanical properties: The characteristics or traits that indicate how flexible and strong a metal will be under stress.

Metallic: Having characteristics of metal, such as ductility, malleability, luster, and heat and electrical conductivity.

Mold line: Lines showing the exact locations of installed structures.

Nonferrous metal: A metal, such as aluminum, copper, or brass, that lacks sufficient quantities of iron to be affected by its properties.

Notch toughness: The impact load required to break a test piece of weld metal measured by a notch toughness test (Charpy V-notch test). This test may be performed on metal below room temperature, at which point it may be more brittle.

Oxide: The scale that forms on the surfaces of metals when they are exposed to air.

Piece part: An individual part.

Postweld heat treatment (PWHT): The process of heat treating the base metal and weld after the weld has been made.

Pulsed transfer: A GMAW welding process in which the welding power is cycled from a low level (background level) to a higher level (welding level), resulting in lower average voltage and current; this is a variation of the spray transfer process.

Quench: To rapidly cool a hot metal using air, water, or oil.

Residual stress: The strain that occurs in a welded joint after the welding has been completed, as a result of mechanical action, thermal action, or both.

Retrofit: (a) Incorporation of a change into a deployed system; (b) modification of an existing device or structure.

Scribe: To trace the contour of a surface with a pointed instrument.

Set marks: Groups of center-punch marks used to show exactly where the edge of a plate should be set.

Shell plates: The outer hull plating of a ship.

Shielded metal arc welding (SMAW): The most common type of welding used on iron and steel, but can also be used on stainless steel; uses a flux-coated consumable electrode.

Short-circuiting transfer: A GMAW welding process in which the electrode wire shorts against the base metal. The welding current repeatedly and rapidly melts the wire and establishes an arc, which is extinguished when the wire again touches the base metal.

Society of Automotive Engineers (SAE): An automotive society responsible for creating an early system for classifying carbon steels.

Spray transfer: A GMAW welding process in which the electrode wire does not touch the base metal, and the filler metal is transferred axially (in a straight line) in the arc as fine droplets.

Stanchion: An upright member used to support plating.

Stickout: The length of electrode that projects from the contact tube, measured from the arc tip to the contact tube.

Stiffeners: An angle bar, T-bar, or channel used to stiffen the plating of a bulkhead, deck, or shell plate.

Stringers: A fore and aft (front to back) member used to provide longitudinal strength.

Tempering: The process of reheating quench hardened or normalized steel to a temperature below the transformation range and then cooling it at any rate desired.

Tensile strength: The maximum stress or force that a material can withstand without breaking.

Underbead cracking: Subsurface cracking in the base metal under or near the weld.

Unified numbering system (UNS): An industry-accepted numbering system for identifying ferrous metals and alloys.

Web: The plate joining the flanges of a girder, joist, or rail.

Work line: A line created by a fitter to serve as a reference line when laying out structures.

Wrought: Formed or shaped by hammering or rolling.

Index

Figures in *italic* refer to the contents of the drawing package included with this book.

B

C

D

E

F

G

M

N

O

P

Q

R

S

T

U

V

W

X

Y

Z

www.ingramcontent.com/pod-product-compliance
Lightning Source LLC
Jackson TN
JSHW062302150926
108205JS00006B/60